Beyond Oil and Gas:
The Methanol Economy

George A. Olah, Alain Goeppert,
and G. K. Surya Prakash

WILEY-
VCH

WILEY-VCH Verlag GmbH & Co. KGaA

The Authors

Prof. Dr. George A. Olah
Dr. Alain Goeppert
Prof. Dr. G. K. Surya Prakash
Loker Hydrocarbon Research Institute
University of Southern California
837 W. 37th. Street
Los Angeles, CA 90089-1661
USA

1st Edition February 2006
  1st Reprint June 2006
  2nd Reprint Oktober 2006

Library of Congress Card No.: Applied for

British Library Cataloguing-in-Publication Data:
A catalogue record for this book is available
from the British Library.

Bibliographic information published by
Die Deutsche Bibliothek
Die Deutsche Bibliothek lists this publication
in the Deutsche Nationalbibliografie;
detailed bibliographic data is available in the
Internet at < http://dnb.ddb.de >.

Printed in the Federal Republic of Germany.
Printed on acid-free paper.

Typesetting   Hagedorn Kommunikation,
Viernheim
Printing   Betz Druck GmbH, Darmstadt
Bookbinding   Litges & Dopf Buchbinderei
GmbH, Heppenheim
Cover   Grafik-Design Schulz, Fußgönheim

ISBN-13:   978-3-527-31275-7
ISBN-10:   3-527-31275-7

Contents

Chapter 1
Introduction *1*

Chapter 2
Coal in the Industrial Revolution, and Beyond *11*

Chapter 3
History of Oil and Natural Gas *18*
Oil Extraction and Exploration *22*
Natural Gas *23*

Chapter 4
Fossil Fuel Resources and Uses *27*
Coal *28*
Oil *32*
Tar Sands *37*
Oil Shale *38*
Natural Gas *39*
Coalbed Methane *46*
Tight Sands and Shales *47*
Methane Hydrates *47*
Outlook *49*

Chapter 5
Diminishing Oil and Gas Reserves *51*

Chapter 6
The Continuing Need for Hydrocarbons and their Products *60*
Fractional Distillation *63*
Thermal Cracking *64*

Beyond Oil and Gas: The Methanol Economy. G. A. Olah, A. Goeppert, G. K. S. Prakash
Copyright © 2006 WILEY-VCH Verlag GmbH & Co. KGaA, Weinheim
ISBN 3-527-31275-7

Chapter 7
Fossil Fuels and Climate Change *72*
Mitigation *81*

Chapter 8
Renewable Energy Sources and Atomic Energy *84*
Hydropower *87*
Geothermal Energy *91*
Wind Energy *94*
Solar Energy: Photovoltaic and Thermal *97*
Electricity from Photovoltaic Conversion *98*
Solar Thermal Power for Electricity Production *100*
Electric Power from Saline Solar Ponds *101*
Solar Thermal Energy for Heating *102*
Economic Limitations of Solar Energy *102*
Biomass Energy *103*
Electricity from Biomass *103*
Liquid Biofuels *104*
Ocean Energy: Thermal, Tidal, and Wave Power *108*
Tidal Energy *109*
Waves *110*
Ocean Thermal Energy *110*
Nuclear Energy *111*
Energy from Nuclear Fission Reactions *113*
Breeder Reactors *118*
The Need for Nuclear Power *119*
Economics *121*
Safety *121*
Radiation Hazards *124*
Nuclear Byproducts and Waste *125*
Emissions *127*
Nuclear Power: An Energy Source for the Future *127*
Nuclear Fusion *128*
Future Outlook *131*

Chapter 9
The Hydrogen Economy and its Limitations *133*
The Discovery and Properties of Hydrogen *133*
The Development of Hydrogen Energy *135*
The Production and Uses of Hydrogen *138*
Hydrogen from Fossil Fuels *140*
Hydrogen from Biomass *141*
Photobiological Water Cleavage *142*
Water Electrolysis *142*
Hydrogen Production Using Nuclear Energy *144*

The Challenge of Hydrogen Storage *145*
Liquid Hydrogen *147*
Compressed Hydrogen *148*
Metal Hydrides and Solid Absorbents *149*
Other Means of Hydrogen Storage *150*
Hydrogen: Centralized or Decentralized Distribution? *150*
Safety of Hydrogen *153*
Hydrogen in Transportation *154*
Fuel Cells *155*
History *155*
Fuel Cell Efficiency *156*
Hydrogen-Based Fuel Cells *159*
PEM Fuel Cells for Transportation *162*
Regenerative Fuel Cells *165*
Outlook *166*

Chapter 10
The "Methanol Economy": General Aspects *168*

Chapter 11
Methanol as a Fuel and Energy Carrier *173*
Properties and Historical Background *173*
Present Uses of Methanol *175*
Use of Methanol and Dimethyl Ether as Transportation Fuels *177*
Alcohol as a Transportation Fuel in the Past *177*
Methanol as Fuel in Internal Combustion Engines (ICE) *180*
Methanol and Dimethyl Ether as Diesel Fuels Substitute in
Compression Ignition Engines *182*
Biodiesel Fuel *186*
Advanced Methanol-Powered Vehicles *187*
Hydrogen for Fuel Cells from Methanol Reforming *187*
Direct Methanol Fuel Cell (DMFC) *191*
Fuel Cells Based on Other Fuels and Biofuel Cells *195*
Regenerative Fuel Cell *196*
Methanol for Static Power and Heat Generation *196*
Methanol Storage and Distribution *197*
Methanol Price *200*
Methanol Safety *201*
Emissions from Methanol-Powered Vehicles *205*
Methanol and the Environment *206*
Methanol and Issues of Climate Change *208*

Chapter 12
Production of Methanol from Syn-Gas to Carbon Dioxide   *209*
Methanol from Fossil Fuels   *212*
Production via Syn-Gas   *212*
Syn-Gas from Natural Gas   *215*
Methane Steam Reforming   *215*
Partial Oxidation of Methane   *216*
Autothermal Reforming and Combination of Steam Reforming
and Partial Oxidation   *217*
Syn-Gas from CO$_2$ Reforming   *217*
Syn-Gas from Petroleum and Higher Hydrocarbons   *218*
Syn-Gas from Coal   *218*
Economics of Syn-Gas Generation   *219*
Methanol through Methyl Formate   *219*
Methanol from Methane Without Syn-Gas   *220*
Selective Oxidation of Methane to Methanol   *221*
Catalytic Gas-Phase Oxidation of Methane   *221*
Liquid-Phase Oxidation of Methane to Methanol   *224*
Methanol Production through Mono-Halogenated Methanes   *226*
Microbial or Photochemical Conversion of Methane to Methanol   *228*
Methanol from Biomass   *229*
Methanol from Biogas   *235*
Aquaculture   *237*
Water Plants   *237*
Algae   *238*
Methanol from Carbon Dioxide   *239*
Carbon Dioxide from Industrial Flue Gases   *242*
Carbon Dioxide from the Atmosphere   *243*

Chapter 13
Methanol-Based Chemicals, Synthetic Hydrocarbons and Materials   *246*
Methanol-Based Chemical Products and Materials   *246*
Methanol Conversion to Olefins and Synthetic Hydrocarbons   *248*
Methanol to Olefin (MTO) Process   *249*
Methanol to Gasoline (MTG) Process   *251*
Methanol-Based Proteins   *252*
Outlook   *253*

Chapter 14
Future Perspectives *254*
The "Methanol Economy" and its Advantages *256*

Further Reading and Information *260*

References *274*

Index *283*

Preface

Humankind, for its continued existence, requires not only such essentials as food, clean water, shelter, and clothing materials, but also large amounts of energy. Ever since cavemen succeeded in kindling fire, our ancestors have used a variety of sources for heating and cooking, ranging initially from wood and vegetation followed by peat moss and other carbon-based fuels. Since the industrial revolution, the major source of energy was coal to which, during the twentieth century, oil and natural gas were added. The latter resources – termed "fossil fuels" – were formed by Nature over eons, but once combusted they are not renewable on our human time scale and are thus increasingly depleted by overuse. Our readily accessible oil and gas reserves may not last much past the twenty-first century, while coal reserve may be available for another century or two. We need, therefore, to find new ways and resources for the future.

This book discusses a new approach based on what we call the "Methanol Economy®". The production of methanol directly from still-available fossil fuel sources, and the recycling of carbon dioxide via hydrogenative reductions, are – we believe – feasible and convenient ways to store energy generated from all possible sources including, alternative energy sources (solar, hydro, wind, geothermal, etc.) and atomic energy. In the short term, new efficient production of methanol not only from still-available natural gas resources (without going through the syn-gas route) but also by the hydrogenative conversion of carbon dioxide from industrial exhausts, offer feasible new routes. In the long term, recycling of the carbon dioxide captured from the air itself will be possible. Air, in contrast to oil and gas resources, is available to everybody on Earth, and its CO_2 content represent an inexhaustible recyclable carbon resource. Methanol produced from this CO_2 (using any energy source to produce the required hydrogen from water), is an excellent fuel on its own for internal combustion engines or fuel cells of the future. It can be also readily converted, via its dehydration to ethylene and propylene, into synthetic hydrocarbons and their products. Consequently, it can free mankind's dependence on our diminishing oil and natural gas (even coal) resources. At the same time, by being able to recycle excess CO_2 we can mitigate or eliminate a major source of global climate change – that is, warming of the Earth – caused by human activities.

Beyond Oil and Gas: The Methanol Economy. G. A. Olah, A. Goeppert, G. K. S. Prakash
Copyright © 2006 WILEY-VCH Verlag GmbH & Co. KGaA, Weinheim
ISBN 3-527-31275-7

We are fully aware that to solve our outlined problems for the future, including energy storage and transportation, non-oil- and gas-based fuels and raw materials for the production of synthetic hydrocarbons and their products (to which we are accustomed in our everyday life) and new approaches are needed. Much has been said about the future in view of our diminishing and non-renewable fossil fuel resources. The outlined "Methanol Economy" is one of the feasible and achievable solutions, which deserves serious further consideration and development. We hope that this book will call more attention to this approach, and spur future activities in the area.

Los Angeles, December 2005

George A. Olah
Alain Goeppert
G. K. Surya Prakash

Acronyms, Units and Abbreviations

Acronyms

AFC	Alkaline Fuel Cell
BP	British Petroleum
BWR	Boiling Water Reactor
CEA	Commissariat à l'Energie Atomique (France)
CEC	California Energy Commission
CIA	Central Intelligence Agency
DMFC	Direct Methanol Fuel Cell
DOE	Department of Energy (United States)
EDF	Electricité de France
EIA	Energy Information Administration (DOE)
EPA	Environmental Protection Agency (United States)
EU	European Union
GDP	Gross Domestic Product
GHG	Greenhouse Gas
IAEA	International Atomic Energy Agency
ICE	Internal Combustion Engine
IEA	International Energy Agency
IGCC	Integrated Gasification Combined Cycle
IPCC	International Panel on Climate Change
ITER	International Thermonuclear Experimental Reactor
LNG	Liquefied Natural Gas
MCFC	Molten Carbonate Fuel Cell
NRC	National Research Council (United States)
NREL	National Renewable Energy Laboratory (United States)
OECD	Organization for Economic Cooperation and Development
OPEC	Organization of Petroleum Exporting Countries
ORNL	Oak Ridge National Laboratory
OTEC	Ocean Thermal Energy Conversion
PAFC	Phosphoric Acid Fuel Cell
PEMFC	Proton Exchange Membrane Fuel Cell
PFBC	Pressurized Fluidized Bed Combustion
PV	Photovoltaics

Beyond Oil and Gas: The Methanol Economy. G. A. Olah, A. Goeppert, G. K. S. Prakash
Copyright © 2006 WILEY-VCH Verlag GmbH & Co. KGaA, Weinheim
ISBN 3-527-31275-7

PWR	Pressurized Water Reactor
R/P	Reserve/Production ratio
SUV	Sport Utility Vehicle
TPES	Total Primary Energy Supply
UNO	United Nations Organization
UNSCEAR	United Nations Scientific Committee on Effects of Atomic Radiation
URFC	Unitized Regenerative Fuel Cell
USCB	United States Census Bureau
USGS	United States Geological Survey
WCD	World Commission on Dams
WCI	World Coal Institute
WEC	World Energy Council
ZEV	Zero Emission Vehicle

Units and Abbreviations

b and bbl	barrel
Btu	British thermal unit
°C	degree Celsius
cal	calorie
g	gram
h	hour
ha	hectare
kWh	kilowatt-hour
m	meter
Mb	megabarrel (10^6 barrels)
ppm	parts per million
toe	tonne oil equivalent
s	second
Sv	Sievert
t	metric tonne
W	watt

Prefixes

μ	micro	10^{-6}
m	milli	10^{-3}
k	kilo	10^{3}
M	mega	10^{6}
G	giga	10^{9}
T	tera	10^{12}
P	peta	10^{15}
E	exa	10^{18}

Conversion of Units

Volume

1 tonne of crude oil = 7.33 barrels of oil
1 gallon = 3.785 liters
1 barrel of oil = 42 U.S. gallons = 159 liters
1 m^3 = 1000 liters
1 m^3 = 35.3 cubic feet (ft^3)

Energy

1 kcal = 4.1868 kJ = 3.968 Btu
1 kJ = 0.239 kcal = 0.948 Btu
1 kWh = 860 kcal = 3600 kJ
1 toe = 41.87 GJ
Quadrillion Btu, QBtu = 1×10^{15} Btu

Chapter 1
Introduction

Ever since our distant forerunners managed to light fire for providing heat, means for cooking and many essential purposes, humankind's life and survival is inherently linked with our ever-increasing thirst for energy. From burning wood, vegetation, peat moss and other sources, to the use of coal, followed by that of petroleum oil and natural gas (fossil fuels), we have thrived using Nature's resources. Fossil fuels include coal, oil and gas, all composed of hydrocarbons with varying ratios of carbon and hydrogen.

Hydrocarbons derived from petroleum, natural gas or coal are essential in many ways to modern life and its quality. The bulk of the world's hydrocarbons is used as fuels for propulsion, electrical power generation, and heating. The chemical, petrochemical, plastics and rubber industries are also dependent upon hydrocarbons as raw materials for their products. Indeed, most industrially significant synthetic chemicals are derived from petroleum sources. The overall use of oil in the world now exceeds 11 million metric tons per day. An ever-increasing world population (presently exceeding 6 billion, projected to increase to 8–10 billion by the middle of the 21st century; Table 1.1) and energy consumption, compared with our finite non-renewable fossil fuel resources, which will be increasingly depleted, are clearly on a collision course. New solutions will be needed for the 21st century to sustain the standard of living to which the industrialized world become accustomed and to which the developing world is striving to achieve.

The rapidly growing world population, which stood at 1.6 billion at the beginning of the 20th century, has now exceeded 6 billion. With an increasingly technological society, the world's per capita resources have difficulty keeping up with demands. Satisfying our society's needs while safeguarding the environment and

Table 1.1 World population (in millions).

1650	1750	1800	1850	1900	1920	1952	2000	Projection 2050*
545	728	906	1171	1608	1813	2409	6200	8000 to 10000

* Medium estimate. Source: United Nations, Population Division.

Beyond Oil and Gas: The Methanol Economy. G. A. Olah, A. Goeppert, G. K. S. Prakash
Copyright © 2006 WILEY-VCH Verlag GmbH & Co. KGaA, Weinheim
ISBN 3-527-31275-7

allowing future generations to continue to enjoy planet Earth as a hospitable home is one of the major challenges that we face today. Man needs not only food, water, shelter, clothing and many other prerequisites, but also increasingly huge amounts of energy. Today, the world uses some 1.05×10^{18} calories per year (120 Petawatt-hours), equivalent to a continuous power consumption of about 13 terawatts (TW), comparable to the production of 13 000 nuclear power plants each of 1 GW output. With increasing world population, development and higher standards of living, this demand for energy is expected to grow to 21 TW in 2025 and to about 30 TW in 2050 (Fig. 1.1).

Our early ancestors discovered fire and started to burn wood. The industrial revolution was fueled by coal, and the 20th century added oil, natural gas and introduced atomic energy.

When fossil fuels such as coal, oil or natural gas (i.e., hydrocarbons) are burned in power plants to generate electricity or to heat our houses, propel our cars, airplanes, etc., they form carbon dioxide and water. They are thus used up, and are non-renewable on the human timescale.

Fossil Fuels

Petroleum Oil, Natural Gas, Tar-Sand, Shale Bitumen, Coals

They are mixtures of hydrocarbons (i.e., compounds of the elements carbon and hydrogen). When oxidized (combusted) they form carbon dioxide (CO_2) and water (H_2O) and thus are not renewable on the human scale.

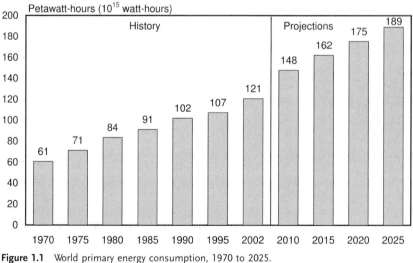

Figure 1.1 World primary energy consumption, 1970 to 2025.
Based on data from: Energy Information Administration (EIA),
International Energy Outlook 2005.

Nature has given us, in the form of oil and natural gas, a remarkable gift. What was created over the ages, however, mankind is consuming rather rapidly. Petroleum and natural gas are used on a large scale to generate energy, and also as raw materials for diverse man-made materials and products such as the plastics, pharmaceuticals and dyes that have been developed during the 20th century. The United States energy consumption is heavily based on fossil fuels, with atomic energy and other sources (hydro, geothermal, solar, wind, etc.) representing only a modest 15% (Table 1.2).

With regard to electricity generation, coal still represents more than half of the fuel used, with about 16% for natural gas and 20% for nuclear energy (Table 1.3).

Other industrial countries, in contrast, obtain between 20% and 85% of their electrical energy from non-fossil sources (Table 1.4).

Table 1.2   United States Energy Consumption (%).

Energy source	1960	1970	1980	1990	2000
Oil	44.1	43.5	43.6	39.8	38.5
Natural gas	27.5	32.1	26.0	22.9	23.7
Coal	21.8	18.1	19.6	22.8	22.7
Nuclear energy	0.002	0.35	3.5	7.3	8.1
Hydro-, Geothermal, Solar, Wind, etc. energy	6.6	6.0	7.2	7.4	6.9

Source: US Census Bureau, Statistical Abstract of the US 2002, Section 19, Energy and Utilities.

Table 1.3   Electricity Generation in the United States (%).

	1990	2000
Coal	52.6	51.8
Petroleum	4.1	2.9
Natural gas	12.5	15.7
Nuclear	19.1	19.9
Hydroelectric	9.7	7.2
Geothermal	0.5	0.4
Wood	1.0	1.0
Waste	0.4	0.6
Other waste	0.076	0.090
Wind	0.099	0.129
Solar	0.020	0.021

Source: US Census Bureau, Statistical Abstract of the US 2002, Section 19, Energy and Utilities.

Table 1.4 Electricity generated in Industrial countries by non-fossil fuels (%, 2001).

Country	Conventional thermal	Hydro-electric	Nuclear	Geothermal, Solar, Wind, Wood and Waste	Total non-fossil
France	8.4	14.1	76.8	0.7	91.6
Canada	28.1	57.8	12.8	1.3	71.9
Korea, South	58.2	1.5	40.1	0.2	41.8
Japan	58.9	8.5	31.0	1.6	41.1
Germany	62.5	3.7	29.7	4.1	37.5
United States	71.6	5.6	20.6	2.2	28.4
United Kingdom	73.6	1.1	23.7	1.6	26.4
Italy	78.7	18.0	0.0	3.4	21.3

Source: Energy Information Administration, International Energy Annual 2002, World Net Electricity Generation by Type, 2001.

Oil use has grown to the point where the world consumption is around 82 million barrels (1 barrel equals 42 gallons, i.e. some 160 L) a day, or 11 million metric tonnes. Fortunately, we still have significant worldwide reserves, including heavy oils, oil shale and tar-sands and even larger deposits of coal (a mixture of complex carbon compounds more deficient in hydrogen than oil and gas). Our more plentiful coal reserves may last for 200–300 years, but at a higher socio-economical and environmental cost. It is not suggested that our resources will run out in the near future, but it is clear that they will become even scarcer, much more expensive, and will not last for very long. With a world population exceeding 6 billion and still growing (as indicated earlier, it may reach 8–10 billion), the demand for oil and gas will only increase. It is also true that in the past, dire predictions of rapidly disappearing oil and gas reserves have always been incorrect (Table 1.5). As a matter of fact, until recently the reserves have been growing, but lately they have begun to level off.

The question is, however, what is meant by "rapid depletion" and what is the real extent of our reserves? Proven oil reserves, instead of being depleted, have in fact almost doubled during the past 30 years and now exceed 150 billion tonnes (more than one trillion barrels). This seems so impressive that many people assume that there is no real oil shortage in sight. However, increasing consumption (due also to increasing standards of living), coupled with a growing world population makes it more realistic to consider per-capita reserves. If we do this, it becomes evident that our known reserves will last for not much more than half a century. Even if all other factors are considered (new findings, savings, alternate sources, etc.), we will increasingly face a major problem. Oil and gas will not become ex-

Table 1.5 Proven Oil and Natural Gas Reserves (in billion
tonnes oil equivalent).

Year	Oil	Natural Gas
1960	43	15.3
1965	50	22.4
1970	77.7	33.3
1975	87.4	55
1980	90.6	69.8
1986	95.2	86.9
1987	121.2	91.4
1988	123.8	95.2
1989	136.8	96.2
1990	136.5	107.5
1993	139.6	127
2002	156.4	157.6
2003	156.7	158.2

Source for 1993–2003: BP Statistical review of world energy,
June 2004.

hausted overnight, but market forces of supply and demand will inevitably start to
drive the prices up to levels that nobody even wants to contemplate presently.
Therefore, if we don't find new solutions, we will inevitably face a real crisis.

Mankind wants all the advantages that an industrial society can give to all of its
citizens. We all essentially rely on energy, but the level of consumption varies
vastly in different parts of the world (industrialized versus developing countries).
At present, the annual oil consumption per capita in China is still only five to six
barrels, whereas it is more than ten-fold this level in the United States. China's oil
use is expected to at least double during the next decade, and this alone equals
roughly the United States consumption – reminding us of the size of the problem
that we will face. Not only the world population growth but also the increasing
energy demands of China, India and other developing countries is already putting
great pressure on the world's oil reserves, and this in turn contributes to price es-
calation.

Even though the generation of energy by burning non-renewable fossil fuels
(including oil, gas and coal) is feasible only for a relatively short period in the future, it is also generating serious environmental problems (*vide infra*). The advent
of atomic energy opened up a fundamental new possibility, but also created dangers and concerns of safety of radioactive byproducts. It is regrettable that these
considerations brought any further development of atomic energy almost to a
stand still, at least in most of the Western world. Whether we like it or not, it

is clear that we will have no alternative but to rely increasingly on nuclear energy, albeit making it safer and cleaner. Problems including those of the storage and disposal of radioactive waste products must be solved. Pointing out difficulties and hazards as well as regulating them, within reason, is necessary and solutions to overcome them is essential and certainly feasible.

If we continue to burn our hydrocarbon reserves to generate energy, and to use them at the present alarming rate, then diminishing resources and sharp price increases by the mid-21st century will lead eventually to a need to supplement these reserves, or generate them by synthetic manufacturing. Synthetic gasoline or oil products will, however, be costlier. Nature's petroleum oil and natural gas are the greatest gifts we will ever have. A barrel of oil still costs only around $50–75, within market fluctuations. It will be difficult for synthetic manufacturing process to come close to this price. We will need to get used to higher prices, not as a matter of any government policy, but as a fact of free market forces over which free societies have very little control.

Synthetic oil is feasible, its production having been proven feasible from coal or natural gas via synthesis-gas, a mixture of carbon monoxide and hydrogen obtained from the incomplete combustion of coal or natural gas (which are themselves non-renewable). Coal conversion was used in Germany during World War II and in South Africa during the boycott years of the Apartheid era. Nevertheless, the size of these operations hardly amounted to 0.3% of the present United States consumption. This route – the so-called Fischer–Tropsch synthesis – is also highly energy consuming, giving complex, unsatisfactory product mixtures, and can hardly be seen as the technology of the future. To utilize still-existing large natural gas reserves, their conversion to liquid fuels through syn-gas is presently developed for example on a large scale in Qatar, where major oil companies including ExxonMobil, Shell or ChevronTexaco, have recently committed over $20 billion to the construction of gas-to-liquid (GTL) facilities, mainly to produce sulfur-free diesel fuel. However, when completed, this will provide a daily total of some 100 000 t compared with present world use of transportation fuels in excess of 6 Mt per day. These figures demonstrate the enormity of the problem that we face. New and more efficient processes are clearly needed. Some of the required basic science and technology is evolving. As will be discussed (*vide infra*), still abundant natural gas can be, for example, directly converted, without first producing syn-gas, to gasoline or hydrocarbon products. Using our even larger coal resources to produce synthetic oil could extend its availability, but new approaches based on renewable resources are essential for the future. The development of biofuels, primarily by the fermentative conversion of agricultural products (derived from sugar cane, corn, etc.) to ethanol is evolving. Whereas ethanol can be used as a gasoline additive or even alternative fuel, the enormous amounts of transportation fuel needed clearly limits the applicability to specific countries and situations. Other plant-based oils are also being touted as renewable equivalents of diesel fuel, although their role in the total energy picture is minuscule.

When hydrocarbons are burned, as pointed out, they produce carbon dioxide (CO_2) and water (H_2O). It is a great challenge to reverse this process and to produce efficiently and economically hydrocarbon fuels from CO_2 and H_2O. Nature, in its process of photosynthesis, of course recycles CO_2 with water into new plant life. The natural formation of new fossil fuels form CO_2, however, takes a very long time, making them non-renewable on the human time scale.

The "Methanol Economy®" [1] – the subject of our book – elaborates an approach of how humankind can decrease and eventually liberate itself from its dependence on diminishing oil and natural gas (and even coal) reserves while mitigating global warming caused by their excess combustion giving carbon dioxide. The "Methanol Economy" is based in the interim on the efficient direct conversion of still-existing natural gas resources to methanol or dimethyl ether, or their production by chemical recycling of CO_2 from the exhaust gases of fossil fuel-burning power plants and other industrial sources. Eventually, atmospheric CO_2 itself can be recycled using catalytic or electrochemical methods. Methanol and dimethyl ether are both excellent transportation fuels on their own for internal combustion engines. Methanol is also an adequate fuel for fuel cells, being capable of producing electric energy by reaction with atmospheric oxygen (air). Fuel cells provide a convenient, efficient source for electric power. It should be emphasized that the "Methanol Economy" is not producing energy. In the form of liquid methanol, it only stores energy more conveniently and safely compared to extremely difficult to handle and highly volatile hydrogen gas, the basis of the "hydrogen economy". Besides being a most convenient energy storage material and a suitable transportation fuel, methanol can also be catalytically converted to ethylene and/or propylene, the building blocks of synthetic hydrocarbons and their products, which are currently obtained from our diminishing oil and gas resources.

The far-reaching implications of the new "Methanol Economy" approach clearly have great potential and societal benefit for mankind. As mentioned earlier, the world is presently consuming more than 80 million barrels of oil each day, and about two-thirds as much natural gas equivalent, both being derived from our declining and non-renewable natural sources. Oil and natural gas (as well as coal) were formed by Nature over the eons in scattered and frequently increasingly difficult-to-access locations (under desert areas, in the depths of the seas, the inhabitable reaches of the polar regions, etc.). In contrast, the recycling of CO_2 from industrial exhausts and eventually from air itself (which belongs to everybody) opens up an entirely new vista. The energy needs of humankind will, in the foreseeable future, come from any available source, including alternate sources of fossil fuels and atomic energy. As we still cannot store energy efficiently on a large scale, new ways are needed. The production of methanol offers a convenient means of energy storage. Even now, our existing power plants, during off-peak periods, could, by the electrolysis of water, generate the hydrogen needed to produce methanol. Other means of cleaving water by either biochemical (enzymatic) or photovoltaic (using energy from the Sun, our ultimate clean energy source) pathways are also evolving.

Initially, CO_2 will be recycled from industrial emissions (fossil fuel-burning power plants, cement factories, etc.) to produce methanol and to derive synthetic

Table 1.6 Composition of air.

Nitrogen	78%
Oxygen	20.90%
Argon	0.90%
Cabon dioxide	0.037%
Water	few % (variable)
Methane Nitrogen oxides Ozone	trace amounts

hydrocarbons and their products. The CO_2 content of these emissions is high and can be readily separated. In contrast, the average CO_2 content of air is very low (0.037%) (Table 1.6), and atmospheric CO_2 is therefore presently difficult to separate on an economic basis. However, these difficulties can be overcome by future developments using selective absorption or other separation technologies. Mankind's ability to chemically recycle CO_2 to useful fuels and products will eventually provide an inexhaustible renewable carbon source.

CO_2 can, as mentioned earlier, even now be readily recovered from flue gas emissions of power plants burning carbonaceous fuels (coal, oil, and natural gas), from fermentation processes, and from the calcination of limestone (cement production), production of steel, or other industrial sources. As these plants emit very large amounts of CO_2 they contribute to the so-called "greenhouse warming effect" of our planet, which is causing grave environmental concern. The relationship between the atmospheric CO_2 content and temperature was first studied scientifically by Arrhenius as early as 1898. The warming trend of our Earth can be evaluated only over longer time periods, but there is clearly a relationship between the CO_2 content in the atmosphere and Earth's temperature.

Recycling CO_2 into methanol (or dimethyl ether) and, through this into useful fuels and synthetic hydrocarbons and products, will not only help to alleviate the question of our diminishing fossil fuel resources, but at the same time help to mitigate global warming caused at least in part by man-made greenhouse gases.

One highly efficient method of producing electricity directly from fuels is achieved in fuel cells via their catalytic electrochemical oxidation, primarily that of hydrogen.

Fuel cell is an electrochemical device that converts free chemical energy of a fuel directly into electrical energy

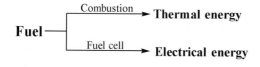

The principle of fuel cells was first recognized by Sir William Grove during the early 1800s, but their practical use was only recently developed. The basis of most fuel cell technologies is still based on Grove's principle – that is, hydrogen and oxygen (air) are combined in an electrochemical cell-like device, producing water and electricity.

The process is clean, giving only water as a byproduct, but hydrogen itself must be first produced in an energy-consuming process, using (at present) mainly fossil fuels. The handling of hydrogen gas itself is not only technically difficult, but also dangerous. Nonetheless, the use of fuel cells is gaining application in static installations or in specific cases, such as space vehicles. Currently, hydrogen gas is produced mainly from still-available hydrocarbon (fossil fuel) sources using reformers, which converts them to a mixture of hydrogen and carbon monoxide. These two gases are then separated. Although this process relies on our diminishing fossil fuel sources, electrolysis or other processes to cleave water can also provide hydrogen without any reliance on fossil fuels. Hydrogen-burning fuel cells, by necessity, are still limited in their applicability. In contrast, a new approach (discussed in Chapter 11) uses directly liquid methanol (or its derivatives) in a fuel cell without first converting it to hydrogen. The direct oxidation liquid-fed methanol fuel cell (DMFC) has been developed in a cooperative effort between the University of Southern California and Caltech-Jet Propulsion Laboratory of NASA (who for a long time worked on fuel cells for the US space programs). In such a fuel cell, methanol reacts with oxygen of the air over a suitable metal catalyst, producing electricity while forming CO_2 and H_2O:

$$CH_3OH \; + \; 1.5O_2 \longrightarrow CO_2 \; + 2H_2O \; + \text{electrical energy}$$

More recently, it was found that the process could be reversed. Methanol (and related oxygenates) can be made from CO_2 via aqueous electrocatalytic reduction without prior electrolysis of water to produce hydrogen in what is termed a "regenerative fuel cell". This process can convert CO_2 and H_2O electrocatalytically into oxygenated fuels (i.e., formic acid, formaldehyde and methyl alcohol), depending on the electrode potential used in the fuel cell in its reverse operation.

The reductive conversion of CO_2 to methanol can also be carried out by catalytic hydrogenation using hydrogen. Hydrogen can be obtained by electrolysis of water (using all kinds of energy sources such as atomic, solar, wind, geothermal, etc.) or other means of cleavage (photolytic, enzymatic, etc.). Methanol is a convenient medium to store energy, and is an excellent transportation fuel. It is a liquid (with a boiling point of 64.6 °C) that can be easily transported using the existing infrastructure. Methanol can also be readily converted to dimethyl ether. Dimethylether has a relatively higher calorific value and is an excellent diesel substitute.

Methanol produced directly from methane (natural gas) without going to syngas or by recycling of CO_2 into CH_3OH or dimethyl ether can subsequently also be used to produce ethylene as well as propylene. These are the building blocks in the petrochemical industry for the ready preparation of synthetic aliphatic and aromatic hydrocarbons, and for the wide variety of derived products and materials, obtained presently from oil and gas, on which we rely so much in our everyday life.

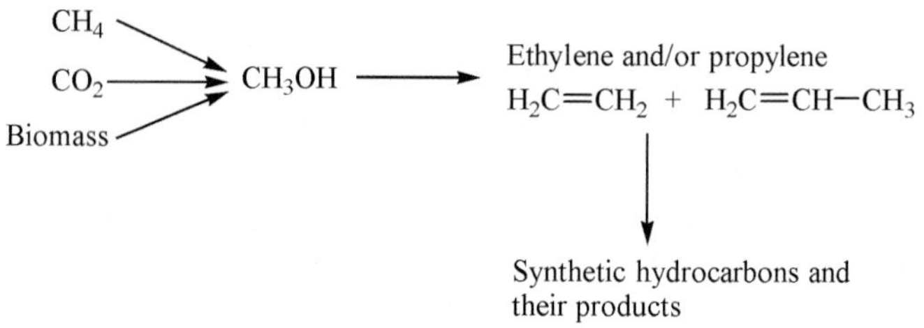

CH_4

$CO_2 \longrightarrow CH_3OH \longrightarrow$ Ethylene and/or propylene

$H_2C{=}CH_2 \; + \; H_2C{=}CH{-}CH_3$

Biomass

Synthetic hydrocarbons and their products

Chapter 2
Coal in the Industrial Revolution, and Beyond

Coal was formed during the carboniferous period – roughly 360 to 290 million years ago – from the anaerobic decomposition of then-living plants. These plants ended up as coal because, upon their death, they failed to decompose in the usual way, by the action of oxygen to form eventually CO_2 and water. As the carboniferous plants died, they often fell into oxygen-poor swamps or mud, or were covered by sediments. Because of the lack of oxygen they only partly decayed. The resulting spongy mass of carbon-rich material first became peat. Then, by action of the heat and pressure of geological forces, peat eventually hardened into coal.

During this process, the plant's carbon content was trapped in coal together with the Sun's energy used in the photosynthesis of plants and accumulated over millions of years. This energy source was buried until modern man dug it up and made use of it. It is only very recently on the Earth's time scale that mankind has started to use coal. Historically, the use of coal began when the Romans invaded Britain. While it was used occasionally for heating purposes, the main use of this "black stone" was to make jewelry, since it could be easily carved and polished. It was only during the late 12th century that coal re-emerged as a fuel along the Tyne river in Britain, especially around the rich coal fields of Newcastle. The widespread use of coal however, would not be significant before the middle of the 16th century. At that time, England's population – and that of London especially – was growing rapidly. And, as the city was growing, the nearby land was deforested such that the wood had to be hauled from increasingly distant locations. Wood was used not only for home heating and cooking purposes but also in most industries, such as breweries, iron smelters and in ship building. As the shortage of wood became increasingly pronounced, its price increased such that the poorest of the population were increasingly unable to afford it. These were particularly hard times because Europe had just entered into a so-called "little ice age" which would last until the 18th century. However, a severe energy crisis never materialized thanks to coal, which became increasingly the country's main source of fuel by the beginning of the 17th century. This was not without problems; coal's thick smoke upon burning, made London's air one of the poorest in all of Europe. On some days, the Sun was hardly able to penetrate the coal smoke, and travelers could smell the city miles before they actually saw it.

Beyond Oil and Gas: The Methanol Economy. G. A. Olah, A. Goeppert, G. K. S. Prakash
Copyright © 2006 WILEY-VCH Verlag GmbH & Co. KGaA, Weinheim
ISBN 3-527-31275-7

What really brought about the power of coal as an energy source came along in the early 18th century with the invention of the steam engine. The steam engine was at the heart of the resulting industrial revolution, and it was fueled by coal. At the time, one of the main problems facing coal mining was water seepage and flooding from various sources. Rainwater seeping down from the surface accumulated in the tunnels, and once the mines reached below the water table the surrounding groundwater also contributed to the problem. Consequently, the

A—AXLES.   B—WHEEL WHICH IS TURNED BY TREADING.   C—TOOTHED WHEEL.
D—DRUM MADE OF RUNDLES.   E—DRUM TO WHICH ARE FIXED IRON CLAMPS.
F—SECOND WHEEL. G—BALLS.

Figure 2.1 Water removal in mines during the middle ages.
From an engraving by Georgius Agricola-De Re Metallica: book
6 ill. 36, 1556.

mines became slowly submerged in water. If the mine was located on a hill, simple draining shafts could be used, but as the mines were pushed deeper into Earth, the water had to be removed by other means. The earliest method relayed on miners hauling the water up in buckets strapped to their backs. This was not really convenient, so various ways were designed to increase the effectiveness of the human labor. Among these were chains of buckets or primitive forms of pumps powered not only by human muscle but also in some cases by windmills, waterwheels (Fig. 2.1), or horse power. However, none of these was very economical or suitable.

One of the most pressing challenges for contemporary England was to find a way to keep its coal mines dry. This led eventually to the introduction of a device invented by Thomas Newcomen, who was not a scholar but a very inventive small-town ironmonger. His device consisted of a piston which moved up by steam generated by heating water with burning coal, and down by reduced pressure resulting from the condensation of steam with cold water. The piston was connected to the rod of a pump used to pump water.

In 1712, one of these Newcomen engines was first used in a coal mine and became an almost immediate hit among mine operators, largely because it was much cheaper to operate than horses and could pump water from a much greater depth than ever before. The drawback was that the engine needed large amounts of coal to generate the steam necessary to keep it going, and therefore found little use outside of the coal mines.

At about this time, James Watt, a carpenter's son from Scotland, improved Newcomen's steam engine dramatically. Watt realized that as steam was injected and then cooled with water, heat was wasted in the constant reheating and cooling of the cylinder. The installation of a separate condenser immersed in cold water connected to the cylinder kept it hot and avoided unnecessary heat losses (Fig. 2.2). This improved the efficiency of the steam engine by at least a factor of four, and allowed it to move out from the coal mines and find its place in factories.

To really move the industrial revolution ahead, however, another technologic advance was needed: the manufacture of iron using coal-based coke. Until that time, the iron needed to build engines and factories was essentially made using charcoal obtained by burning huge amounts of wood, which was increasingly becoming scarces in Britain. Charcoal provided both the heat and the carbon needed for the reduction of the iron ore. The use of coal to smelt iron was hindered by the impurities it contained, which made it unsuitable. After more than a century of experimentation, however, the key to making iron using coal was found. In the same way that wood was turned into charcoal, coal had first to be baked to drive off the volatiles and form coke. By the 1770s, the technology had advanced to the point where coke could be used in all stages of iron production. With this breakthrough, Britain – rather than being dependent upon iron imports – became, in just a few years, the most efficient iron producer in the world. This allowed it to build its powerful industries at home and its vast empire abroad.

The "coal economy" resulted in a concentration of the ever-larger and mechanized factories, as well as their workforces, into urban areas, making them more

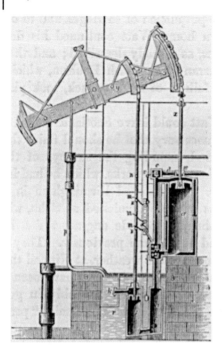

Figure 2.2 Watt's engine, 1774.

Figure 2.3 Stephenson's locomotive, *The Rocket*, 1829.

efficient. The epicenter of this industrial revolution was Manchester, which became the premier center of manufacture in England. The city also became home of the first steam locomotive-driven public railway, the Liverpool and Manchester railway, which opened in 1830 (Fig. 2.3). The "father of the railways" was George Stephenson, who first envisioned moving large quantities of coal over land. It was through the steam locomotive that this became possible, although this invention would in time have revolutionary consequences far beyond the coal industry.

The Liverpool and Manchester Railway became a huge success, transporting hundreds of thousands of passengers during the first months of operation.

This established a bright future for railway as a transportation system and triggered massive investment in this industry. Although other European nations followed its example, Britain had a good 50 years head start in industrialization, and maintained its lead for most of the 19th century. In 1830, Britain produced 80% of the world's coal and, in 1848, more than half of the iron of the world, making the nation the most powerful on Earth until the end of the 19th century. Across the Atlantic, however, the United States – having even more coal and other resources than England – also began to undergo an even faster industrial transformation.

Historically, coal has probably been the most important fossil fuel as it triggered the industrial revolution that led to our present-day modern industrial society. During the 20th century, coal has been supplemented and displaced progressively by oil and natural gas, as well as nuclear power, for electricity generation. Coal was increasingly considered as a "dirty fuel" of the past, and was deemed to have a limited future. Only with the energy crisis in the 1970s and the growing concerns about the safety of nuclear energy, did coal again become an attractive energy source especially for electricity production. Because the reserves of coal are geographically widespread and coal is a heavy and bulky solid which is costly to transport, it is mainly utilized close to its source. The economically recoverable proven coal reserves are enormous, and estimated as being close to one trillion tonnes [2, 3] – enough at current rates of consumption to supply our needs for more than 170 years. This Reserve over Production (R/P) ratio is about three times as large as the one for natural gas, and more than four times as high as that for oil. Unlike oil and natural gas, our coal resources should last at least for the next two centuries. The total coal resources are estimated to be more than 6.2 trillion tonnes [4]. The main reason why the R/P ratio for coal is not even higher is the limited incentive to find new exploitable reserves, given the size of already-known reserves. Production has increased ten-fold over the past 100 years, without any significant increase in coal price. In contrast, the implementation of advanced mining technologies has improved, and will continue to improve productivity and steadily lower the cost of coal extraction and treatment. Higher efficiency in coal transportation, which can represent as much as 50% of the import cost into Europe or Japan, are also improving [5]. Furthermore large reserves, coupled with competition between coal-producing countries, makes a sustained price increase unlikely and should result in relatively flat coal prices in the foreseeable future. In the case of coal, neither the abundant resources nor the competitive prices are determining factors in the fuel's future. The rate of extraction of coal is presently only a function of its relative limited demand. In industrialized countries, where coal is used mainly to generate electricity, the demand will be governed by the ability of coal to compete with natural gas, not only from an economic point of view but also increasingly from environmental considerations. One of the reasons why we no longer rely heavily on coal is that, from an environmental aspect, it is the most polluting fossil fuels in comparison to oil and gas. It usually emits significant levels of pollutants, especially sulfur dioxide, nitrogen oxides, and particulates. Heavy metals such as mercury, lead, arsenic or even uranium are difficult to remove from coal, and are generally released into the air upon combustion. In a continuous effort to diminish the environmental impact of coal burning, the development and progressive introduction of new separation technologies applied to existing or new power plants can greatly reduce or nearly eliminate the emissions of SO_2, NO_x and particulates. Emission regulations for mercury and other impurities present in coal are under evaluation in several nations, including the United States. However, at the present time there are no CO_2 emission capture technologies operating in a large-scale power plants. Given the growing concerns about global warming, coal-burning power plants face a major challenge. Compared to oil and gas, coal is the fuel that produces the most CO_2 per unit of energy released. To tackle this problem, so-called "clean coal technologies" are being developed to improve the thermal efficiency and reduce emissions and, consequently, the environmental impact of coal-fired power plants [4,6]. Among these technologies, some are already commercially available.

In the Atmospheric Fluidized Bed Combustion (AFBC process), coal is burned in a fluidized bed at atmospheric pressure and the heat recovered to power steam turbines. An improved version of that system, Pressurized Fluidized Bed Combustion (PFBC), in which gas produced by the combustion of coal is used to drive directly a gas turbine, is currently under development. Supercritical and ultra-Supercritical power plants operate under supercritical conditions at steam pressure above 22.1 MPa and 566 °C, where there is no longer any distinction between the gas and liquid phases of water as they form a homogeneous fluid. Such plants are well established and operate routinely at pressures up to 30 MPa with efficiencies above 45%. The introduction of special metal alloys that are more resistant to corrosion (but also are more expensive) to increase the operation pressure to 35 MPa and the thermal efficiency of power generation to over 50% are under development.

Integrated Coal Gasification Combined Cycle (IGCC) is an other emerging technology that has been demonstrated on a commercial scale, but not yet widely deployed. In this case, the coal is first gasified to produce syn-gas, which is then combusted under high pressure in a gas turbine to generate electricity. The hot exhaust gas from this turbine is used to generate steam that can produce additional electricity using a steam turbine. The goal, for the United States using this technology, is to reach a 52% efficiency by 2010. Nevertheless, even with this level of efficiency, an IGCC plant would still produce twice as much CO_2 per kWh generated than a combined cycle natural gas turbine [5], the current favored option for power generation. However, the process has also a longer-term strategic importance because it is the first electric power-generating technology to rely on gasified coal. It could therefore act as a bridge to more advanced coal gas-based power plants such as the ambitious "FutureGen", a $1 billion dollar project initiated by the U.S. Department of Energy to develop an economically viable plant which would have zero emissions [7,8]. Beside electricity, this plant would also produce hydrogen from the syn-gas generated during coal gasification, whereas CO_2 would be captured and sequestered underground (see also Chapter 12, the use of CO_2 and H_2 in producing methanol).

From an energy perspective, coal has a major advantage as its resources are still vast and are widely distributed around the world. Furthermore, the outlook for coal supply and prices are subject to less fluctuation than are those for oil and gas. However, coal could be penalized for its high carbon content, and the key uncertainty affecting the future of coal is the impact of environmental policies. Longer-term prospects for coal may therefore depend on the development and introduction of clean coal technologies that would reduce or even eliminate carbon emissions. In any case, coal resources will not last for more than two or three centuries – longer than oil and gas, but still a short period on the time scale of humanity.

Chapter 3
History of Oil and Natural Gas

Oil and gas, in similar manner to coal, are generally the result of the degradation of organic materials, primarily plankton, which settled on the seafloor millions of years ago. This process occurred in the so-called "source rock", where the biomass was trapped along with other sediments. Depending on the depth at which this source rock was buried during its existence, the biomass will form either oil or gas. If this source rock was buried for a sufficient time between 2500 and 5000 m depth, where the temperature is around 80 °C, the hydrocarbon chains will break to form mainly oil. At depths exceeding 5000 m, however, usually no oil will be found, and at temperatures around 145 °C at that depth, over the geologic time, all carbon–carbon bonds will break to form the dominant component of natural gas: methane. The formations from which oil and gas are extracted are generally different from the source rock in which they were originally formed, however. In fact hydrocarbons, once liberated from the source rock, can migrate upward to form shallow oil or gas fields called "reservoirs", or even appear as surface oil seeps, for example as in the Los Angeles Basin at the La Brea Tar Pits in Southern California.

Natural seepage of oil has been used since ancient times in locations in the Middle East and the Americas for a variety of medicinal, lighting, and other purposes. Petroleum was referred to as early as the Old Testament. The word petroleum means "rock oil" from the Greek *petros* (rock) and *elaion* (oil). Uses of petroleum oil, however, were very limited and it was not before the mid-19th century that wide use and the real potential of these natural resources began to evolve.

America's first commercial oil well was drilled in 1859 by Colonel (titular) Edwin Drake near Titusville in the State of Pennsylvania (Fig. 3.1), yielding about 10 barrels of oil per day [9]. Drake's single well soon surpassed the entire production of Romania, which was at that time a major source of oil for Europe [10]. The area was known previously to contain petroleum, which seeped from the ground and was skimmed from a local creek's surface called therefore "Oil Creek". Drake was a former railway conductor who, because of ill health, was urged by his doctor to move from the east coast to a more rural area. The "Colonel" label was given to him not as a result of any military service but by the Seneca Oil Company, which hired him and believed that such a title would help Drake to get the assistance of the local people. The efforts to find oil grew

Beyond Oil and Gas: The Methanol Economy. G. A. Olah, A. Goeppert, G. K. S. Prakash
Copyright © 2006 WILEY-VCH Verlag GmbH & Co. KGaA, Weinheim
ISBN 3-527-31275-7

out of technology evolution and the need for lubrication and illumination products. Without evolving markets and processes for such products, Drake would have never been sent to the Pennsylvania hinterland to prospect and develop his oil operation.

In the mid-19th century, the need for illumination in cities such as Boston or New York led to the development of gas lighting. This was not done using natural gas, but by employing an illuminating gas that was produced by heating coal at gasworks located at the edge of the towns. Where gas was not available, and before the discovery of oil, whale oil more than any other product was used to fulfill the demand for clean and efficient illumination. Operating out of the North-Eastern United States, and later Hawaii and the North-Western States, the American whaling fleet mainly searched for sperm whales which contained large amounts of high-quality oil. However, with a steadily declining whale population, coupled with supply problems during the civil war, and ever-higher prices, the need for less expensive or more easily available alternatives was growing. The advent of petroleum oil signaled the end of the use of whale oil, these lamp oils were replaced by the more convenient and easily obtained, and seemingly inexhaustible, kerosene.

Kerosene, which was derived initially from bituminous coal and also known as "coal oil", proved to be a competitive illuminant during the 1850s, although its

Figure 3.1 Edwin Drake (right) in front of his well in Titusville, Pennsylvania, 1866. (Source: Pennsylvania Historical & Museum Commission, Drake Well Museum, Titusville, Pennsylvania.)

foul odor kept the rate of development relatively slow. The name kerosene was quickly extended to all illuminating oils made from minerals, however. For the production of kerosene, petroleum oil could be used instead of coal in the same distillation process, and the product was distributed over the same existing network, thereby facilitating its development as a fuel. On a positive point, these fossil fuel light sources put an end to sperm whaling for oil, just as the use of coal had helped to save the remaining forests.

Kerosene was the first petroleum product to find a wide market, allowing John D. Rockefeller Sr., a Cleveland (Ohio) entrepreneur, to build his Standard Oil Company into a vast industrial empire that would, during the subsequent rise in the use of gasoline and the internal combustion engine, enjoy a virtual monopoly over the production and distribution of oil in the United States. In 1911 however, Standard Oil, due to antitrust regulations, was broken up into a number of separate companies, including those which became Chevron, Amoco, Conoco, Sohio and, of course, Mobil and Exxon. The later two giants recently merged again without causing much public concern. "Tempora mutantur" – such is the effect of changing times.

The discovery and uses for petroleum has paralleled and, to a large extent been responsible for, the growth in oil production. Inspired by the invention of the gasoline-burning engine by Nikolaus Otto in 1876, the combination of Gottlieb Daimler's engine, Carl Benz's electrical ignition and Wilhelm Maybach's float-feed carburetor resulted in the 1890s in the first successful commercially produced internal combustion automobile (Fig. 3.2). Henry Ford's mass production methods soon made it widely available and changed mankind's life in the 20th century (Fig. 3.3).

The use of oil began to increase dramatically to produce the large quantities of gasoline needed to fuel automobiles. However, the amount of gasoline that could be obtained from crude oil was low, anywhere from 10 to 20%. The original production process was based on simple distillation (fractionation), separating hydrocarbons through differences in their boiling points. Later, due mainly to growing demand, the refining of crude oil to yield a range of liquid fuels suited to a variety

Figure 3.2 Daimler and Maybach in their first 4-wheel automobile.

Figure 3.3 The Ford's first assembly line.

of specific applications that ranged (eventually) from massive diesel locomotives to supersonic airplanes, was transformed by the introduction of cracking and other refining processes. Thermal cracking in combination with high pressure was introduced in 1913. The high temperature and pressure reproduced, on a short time scale, the naturally occurring process in breaking larger molecules into smaller ones. This process was further improved by the introduction of catalytic cracking in 1936. World War II saw also the introduction of high-octane gasoline produced by alkylation and isomerization. Without these processes it would be impossible to produce, inexpensively, the required large amounts of more valuable lighter fractions from the intermediate and heavy, higher molecular weight compounds of the crude oils. Furthermore, with these processes, the route to petrochemicals was opened up, since cracking provides the ability to produce unsaturated hydrocarbons – molecules which, in contrast to saturated hydrocarbons (paraffins), the main components of oil, can be readily used and further transformed in chemical reactions to yield products such as lubricants, detergents, solvents, waxes, pharmaceuticals, insecticides, herbicides, synthetic fibers for clothing, plastics, fertilizers, and much more. Today, our daily lives would be unthinkable without all these products.

Since the first "black gold" rush initiated by Drake in Pennsylvania, the search for new oil fields worldwide has never stopped. The most intensive exploration occurred in the United States, where vast quantities of oil were found in a number of States including Oklahoma, California, Texas and, more recently, Alaska. In Eurasia, the earliest oil exploitations occurred around the Black Sea and on the Caspian Sea near Baku in the then Russian Empire (now Azerbaijan), where the Swedish Nobel family had for some years held the concession for oil production. More recently, in the harsh regions of Siberia, vast reserves of oil were found and developed. In Sumatra, Java and Borneo, petroleum resources have been exploited since the second half of the 19 century.

In the Middle East, exploration for oil began during the 1910s, and this led to the discoveries of the world's largest oilfields; al-Burgan in Kuwait, the world's second largest oilfield in 1938 and Saudi's al-Ghawar, the world's largest super-giant oilfield containing almost 7% of the world's oil reserves (according to estimates in 2000), a decade later. Other major oil production areas include South America, South-East Asia, Africa and, more recently, the North Sea.

Oil Extraction and Exploration

Oil extraction begins with the drilling of a well. Edwin Drake's first oil well was only around 20 m deep. With rotary drilling techniques, used for the first time at the Spindeltop well in Beaumont, Texas in 1901, oil wells surpassed 3000 m in the 1930s, and drilling production wells deeper than 5000 m are now possible and used in several hydrocarbon reservoirs. A relatively recent innovation in drilling technology has been the routine use of directional and horizontal drilling. For the same reservoir, horizontal wells can produce several times as much oil as a traditional vertical wells. The longest horizontal wells are now around 4000 m in length.

In the early days of oil exploration, drilling took place exclusively on land, but moved to off-shore locations as the land deposits became less abundant and the necessary technologies were developed. The first off-shore well was drilled at Summerland, south east of Santa Barbara, California in 1897, and the first deep-water off-shore oil well along the Gulf coast of Louisiana in 1947. Today, some of these off-shore platforms are working in waters 2000 m deep or more.

Since oil reservoirs are unevenly distributed and often far away from major consumption centers, the crude oil must be transported over long distances, sometimes thousands of miles. For the long-distance transport of oil products on land, pipelines and railway tank cars are used. Pipelines are expensive to build and maintain, and breaks along the line can cause severe oil spills. However, they are the most energy-efficient means of transporting oil overland.

When oil must be transported overseas, for example from the Persian Gulf to North America or Europe, it is carried in specialized oil tankers. The need to carry ever-increasing quantities of oil has resulted, since the 1970s, in the construction of so-called "supertankers"; these are the largest ships afloat in the world, and larger even than aircraft carriers (Fig. 3.4). However, if a tanker is damaged in an accident it can cause severe environmental problems. Some of the oil spills resulting from such accidents have become very famous (or "infamous"). For example, the *Amoco Cadiz*, which in 1978 broke up off the coast of France spilling 1.6 million barrels of crude oil damaged not only the ecosystem but also the lucrative French tourist industry. In 1989, the *Exxon Valdez* spilled almost 270 000 barrels off the coast of Alaska. Despite the increasing oil quantities transported overseas, the amount of oil spilled has decreased over the years, thanks to the development of new technologies and infrastructures such as double-hulled tankers or deep-water "superports".

Figure 3.4   An oil supertanker.

The 20th century saw a tremendous increase in the use of oil for a variety of purposes. With the development of the automobile and diesel locomotives, oil replaced coal as the main fuel for transportation. Oil also replaced coal as the primary home heating fuel because of its greater convenience. By the mid-20th century oil had become the world's primary energy source, but after the two oil crises of the 1970s and concerns about the reliability of our oil sources, another fossil fuel was gaining wider use, namely natural gas.

Natural Gas

Natural gas used to be regarded as an undesirable byproduct of petroleum production, and was simply burned, or "flared" at the oil wells. This is still done in some parts of the Middle East or some off-shore platforms around the world, where no nearby markets exist for natural gas and transportation to more distant markets is not yet economically feasible. Most commercially sold natural gas comes from wells, which are used solely for natural gas production.

Before the nature of natural gas was understood, it posed a mystery to mankind. Sometimes, lightning strikes would ignite natural gas that was escaping from the Earth's crust; this would create a fire coming from the Earth, burning the natural gas as it seeped out from underground. These fires puzzled most early civilizations, and were the root of much myth and superstition. One of the most famous of these flames was found in ancient Greece, on Mount Parnassus, approximately around 1000 BC. According to the legend, a goat herdsman came across what looked like a "burning spring", a flame rising from a fissure in the rock. The Greeks, believing it to be of divine origin, built a temple over the flame. This temple housed a priestess, who was known as the Oracle of Delphi, giving out prophecies she claimed were inspired by the flame.

Such flames became prominent in the religions of India, Greece, and Persia. Since the origin of these fires could not be explained, they were often regarded as divine, or supernatural. It was not until about 500 BC that the Chinese discovered the potential to use such fires to their advantage. After first finding places where gas was seeping to the surface, the Chinese built crude pipelines from bamboo stems to transport the gas, which was then used to boil sea water, separating the salt and making the water potable.

In America, naturally occurring gas was identified as early as 1626, when French explorers discovered natives igniting gases that were seeping from the ground around Lake Erie. Indeed, the beginnings of the American natural gas industry arose in this area. Actually, the very same first well dug by Colonel Drake in 1859 was producing not only oil but also natural gas. At the time, a 5-cm (2-inch) diameter pipeline was built, running some 9 km from the well to the village of Titusville, Pennsylvania. The construction of this pipeline proved that natural gas could be brought safely and relatively easily from its underground source to be used for practical purposes. For this reason, most consider this well as the beginning not only of the oil industry but also of the natural gas industry in America.

In 1821, the first well specifically intended to obtain natural gas was dug in Fredonia, New York, by William Hart. After noticing gas bubbles rising to the surface of a creek, Hart dug a well about 10 m deep to obtain a larger flow of gas to the surface. Hart is regarded by many as the "father of natural gas" in America. Expanding on Hart's work, the Fredonia Gas Light Company was eventually formed, becoming the first American natural gas company.

In 1885, Robert Bunsen invented what became known as the Bunsen burner (Fig. 3.5). He created a device that mixed natural gas with air in the correct proportions, creating a blue flame that could be safely used for cooking and heating. The invention of the Bunsen burner opened up new opportunities for the use of natural gas. The invention of temperature-regulating thermostatic devices made it possible to better use the heating potential of natural gas, allowing the temperature of the flame to be adjusted and monitored.

Without any means of transporting it effectively, natural gas discovered prior to World War II was usually just allowed to vent into the atmosphere, or burned when found together with coal and oil, or simply left in the ground. Transportation by pipelines developed only gradually. One of the first natural gas pipelines of considerable length was constructed in 1891; this was 200 km long, and carried natural gas from wells in central Indiana to the booming metropolis of Chicago. This early pipeline was very rudimentary and used no artificial compression, relying completely on the natural underground pressure. As might be imagined, the pipeline was not very efficient in transporting natural gas. It was not until the 1920s that any significant effort was put into building a suitable pipeline infrastructure. However, it was after World War II that newly developed welding techniques, pipe rolling, and metallurgical advances permitted the construction of reliable pipelines. The post-war pipeline construction boom lasted well into the 1960s, and allowed many thousands of kilometers of pipeline to be constructed in America and worldwide.

Once its transportation was made feasible and safe, many uses for natural gas were developed. These included its use to heat homes and operate appliances such as water heaters, cooking oven ranges, and clothes dryers. This was made

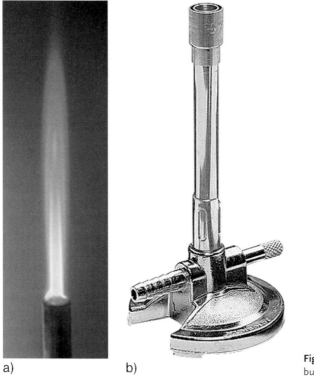

a)   b)

Figure 3.5 The Bunsen burner.

relatively easy because customers were already comfortable with the town gas which was generated from coal and had been widely used for close to a century.

Industry also began to use natural gas extensively in manufacturing and processing plants. Natural gas was also increasingly used to generate electricity. Its transportation infrastructure had made natural gas easy to obtain, and it became increasingly a most popular and efficient form of energy.

Moving large amounts of natural gas by pipeline is relatively easy, but the process is not adaptable to transport natural gas from remote locations far from consumer markets such as Saudi Arabia to North America or Europe. Because of the gaseous state of natural gas, it occupies enormous volumes compared to liquid or solid fuels. For intercontinental transport, across oceans, natural gas is usually liquefied to yield liquefied natural gas (LNG). This process is, however, energy-intensive, and requires specially designed and highly expensive tankers that keep natural gas at or below its boiling point (−162 °C) in highly insulated, double-walled tanks. LNG can also be dangerous. In 1944, the explosion of an LNG storage plant in Cleveland, Ohio, killed 128 persons and injured several hundreds. More recently, in 2003, 27 people were killed in the explosion of a natural gas liquefaction plant in Algeria [11]. A study by Sandia National Laboratory found that an explosion from a LNG tanker leak could result in major injuries and significant damage up to 500 m away from the leak, while people up to 2 km away would suffer second-degree burns [12]. Besides accidents, LNG facilities and tankers are also potential targets for terrorists. For these reasons, LNG terminals are usually not welcomed near major cities or population centers and are increasingly located off-shore.

Today, natural gas is extensively used because of its clean-burning properties and convenience, for space heating and generation of electricity to replace older and more polluting coal-fired power-plants. From an environmental point of view, natural gas is also advantageous because it produces the least amount of CO_2 greenhouse gas per unit of energy generated.

As the 19th century had been the golden age of coal, the 20th century saw the rise of petroleum oil and natural gas as the new "kings" of fuels. Advances in the petrochemical industry, which provide us with many of the necessities of modern life, may be credited, more than anything else for the high standard of living that we enjoy today. As we advance into the 21st century, the use of oil and natural gas will continue. The unquestionable fact, however, is that these non-renewable resources are limited. With their increasing use and the growing population of the Earth, they will become increasingly scarce and expensive. We have no choice but to develop new sources and technologies in order eventually to replace fossil fuels. The time to do this is now, when we still have extensive sources of fossil fuels available to make the inevitable changes gradually, without major disruptions or crises.

Chapter 4
Fossil Fuel Resources and Uses

Today's enormous energy demands are mainly fulfilled by the use of fossil fuels. In 2000, they contributed to 85% of the energy consumption in the United States, and to 86% worldwide. The fossil fuels used are all different forms of hydrocarbons: coal, oil, and natural gas, all differing in their hydrogen to carbon ratio. To liberate their energy content and to power our electric plants, heat our houses, and to propel our cars and airplanes, the fuels must be burned, forming CO_2 and water. Consequently, they are non-renewable. This means that on the human time-scale they cannot be naturally regenerated and are only available in a finite amount on Earth. The recurring question is: how much of these reserves are still available? Most estimates put our overall worldwide fossil fuel reserves as lasting not more than 200 to 300 years, of which oil and gas would last for less than a century. These forecasts however, are based on our present state of knowledge. The dynamic nature of the discovery and consumption rates of fossil fuels makes it difficult to provide any definite estimate. In fact, estimates have changed over the years, as new large oil deposits have been found, although the probability of finding new oil deposits is diminishing. At the same time, consumption and future needs are growing significantly. As our readily recoverable resources are finite, it is certain that we will face a real and unavoidable major problem.

In order to better understand how the available amount of supply of a given energy is determined, the notions of reserve and resource must be defined. In geological terms, and applied to fossil fuels, the *reserve* is the amount of material that can be recovered economically with known technology. A *resource* is the entire amount of the material known or estimated to exist, regardless of the cost or technological development needed to extract it. The amounts of energy contained in these categories change over time. Continued exploration and the discovery of new energy fields or advances in technology could transfer an energy source from the resource to the reserve category. The criteria of profitability can also change as market conditions change. This is especially well demonstrated in the case of oil price increases. Much oil that was formerly considered as "unrecoverable" because of high extraction costs can be profitably recovered when the price goes up and remains above a certain level and thus becomes categorized as "reserve".

Beyond Oil and Gas: The Methanol Economy. G. A. Olah, A. Goeppert, G. K. S. Prakash
Copyright © 2006 WILEY-VCH Verlag GmbH & Co. KGaA, Weinheim
ISBN 3-527-31275-7

Based on the foregoing, resource exhaustion is rather a question of unacceptable costs than necessarily actual physical depletion. It is also important to note that the reported reserves may vary greatly depending on the country and the prevailing political and economical conditions. The best example that can be given are the huge and abrupt reported increases in oil reserves in the late 1980s for several OPEC nations [13]. This occurred almost overnight, and was not the result of any major oil discovery at that particular time. Until then, OPEC assigned a share of the oil market based on each country's annual production capacity. However, during the 1980s OPEC changed the rules to also take into account the oil reserves of each country, leading most of them to promptly increase their reserve estimate in order to obtain a larger share of the market. On the other hand, the former Soviet Union used to publish wildly optimistic estimates of its reserves, probably to impress or mislead its cold war adversaries. Major oil companies also tended to exaggerate their reserves. Much re-evaluation is consequently taking place on a regular basis.

Coal and gas – due to their different availability and utility – have different markets, and their reserves also vary accordingly.

Coal

Historically, coal was the first fossil fuel to be used on a massive scale, and still represents today 23% of our primary energy source worldwide [14] (Fig. 4.1). In most industrialized countries coal has been replaced by oil, gas or electricity in household uses. Electricity generation is the one area where coal still plays a major role as an energy source in the industrialized world. It is the premier fuel, its share having remained almost unchanged for the past four decades. Worldwide, almost 40% of electric power is generated using more than 60% of

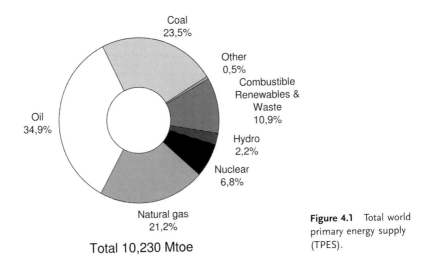

Coal
23,5%

Other
0,5%

Combustible
Renewables &
Waste
10,9%

Hydro
2,2%

Nuclear
6,8%

Oil
34,9%

Natural gas
21,2%

Total 10,230 Mtoe

Figure 4.1 Total world primary energy supply (TPES).

the global coal production. Many countries are heavily dependent on coal for electricity production, including Poland (95%), South Africa (93%), Australia (77%), India (78%), and China (76%) [15]. In the coal-rich United States, 92% of the domestic coal production is used to generate 51% of the country's electricity needs in giant coal-burning power-plants [16]. The heat created by coal combustion is used to vaporize water which, under high pressure and temperature, drives turbines connected to generators, the modern steam engines. Beside steam generation for electricity, coal is used mainly for industrial applications especially in steel and cement manufacturing.

In 2004, the world coal production (hard coal and brown coal) amounted to 5.5 billion tonnes, China being the main producer with almost 2 billion tons, representing about 35% of global production, followed at some distance by United States, but far ahead of India, Australia, and others [17]. The largest future increase in coal production will come from increasing electricity demand in China, India, and other developing countries in Asia.

Our reserves of coal are still large and geographically widely dispersed. Economically recoverable proven reserves are close to one trillion tons, representing about 170 years of supply at the current rate of consumption. The largest deposits are concentrated in the United States (27% of the proven reserves), Russia (17%), China (13%), India (10%) and Australia (9%) [17] (Fig. 4.2). The amount of coal in the world is so huge that even nations with a small percentage of the proven reserves can be major coal exporters. This includes South Africa, Indonesia, Canada, and Poland. The global coal resources are estimated at more than 6.2 trillion tons [4] (Fig. 4.3), but given the size of already-known reserves there is for now, little incentive to engage in extensive exploration to find new exploitable reserves.

The quality of coal and the geological characteristics of coal deposits have as much importance as the actual size of a country's reserves. Quality of coal can vary widely from one region to another. The classification depends mainly on caloric value, carbon and water content, which itself depends on the degree of maturity of a given coal deposit. Initially, peat (the precursor of coal) is converted into lignite or brown coal – both coal-types with low organic maturity. With exposure to high pressure and temperatures over millions of years, lignite is progressively transformed into sub-bituminous coals. Lignite and sub-bituminous coal are lower grades of coals which contain higher concentration of water. They are softer, friable materials and have a low energy content, between 5700 kcal kg^{-1} and 4165 kcal kg^{-1} for sub-bituminous coal, and less for lignite. They burn with smoky flames at a lower temperature. As the process of maturation continues, these coals become harder, forming bituminous or hard coal. Under the right conditions, further increase in maturity can lead to the formation of anthracite, the rarest and most desirable form of coal, representing less than 1% of the known coal reserves. Anthracite and bituminous coal, the highest grades of coals, have the lowest water content. They burn at high temperatures with no or little smoke and ash formation, have a high caloric value (>5700 kcal kg^{-1}), and are suited for industrial coke production.

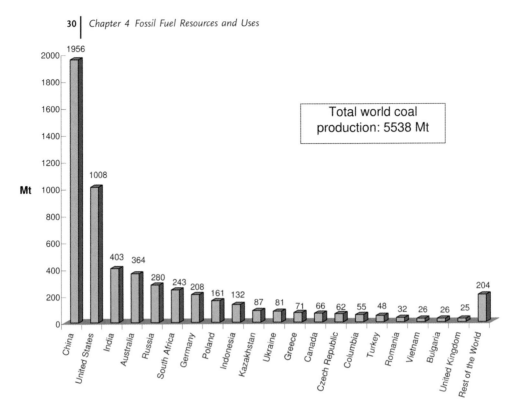

Figure 4.2 World coal production in 2004. Source: BP Statistical Review of World Energy 2005 and IEA coal information 2005.

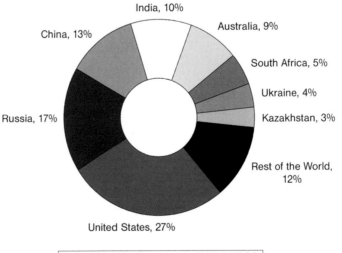

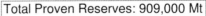

Figure 4.3 Proven coal reserves distribution. Source: BP Statistical Review of World Energy 2005.

Beside the caloric value and water content, the amount of impurities such as sulfur plays an increasingly important role as environmental regulations become more stringent. In the United States for example, many electricity generators switched to lower grade sub-bituminous coal from the Rocky mountains (which contains up to 85% less sulfur than higher-grade bituminous deposits in Eastern States) to reduce acid rain-inducing SO_2 emissions and comply with the new environmental laws [16]. Because coal consumption (in tonnes) per kilowatt-hour generated is higher for sub-bituminous than for bituminous coal, the shift to Western States coal is projected to increase the tonnage consumed per kilowatt-hour of electricity generation. Despite this drawback, and the fact that coal must be transported over longer distances, the switch was also made easier because of the much higher productivity and low cost of coal extraction in Western coal mines, which are operated as open-pit surface mines as compared to underground mining in the Eastern United States. Surface mining, which accounted in 2000 for 65% of the coal extracted in United States and about one-third in the rest of the world, is also safer (in China alone, yearly, thousands of miners perish in coal mine accidents). However, it is only economical when the coal seam is near to the surface and the amount of overburden that must be removed is limited. A majority of the coal (especially hard coal) reserves are too deep and must be mined underground. Increasingly, highly mechanized processes called "room-and-pillar" and "longwall" mining are used, which have nonetheless a lower productivity than surface mining. Worldwide, about two-thirds of the coal is still extracted from underground mines. Coal mining is a dangerous and physically demanding occupation which also involves many health hazards. The socio-economics of underground coal mining are also a major factor in how much coal resources will be indeed recoverable in the future.

Once the coal has been extracted from the mine, it must be moved to the power plant or other location of use. Because it is a bulky commodity, it is expensive to transport. The world coal industry is dominated by production for local use; more than 60% of the coal used for power generation is consumed within 50 km of the mining site. This is made possible by a widespread geographical distribution of coal. Currently, only about 12% of coal is traded internationally.

During the past 20 years, coal prices have declined steadily as productivity for coal extraction and transportation has greatly improved. The nominal price for coal used in power plants has fallen from $50 per ton in 1980 to less than $35 in 2000 [5]. At the same time, environmental consideration for increasingly "clean" coal can become expensive for the end-user. Due to the large reserves and large number of producing countries the coal market will, however, remain very competitive.

Though coal is generally seen as a major fossil-fuel source for the long term, its use presents significantly greater environmental challenges than oil or gas. Upon combustion, it produces significant levels of pollutants such as sulfur dioxide, nitrogen oxides and particulates, as well as heavy metals such as mercury, arsenic, lead, and even uranium. This is the reason why coal bears the image of being a "dirty" and polluting fuel, bringing to mind images of black smoke rising from chimney stacks. Today however, this view is slowly changing, and with proper processes and treatments, coal can be burned quite cleanly and efficiently. Problems arising from emissions were initially alleviated by building tall chimney stacks to improve dispersion following the old diction: "The solution for pollution is dilution". This is a view of the past. Technologies to reduce to acceptable levels or practically eliminate emissions – especially particulates responsible for smoke and dust as well as SO_2 and NO_x inducing acid rains – are now commercially available and progressively applied, mainly in developed countries, and this has led to a dramatic decrease in pollution levels.

The emission that is now at the center of the concerns is CO_2, because of its potential as a greenhouse gas. Compared to oil and gas, coal produces the highest amount of CO_2 for the unit of energy generated. To reduce this emission, efforts are under way to improve the thermal efficiency of coal-fueled power-plants (*vide supra*). Ways to capture and sequester CO_2 from coal-burning power-plants and other industrial sources in subterranean cavities, various geological formation or in the seas, are also being developed. As discussed later (Chapters 10–14) instead of sequestering, chemical recycling of CO_2 to methanol offers a new feasible, long-range solution. This also provides economic value to the recycled carbon source for fuels and feedstocks. Complying with environmental and health regulations is necessary, but should not increase substantially the overall cost of power generation from coal to the point that it becomes prohibitive and uncompetitive in the market place.

Oil

Petroleum oil-derived gasoline and diesel fuel are the fossil fuel products most widely used in our daily lives, and which we are most familiar with. Oil production increased rapidly after World War II, making it the dominant energy source over the past half-century (Fig. 4.4). Close to 30 billion barrels of oil were produced in 2004, representing 35% of the world's total primary energy supply (TPES) [14] – far more than coal (23%) and natural gas (21%). Renewable energy sources accounted for about 14% of the global energy supply, with nuclear energy providing the remaining 7%. Today, the world is using a staggering 82 million barrels of oil every day, representing the content of more than 50 giant supertankers each the length of three football fields. In a span of two days, we consume today as much oil as the yearly oil production in 1900. The U.S. itself consumes about 25% of that oil. The amount of crude oil produced has increased from 58 to 80 million barrels a day between 1973 and 2002. However, the share of oil in the TPES after the two oil crises of the 1970s has decreased from 45% to 35%, mainly in favor of natural gas and nuclear power. Regardless, we are still extremely dependent on oil as the economic prosperity of our modern society is closely related to the availability of abundant and relatively cheap oil. Oil – and all products derived from it – are also essential for our daily lives. Worldwide, about 6% of oil is used as the feedstock for the manufacture of chemicals, dyes, pharmaceuticals, elasto-

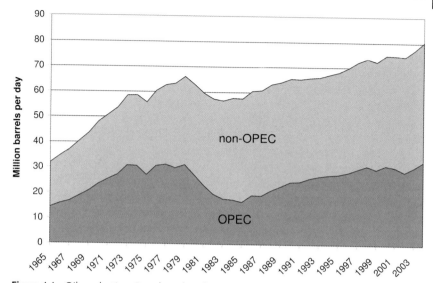

Figure 4.4 Oil production. Based on data from BP Statistical Review of World Energy 2005.

mers, paints and a multitude of other products. These petrochemicals provide many of the necessities of modern life to which we became so accustomed that we do not even notice our increasing dependence on them. The consumption of petrochemicals is still growing at a significant pace. The bulk amount of oil however, is used as fuel for heating, generation of electricity, industrial uses and predominantly as transportation fuels (Fig. 4.5). Transportation, which accounts for about two-thirds of the oil consumed [18] in the United States and almost 60% worldwide [14], is by far the most oil-dependent sector in our modern economy. Automobiles, trucks, buses, locomotives, agricultural machinery, and ships all rely on gasoline or diesel fuel. Air transport would be unthinkable without refined jet and aviation fuels. More than 95% of the energy used in transportation comes from oil. Its use in the transportation sector is growing inexorably, with little immediate prospect for short-term alternatives. In the developed countries, the transportation sector accounts for virtually all new growth for oil demand, while power generation and domestic and industrial uses are being switched increasingly to natural gas and to some degree to alternative fuels. Global demand for oil is currently rising at about 2% per year. From 77 mb per day in 2000, forecasts predict that the world oil demand in 2010 will be 96 Mb per day, reaching 115 Mb per day in 2020 [5]. If these estimates are correct, some 700 billion barrels of oil would be necessary from 2000 to 2020 to satisfy the growing demand. Considering that a cumulative total of about 900 billion barrels have been produced between the beginning of oil exploitation in the 1850s and today, this is a staggering amount. The question that must be raised is: will we have enough resources to cover the demand, and from where do we obtain all this oil?

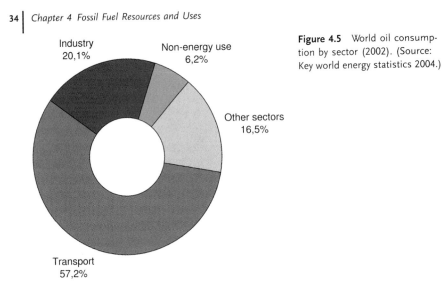

Industry
20,1%

Non-energy use
6,2%

Other sectors
16,5%

Transport
57,2%

Figure 4.5 World oil consumption by sector (2002). (Source: Key world energy statistics 2004.)

The good news is that there are still sufficient oil reserves to satisfy the demand, at least during the next few decades. However, most of this oil will come from the politically unstable Middle-East and other areas. There is considerable uncertainty about the amount of oil that exists worldwide and the proportion of this resource that can be economically recovered. Numerous estimates made by various assessors such as the World Energy council, IHS Energy, Organization of Petroleum Exporting Countries (OPEC), United States Geological Survey (USGS), *Oil and Gas Journal* and others reported current proven oil reserves ranging from 950 to 1100 billion barrels [5]. Current global oil production is about 30 billion barrels per year, so known reserves in 2004 represent only about 30 to 40 year supply. In practice, there has been little decline in estimated reserves, with most assessments showing indeed increases. Proven reserves, instead of being depleted, as a matter of fact, have almost doubled during the past 40 years. This is due to the fact that the nature of oil reserves is very dynamic, depending on many factors. New oil fields were discovered, increasing the amount of proven oil reserves, while rising oil prices make formerly uneconomical oil sources economical. Technological advances can increase the amount of economically recoverable oil. On the other hand, with a growing world population, which has now exceeded 6 billion and is still growing (it may reach 8–10 billion by the mid-21st century), and increasing standards of living around the world, the demand for oil will continue to expand. Oil will not be exhausted overnight, but market forces will inevitably drive prices up as demand will surpass supply, creating a true and lasting oil crisis.

In recent history, there have been three distinct periods in oil supply versus demand involving oil-producing and oil-consuming countries. The period from 1960 to 1973 was one of rapid economic growth and burgeoning oil demand. Wealth in developed countries grew by 90%, energy demand by a similar amount, and oil demand by 120%. The transportation sector boomed and oil also cut deeply into coal's markets as a heating fuel. Worldwide, oil demand rose from some 20 million barrels per day to almost 60 million barrels per day. Many developed countries produced little primary energy or had static or declining production. They became heavily dependent on imports mainly from OPEC countries, especially from the politically unstable Middle East.

The oil price shock resulting from the 1973 crisis reinforced by the Iran/Iraq war of the late 1970s, had profound effects. It abruptly ended a period of extremely rapid growth and prosperity. The world was stricken by high inflation, trade and payment imbalances, high unemployment, weak business climate, and low consumer confidence. The 1973 crisis introduced a new period of oil market development, which lasted until the mid-1980s. It was characterized by vigorous efforts by the importing countries to reduce their dependence on oil. During much of this period these efforts were governed by high oil prices. In the first half of the 1980s, however, prices responded to the weakening market as oil supplies increased and demand continued to reflect the oil-savings measures achieved since the mid-1970s. New oil fields also went into production in Alaska and under the North Sea, weakening the monopoly of the OPEC cartel. Nuclear energy, natural gas and coal replaced much oil in electricity generation and energy-savings measures were widely introduced.

The mid-1980s brought an end to the decrease of oil imports. Until the end of the 1990s, with low oil prices for much of the period and steady economic growth, oil consumption has risen again and net imports of oil-importing countries are now much higher than they were in the early 1970. The bulk of additional imports continues to come from the Middle-East, Russia, and other producers.

Oil prices increased sharply in recent years, reaching the historical height of more than $70 per barrel in 2005, fluctuating presently in the range of $50 to $75. It can, however, in the foreseeable future reach even much higher levels. While experiencing temporary dips, with diminishing resources and increasing demand, the future trend seems inevitable. Several factors, including increased oil consumption in Asia, especially from the fast-growing economies of China and India, reduced spare capacity in producing countries as well as political uncertainties threatening oil production in the Middle-East, Venezuela, Nigeria and others, have contributed to this and will continue to affect the market significantly.

As can be seen in every market, the supply and demand mechanism regulates the price of oil. The supply can be artificially manipulated by decreasing or increasing the oil production and to affect prices (as OPEC does frequently). This has serious consequences on the economies of oil-importing countries, but the process is reversible. It also accelerates the importing countries' efforts to find alternative solutions to the use of oil, which is contrary to the short-term interest of many oil-producing countries. Regardless, the inevitably permanent and irreversible decline in world oil production due to depletion of oil resources will have a lasting and devastating effect on our modern societies for which we must plan and prepare ourselves and find solutions to overcome the problems.

Presently, the Middle-East is the epicenter of many of the "oil battles", because it has the largest oil reserves in the world. Except for Russia, the top five countries in terms of oil reserves are in this region. They are Saudi Arabia, Iraq, Iran, United Arab Emirates, and Kuwait. Although the Middle-East contains about 62% of the world's oil reserves [17], it produces at present only about one-third of the oil. Oil is cheap to extract in the Middle-East, and production costs are among the lowest in the world (i.e., less than $2 per barrel). This compares with a cost of some $8–10 to produce a barrel of oil in the North Sea [5]. Saudi Arabia has the world's largest oil reserves, and is the largest oil producer. Given its proven reserves of some 260 billion barrels, Saudi Arabia could produce 8 million barrels per day for 75 years without the discovery of any additional reserves. The largest oil field ever found is the Ghawar field in Saudi Arabia, which contains about 7% of the known world oil reserves. There are almost 60 oil fields in Saudi Arabia, but oil is being produced from only a few. Together, all of the Saudi oil fields contain about one-fourth of the world's oil reserves. Not only are these oil fields huge, they are also highly productive. Saudi state-controlled Aramco, the world's largest integrated oil company, accounted for 11.7% of the world's oil production in 2000. In contrast, the largest western oil company, ExxonMobil accounted at that time for only 3.4% of world oil production [5]. Having more than half of the world's remaining oil reserves concentrated in a relatively small geographical area in the Middle-East has significant economic and geopolitical consequences (Fig. 4.6). The market share enjoyed by the Middle-Eastern countries is even increasing because fewer significant new oil fields are discovered elsewhere. Some predictions suggest that, by 2010, this region could supply as much as 50% of the world's demand.

The rate of new oil discoveries has dropped mainly because most of the areas of the world have already been explored. From the hot Arabian deserts to the freezing arctic circle and the depth of the seas, exploration has gone to the most re-

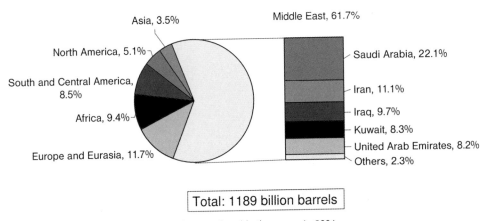

Figure 4.6 Regional distribution of world oil reserves in 2004. Based on data from BP Statistical Review of World Energy 2005.

mote and inhospitable parts of the world in search of oil. The last major oil discoveries have been in the North Sea and Alaska's North Slope. Today, it seems unlikely that there will be other similar new large oil discoveries. One of the remaining promising regions is the South China Sea which, because of territorial dis-
putes over ownership between several countries, has not been fully explored.
However, it seems highly unlikely that it will replicate another Middle-East.
The Caspian Sea region has also large reserves, although their amounts may
have been overestimated and they are increasingly exploited.

The future way to increase the oil reserves is to use unconventional oil sources
which will play an increasingly important role. Oil is considered unconventional
if it is not produced from underground hydrocarbon reservoirs by means of production wells, or if it needs additional processing in producing synthetic crude.
Unconventional oil sources include: oil shales, heavy oils, tar sands, coal-based
liquids, biomass-based oils, processed gas liquid (GTL), and oil obtained from
the chemical processing of natural gas.

Tar Sands

Among the unconventional oil sources, recovery of oil from tar sands is being expanded the most rapidly. Tar sands, as the name suggests, are deposits of sands
containing a high-sulfur tar known as bitumen or very viscous heavy oils. They
are formed when, by erosion, an oil field is brought to the surface and the lighter
components evaporate, leaving an almost solid tar behind. Therefore, in a way it
can be considered as a "dead" oil field. There are tar sands deposits all around the
world, but the larger ones are located in Canada and Venezuela. Canada's Alberta
province contains two enormous deposits: Athabaska and Cold Lake (Figs. 4.7 and
4.8). Of the 2.5 trillions barrels of crude bitumen at these locations, about 12% (or
300 billion barrels) is thought to be recoverable. This is an amount comparable to
the present proven reserves of Saudi Arabia. In Venezuela, over 1.2 trillion barrels
of bitumen is thought to exist in the Orinoco belt and Maracaibo sedimentary
basin, of which 270 billion barrels are presumed to be economically recoverable
with current technology [5]. The potential for unconventional tar sand oil depends

Figure 4.7 Oil sand mining
in Alberta (Courtesy of
Suncor Energy Inc.).

Figure 4.8 Suncor oil sand plant in Alberta (Courtesy of Suncor Energy Inc.).

largely on production costs. In order to thermally extract oil from these heavy tars, large amounts of natural gas are needed. In Canada, due to massive investments and technologic innovations, the operating costs (using *in-situ* recovery) fell from $22 to less than $10 per barrel between 1980 and 2003, making this unconventional oil supply competitive with the conventional oil [2]. In 2004, one million barrels of oil were processed daily in the Athabaska region. With ongoing investments, the production should reach 2 million barrels per day within a decade. Decreasing cheap natural gas resources in that region, however, represent increasing concern.

Oil Shale

If a tar sand deposit can be considered as a "dead" oil field, then oil shale beds can be considered as "unborn" oil fields. Oil shale is a source rock that never sank deep enough for the organic matter to produce petroleum oil. Shale oil indeed neither contains oil nor shale, and the name was probably given in order to attract investments. It contains kerogen, a solid bituminous material that can be used as a substitute for crude oil. When kerogen is heated at high temperature, thermal cracking occurs producing oil which can be refined to petroleum distillates.

In small quantities, oil shale has been extracted for a long time. By the 17th century, oil shales were being exploited in several countries in Europe. In Sweden, oil shales were roasted over wood fires, not to extract the oil but to obtain potassium aluminum sulfate, a salt contained in these deposits and used in tanning leather and for fixing colors in fabrics. During the late 1800s, the same deposits were used to produce hydrocarbons. An oil shale deposit at Autun, France, was exploited as early as 1839. The industrial-scale extraction of oil shale, however, began in Scotland around 1859. As many as 20 oil shale beds were mined at different times and, in 1881, oil shale production had reached 1 million metric tons per year. With the exception of the World War II years, between 1 and 4 million metric tons of oil were mined yearly from 1881 to 1955, when production started to decline and finally ceased in 1962. Other commercial operations exist in Brazil, Estonia, Russia, and China. From the early 1930s, production of oil shale increased until peaking at 46 million metric tons in the early 1980s, with Estonia producing 70% of that oil shale. Production of oil shale subsequently steadily declined to a low of 15 million metric tons in 1999. It was not because of a diminishing supply but rather due to the fact that oil shale could not compete economically with petroleum oil. The world's potential shale oil resources are enormous. Estimated at 2.6 trillion barrels, they have barely been touched. By far the largest known deposit is the Green River oil shale in the Rocky Mountains of the Western United States, that contains a total estimated resource of 1.5 trillion barrels of oil. In Colorado alone, the resource reaches 1 trillion barrels of oil. This represents only those resources that are rich enough in oil and close enough to the surface to be open-pit mined and therefore thought to be economically recoverable. Numerous pilot plants using different extraction technologies have been operated over the past decades in this region by oil companies such as Shell, Exxon, Amoco, Unocal, and Occidental. These projects however, were found to be uneconomical and were terminated. Unocal (now part of Chevron) operated the last large-scale experimental mining facility from 1980 until its closure in 1991, producing 4.5 million barrels of oil from oil shale, and averaging 34 gallons of oil per ton of rock extracted. The future development of the oil shale industry is clearly dependent on the availability and price of petroleum crude oil. When the price of extraction of oil from oil shale will be competitive with the price of oil from conventional sources either by technological improvements or higher oil prices, then shale oil will find a place in the fossil fuel energy market. The present level of oil prices is already prompting a renewed interest in oil shale. In Utah for example, Oil Tech Inc. has recently built a demonstration facility to extract shale oil from underlying deposits, claiming a production cost of less than $20 per barrel [19, 20]. Of course, very significant problems must be overcome, such as the large amounts of heat needed to process the oil shale and the great amounts of expanded shale rock left after extraction.

The highly successful development of the Athabaska tar sand resources of Canada could also serve as a model for the initiation and growth of the shale oil industry especially in the Western United States, as the characteristics and technologies for the exploitation of these two resources are similar.

Natural Gas

Global demand for natural gas has grown much faster than that for oil and coal over the past three decades. In 2004, the World natural gas consumption reached 2.7 trillion cubic meters, representing 21% of the world's total primary energy source (TPES) compared to 23% for coal and 35% for oil (Fig. 4.9). The United States alone accounts for about one-fourth of the global natural gas demand,

and is the second largest gas producer after Russia. However, because of its huge demand, the USA is also the largest natural gas importer. The global market for natural gas is expected to continue to expand rapidly due to its still ample availability, cost-competitiveness, and environmental advantages over other fossil fuels. Natural gas is considered as a premium fuel. It has a high caloric value and is clean-burning and relatively easy to handle. For these reasons it is an excellent fuel for domestic use and heating. The bulk increment of natural gas demand will, however, come from new power plants. In the United States, from 2000 to 2002, virtually all the new electricity generating capacity constructed uses natural gas as a fuel. Natural gas has especially become the preferred fuel for the so-called "peak-load" power plants which use combustion turbines to operate the generators, and can be rapidly switched on or off following the vagaries of demand. Given the significant improvement in the efficiency of gas-fired electricity generation, the *de facto* natural gas consumption in electric plants has however, increased less rapidly than the amount of electricity produced. Beside its flexibility and cleanness, natural gas also has the advantage of adding the smallest amount of CO_2 to the atmosphere per unit of released energy. Typical emissions are about 105 kg carbon Gcal^{-1} for bituminous coal, 80 kg carbon Gcal^{-1} for refined fuel, and less than 59 kg carbon Gcal^{-1} for methane (the major component of natural gas). The reason is that methane has a H:C ratio of 4, the highest of all hydrocarbons, and contains only 75% carbon by weight compared to some 85%

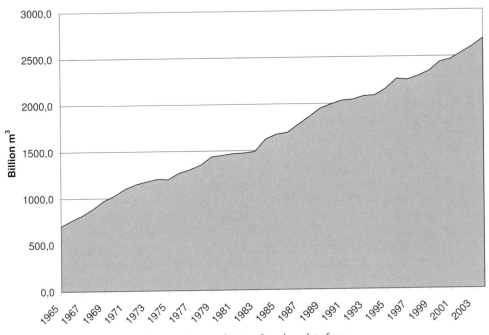

Figure 4.9 World natural gas production. Based on data from BP Statistical Review of World Energy 2005.

for crude oil and more than 90% for coal. Considering the growing concerns about global warming, a reduction of CO_2 emissions by using natural gas is certainly a step in the right direction. However, one should bear in mind that methane itself is a very potent greenhouse gas, which has a global warming potential 23 times higher than CO_2 over a 100-year period. Methane accounted for about 20% of the cumulative greenhouse effect from 1750 to 2000, the largest sources of anthropogenic methane being derived from livestock, rice fields, and landfills [21]. Oil and gas production were responsible for some 15% of the emissions, due mainly to leaks during natural gas transport. Pipelines in developed countries do not lose more than 1% of the carried gas. In less-developed countries, however, loses of 5% or more are quite common. With rising natural gas extraction and long-distance transport, these leakage problems should be resolved, not only to reduce the contribution of methane to global warming but also from an economic point of view, to make the best use of this valuable resource.

Gas is an abundant source of energy, with its proven reserves estimated at some 180 trillion cubic meters in 2004 [17] (Fig. 4.10), exceeding the energy content of the world's proven reserves of oil. Despite ever-increasing production, the reserves of natural gas tripled in the past 30 years, because important new fields were discovered in many parts of the world and technological developments allowed existing reserves to be upgraded. At the actual rate of consumption in 2004, these re-

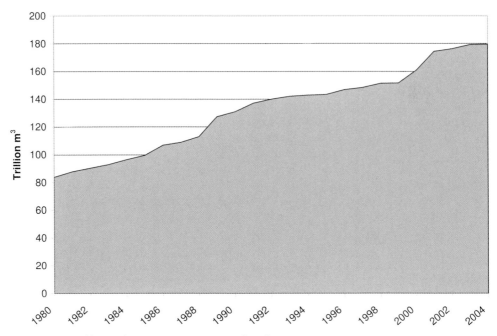

Figure 4.10 World natural gas proven reserves. Based on data from BP Statistical Review of World Energy 2005.

serves would last close to 70 years compared to less than 40 years for oil. Our over-all remaining gas resources are estimated at 450 to 530 trillion cubic meters. Since the beginning of fossil fuel exploration, only slightly more than 10% of the world's gas resources have been consumed, compared to 25% of the estimated world's oil resources (Fig. 4.11).

As in the case of oil, a few countries dominate the natural gas reserves. The largest known accumulations of gas are in Western Siberia and the Persian Gulf. In 2002, more than half of the global gas reserves were found in only three countries: Russia with 26%, and Iran and Qatar each accounting for about 15% of the reserves (Fig. 4.12). North America and Europe represent only about 4% each. Like oil, gas reserves are concentrated in a small number of giant fields. About 190 giant reservoirs contain 57% of the global gas reserves, with some 28 000 other reservoirs holding 28%. The remaining 15% represent marginal fields [4]. Although proven reserves are abundant, new resources tend to be discovered in more remote areas far from consuming centers, in difficult terrain, or as small fields. Even today, the geographical distribution of natural gas-producing regions is thus far from even. Most of the demand is concentrated in North America and Europe, but these regions have relatively limited sources of natural gas, in part because they have already exploited significant amounts of their initial reserves. In Europe, countries such as Norway, The Netherlands

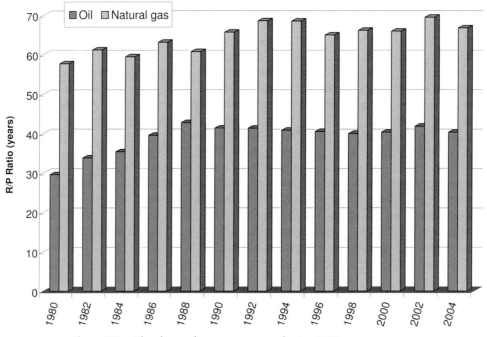

Figure 4.11 Oil and natural gas reserve to production (R/P) ratio. Based on data from BP Statistical Review of World Energy 2005.

and United Kingdom still have significant Northern Sea reserves, but a growing part of Europe's natural gas consumption has to be imported from the rich but remote Western Siberian fields in Russia as well as from Algeria. North America is still largely self-sufficient in natural gas, the larger part of gas imports in the United States being supplied by pipelines from Canada. However, due to stagnating or declining rates of production as well as to a steadily increasing consumption, North America will have to rely more on imports mainly in the form of liquefied natural gas (LNG) from countries such as Algeria, Trinidad and Tobago, Nigeria, and the Middle-East. LNG, however, as discussed subsequently has its own limitations and hazards.

In order to be shipped overseas as LNG, natural gas must first be liquefied at a port-side liquefaction plant in the producing country. Liquefaction removes almost all impurities (CO_2, sulfur compounds, water, etc.) and the gas becomes almost pure methane. Once liquefied, LNG is piped onto giant ships, the size of three football fields, with a carrying capacity of 150 000 to 200 000 m^3, where it is stored in highly insulated, double-walled tanks (Fig. 4.13). After a trip of 10 to 15 days, the LNG is transferred from the ship to a regasification terminal where the liquid is heated to gas ready for distribution. The construction of a LNG infrastructure is very capital intensive, with each liquefaction and regasification plant costing between $1 and $2 billion [11], or even more if located offshore. Nevertheless, ExxonMobil and Qatar Petroleum recently signed two $12 billion deals to build LNG infrastructure to deliver 100 000 m^3 natural gas per

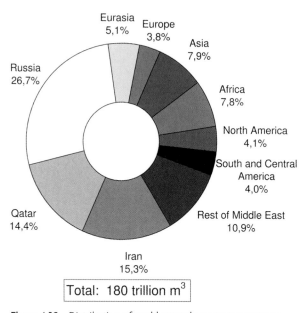

Figure 4.12 Distribution of world natural gas proven reserves in 2004. Based on data from BP Statistical Review of World Energy 2005.

Figure 4.13 Liquid natural gas (LNG) tanker for transportation of natural gas across the seas.

day to both the United States and the United Kingdom. In the United States, the construction of some ten new LNG regasification terminals is projected within a decade, compared to only a half-dozen operating today, all of which were built more than 20 years ago. Because of the safety problems associated with the transport and handling of LNG, most future LNG terminals are expected to be built offshore. In the US and other major countries, much discussion and planning is taking place to build new LNG terminals. Because of the politics involved, the trend is to locate away from major population centers.

Japan, South Korea, and Taiwan, all of which have no gas reserves, imported together more than 70% of the world's LNG production, mainly from Indonesia and Malaysia. For them, LNG is the only economical solution because they are far from gas-producing areas and the costs to construct long-distance off-shore pipelines would be prohibitive. Undersea links are usually only practical when the distances are relatively short and sea is not too deep, as in the case of the North Sea gas network. During the past decade however, development in off-shore pipelines technology has contributed to lower the costs and make possible deep-water projects that were previously impossible such as the Bluestream pipeline under the Black Sea, which reached a maximum depth of 2150 m. Depths of 3000 to 3500 m will probably be feasible in years to come with costs competitive with LNG. The natural gas liquefaction chain, which is an expensive and very energy-intensive process, includes liquefaction, shipping, regasification, and storage. The liquefaction step, during which the gas must be cooled to −162 °C, accounts for about half of the total cost, and shipping and regasification for about 25% each. Liquefaction is usually chosen when the gas has to be transported overseas and/or distances overland longer than 4000–5 000 km, in which case the construction cost of a pipeline would be higher than for LNG. This is especially the case for the Middle-East, which accounts for 40% of the world's natural gas reserves, but is too far from the major consuming centers of Europe or the United states. In 2001 this region represented only 21% of the LNG export market and 4.5% of the natural gas export market. Of course, as mentioned previously there are also significant safety problems associated with LNG transportation and distribution. Besides accidents, LNG tankers and facilities are also potentially vulnerable to terrorist attacks.

Another way of transporting and making use of remote reserves is to convert natural gas into a liquid product, near the gas well, that can be more easily shipped in regular tankers; this is referred to as gas-to-liquid (GTL) technology. Advances in technology and increasing oil prices are stimulating new interest in these GTL processes. Within this broad strategy, two approaches seem to be the most feasible: (i) the conversion of natural gas to a liquid product resembling the light hydrocarbons present in petroleum; or (ii) its transformation to methanol. These processes have so far been based on the Fischer–Tropsch technology using synthesis gas (syn-gas) which was originally developed, using coal as a feedstock, in Germany during the 1920s. Coal is first converted into syn-gas (a mixture of hydrogen and carbon monoxide), which subsequently will yield, depending on the catalyst and reaction conditions, liquid hydrocarbons. Because of the high energy needs and relatively high production costs, the syn-gas-based production of fuels, derived originally from coal, but later from natural gas, was in the past limited to special situations such as Germany during the World War II and South Africa during the Apartheid era restriction on oil imports. Recent technical advances, including reactor design and improved catalysts, have significantly increased yields and thermal efficiency and reduced construction and operating costs. South Africa in particular, due to its decades of "forced" experience in this field, has in the 1990s opened the Mossgas GTL plants yielding 30 000–60 000 barrels per day of mainly diesel fuel and gasoline. GTL plants are also complex, expensive to build and very energy-intensive, consuming up to 45% of the gas feedstock, thereby raising concerns about high CO_2 emissions. In most of the world, such as in New Zealand and Malaysia, syn-gas-based plants use natural gas (methane). Methane conversion to products in the diesel fuel or gasoline range is currently not used to any significant degree. With the anticipated depletion of crude oil reserves, however, a number of oil companies are presently investing massively into GTL technologies. Sasol, in a joint venture with Qatar Petroleum, has begun the construction of a 34 000 barrel per day unit in Qatar, near the giant North Field gas reservoir (containing some 15% of the world's gas reserves) [22]. Other companies, including ExxonMobil, Shell, Marathon Oil, and ChevronTexaco, have committed over $20 billion to the construction of GTL facilities in Qatar [23]. These projects will each have a production capacity ranging from 120 000 to 180 000 barrels per day, and cost between $5 and $7 billion per plant. They will produce mainly diesel fuel and small amounts of other products, including lubricants and motor oil. Due to the economy of scale of such large installations, the cost to produce a barrel of GTL diesel is expected to be in range of $15–$20, which is extremely competitive with recent oil prices. Furthermore, diesel fuel obtained from GTL processes has the advantage of being almost free of most impurities (especially sulfur) contained in regular diesel fuel, and is thus less polluting.

In order to be more easily handled and shipped over long distances, natural gas (i.e., methane) can also be converted to methanol, a convenient liquid. Today already, the production of most of the methanol in the world comes from the conversion of methane first to syn-gas and then to methanol. The direct oxidative conversion of methane to methanol, without going through syn-gas, is however the highly desirable technology of the future and is being developed as part of the "Methanol Economy" (see Chapter 12).

The amount of natural gas reserves referred to generally pertains to conventional natural sources – that is, gas which accumulates under non-permeable rocks alone or in association with oil. There are also considerable resources of so-called unconventional natural gas sources; some of these are already being recovered. They involve gas extracted from coalbeds (coalbed methane), from low-permeability sandstone called "tight sands" and shale formations. Gas hydrates, primarily methane hydrate present at the continental shelves of the seas or underlying the tundra of arctic areas such as Siberia, are also abundant and very significant. Although no suitable technology to utilize these effectively exists today, they represent a very large untapped unconventional gas source. In the United States, during the past 20 years, the use of unconventional gas has grown rapidly, helped significantly by tax incentives. Mainly coalbed methane and tight sand gas account now for about one-fourth of the total U.S. domestic production, helped by the decreasing production of conventional gas and higher natural gas prices. However, they remain more expensive to produce than conventional gas and are presently only of modest interest in the rest of the world, due to the still-abundant sources of conventional gas.

Coalbed Methane

Coalbed methane, as its name indicates, is derived from coal, which acts in both the role of source rock and reservoir for methane. During the geological formation of coal under high temperature and pressure, volatile substances such as water and methane are liberated. Some gas may escape the coal. However, if the pressure of the formation is sufficient, significant quantities of methane are retained in the pressurized coal matrix in an adsorbed form. Because of its large surface area, coal can store up to six or seven times more natural gas than the equivalent rock volume of a conventional reservoir. Gas content generally increases with depth of burial of the coalbed and reservoir pressure. Fractures (or cleats as they are called in this case) that permeate the coalbeds are usually filled with water. The deeper the coalbed, the less water is present, but the more saline it becomes. In order to produce methane, wells are drilled through the coal and the pressure is reduced by pumping out water. This allows the methane to desorb and pass into the gaseous phase, so that it can be produced in a conventional manner and transported by pipeline. While economic recovery of methane is possible, water disposal from coalbeds exploitation raises environmental concerns. Coalbed accumulations are widespread and characterized by their large sizes. Because of the heterogeneous nature of coalbeds, the production rates are, however, highly variable even within a small area. The USGS estimates worldwide resources at up to 210 trillion cubic meters, but this number is uncertain because of the scarcity of basic data on coalbed resources and their gas content. The largest resources are located in regions rich in coal such as the United States,

China, Russia, Australia, Germany, and Canada. Numerous pilot projects are exploiting these resources, and many countries have active coalbed methane wells, but the United States is currently the only country where commercial production on a large scale is taking place. Production from this source now represents about 10% of the methane production in the United States.

Tight Sands and Shales

Tight sands and shales are low-permeability geological formations which sometimes contain large accumulation of natural gas. This means that the underground rock layers holding the gas are very dense, and thus the gas does not flow easily toward wells drilled to collect it, resulting in low recovery rates. This can be enhanced by horizontal drilling and fracturing of the rocks with explosives, or with hydraulic pressure to provide pathways for the trapped gas to flow more easily to the well-bores to be pumped to the surface. In the United States, the exploitation of these sources now accounts for more than 15% of domestic methane production, with the development of tight gas sands representing most of the production. Despite the fact that gas resources from tight sands and shales are immense, there are many uncertainties about the cost of their production. In 2002, the U.S. Geological Survey estimated the total technically recoverable natural gas from these sources in the United States, to be around 10 trillions cubic meters, or about 15 years of natural gas consumption.

Methane Hydrates

Another unconventional gas source that has attracted increasing attention in recent years are methane hydrates (Fig. 4.14). Gas hydrates are naturally occurring crystalline, ice-like solids in which the gas molecules are trapped by water in cage-like structures called clathrates. Although many gases can form hydrates in nature, methane hydrate is by far the most common. Various clathrate compounds

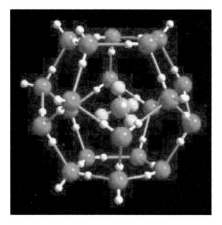

Figure 4.14 Methane hydrate. A molecule of methane is trapped in a cage made out of water molecules (Source: NETL).

were discovered and studied in laboratories beginning in the 1800s. However, as no natural occurrences were known, the subject remained a purely academic curiosity until the 1930s, when solid hydrates were found to be plugging natural gas pipelines. In the 1960s, "solid natural gas" or methane hydrate was observed as a naturally occurring constituent in Siberian gas fields. Since then, gas hydrates have been found worldwide in oceanic sediments of continental and insular slopes, deepwater sediments of inland seas and lakes, and in polar sediments on both continents and continental shelves of the seas (Fig. 4.15). In permafrost Arctic regions, gas hydrates can be present at depths ranging from 150 to 2000 m, but most occurs in oceanic sediments hundreds of meters below the sea floor at water depths greater than 500 m.

Whilst the exact size of the world's methane hydrate deposits remains the subject of debate, largely due to the lack of field data, there is a consensus that the overall amount is huge. The estimates of methane hydrates, at the present time, are in the order of 21 000 trillion cubic meters (Fig. 4.16). This represents more than 100 times our conventional proven gas reserves! However, in the consideration of methane hydrates as a future energy source, knowing the overall amount of methane hydrates present is only a part of the question. As is the case for the production of oil from the tar sands in Alberta, very large investments in research and development will be necessary to resolve significant technical issues before methane hydrates can be considered as an affordable methane source. Furthermore, the gas hydrate reserves are likely to represent only a small fraction of the gas hydrate resources because the deposits are too dispersed or the gas concentration is too low to be economically recovered. Regardless, the amounts of

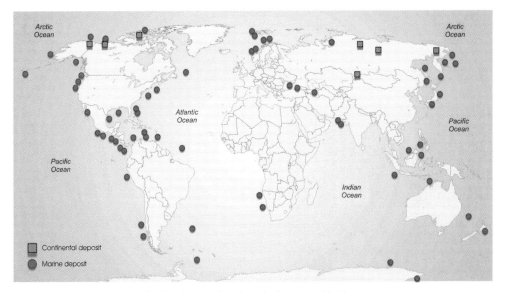

Figure 4.15 The distribution of methane hydrates, worldwide.
Source: Lawrence Livermore National Laboratory.

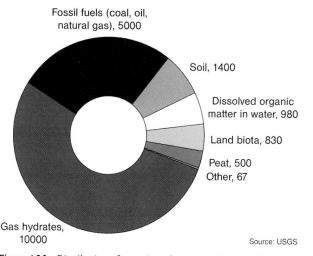

Fossil fuels (coal, oil, natural gas), 5000

Soil, 1400

Dissolved organic matter in water, 980

Land biota, 830

Peat, 500
Other, 67

Gas hydrates, 10000

Source: USGS

Figure 4.16   Distribution of organic carbon on Earth reservoirs (Gigatons). (Source: USGS).

natural gas hydrates seems to be so enormous, that even the exploitation of a small part of the resources could provide all of our energy needs for centuries to come. Japan and India – two countries with large energy needs but limited energy resources, as well as the United States and Canada – are pursuing methane hydrate recovery research programs. For the time being, methane hydrates remain however, a potential but distant energy source.

Outlook

At current production rates, there are sufficient proven reserves to meet oil demand for at least some 40 years, and gas demand for more than 60 years. These reserves indeed have been steadily growing over time and are now higher than ever before in history. However, worldwide consumption is increasing due to growing population and per-capita consumption. The rate of new oil and gas discoveries is now nearing the point where consumption exceeds new discoveries (i.e., replacement) – the so-called Hubbert's peak – meaning that we will start to use more fossil fuels (except coal) than the amount we can add to our reserves. Unconventional sources of oil and gas, such as the tar sands and heavy oil of Canada and Venezuela, oil shales of the Western United States, or the future development of gas hydrates, will play an increasingly important role in extending our resources. Given the higher abundance of coal, its future is not only a matter of resource availability or production costs but also one of environmental acceptability. New combustion and emission control technologies have therefore to be introduced or improved. One of the greatest challenges, not only for coal but also for other carbon-based fossil fuels is the reduction of CO_2 and other emissions responsible for a significant part of mankind-caused greenhouse gases and resulting climate changes. How to manage and prepare ourselves for the time beyond readily available oil and gas, is a challenge that is discussed in the subsequent chapters of this book.

Chapter 5
Diminishing Oil and Gas Reserves

The World is entirely dependent on oil. The transportation sector in particular, so vital in our society for carrying people, goods, food and materials, relies for more than 95% of its needs on gasoline, diesel and kerosene derived from petroleum, and consumes about 60% of the oil produced. We also depend on oil for the large variety of petrochemicals and derived products such as plastics, detergents and synthetic fabrics that today are so ubiquitous in our daily lives that we can hardly think about where they come from.

Oil is so important because it is the most versatile among our three primary fossil fuels. It has a high energy content, is easy to transport, and is relatively compact. Coal is heavier, more bulky and more polluting. Natural gas on the other hand is cleaner burning but bulky in volume and requires pipelines or expensive liquefaction for transportation, which raises safety concerns.

Unlike the case of coal and gas, daily publications of oil prices, reserves and production rates are highly publicized and have immediate effects on our economies and lives. The outcome of every OPEC meeting is covered by the World media like a national or presidential election.

In Chapter 4 the point was discussed that proven oil reserves are estimated at between 950 and 1100 billion barrels [5]. Dividing that figure by the annual production rate of 27 to 30 billion barrels suggests an availability of some 40 to 50 years. This number is called a Reserve/Production or R/P ratio. It is commonly used, especially in industrial reports issued by major oil companies, and can give a useful warning sign of resource exhaustion when the number starts to decline. Today, the R/P ratio is still actually higher than it stood at any time in the past 50 years. At the same time, oil prices are fluctuating rapidly, but even with the cost of a barrel at $70 as in the recent past, it is still cheaper than at the heights of the oil crisis in the early 1980s, when a barrel sold for well over $80 (inflation adjusted) [17]. This frequently leads to forecasts that anticipate a still long and comfortable era of relatively low-priced and abundant oil allowing a smooth transition to other energy sources. This optimistic vision of our oil future, supported by some economists, has been championed by those who believe in the crucial role of prices, continued improvements in exploration, drilling and production technologies to provide adequate oil production supplies for many years to come. Accordingly, it is assumed that whereas there is a finite amount

Beyond Oil and Gas: The Methanol Economy. G. A. Olah, A. Goeppert, G. K. S. Prakash
Copyright © 2006 WILEY-VCH Verlag GmbH & Co. KGaA, Weinheim
ISBN 3-527-31275-7

of accessible petroleum oil on planet Earth, this is basically irrelevant because oil extraction will cease long before its actual physical exhaustion. The point that matters is the cost to find and exploit new reserves. When this cost of exploration and exploitation becomes too high, then oil will be replaced by some other source of energy, leaving part of the oil left in Earth's crust. The challenge is to find acceptable substitutes before oil becomes so expensive to produce that it would disrupt the economic and social fabric of our society. It is argued that similar transitions have taken place in the past when wood was replaced by coal, or coal by oil. Based on this argument, running low on oil will not have major direct relevance. This view, however, is not realistic. A more appropriate evaluation is the inevitable depletion and therefore ending of the era of relatively cheap accessible oil. We are probably already entering into an irreversible decline in oil production following its peak. This prediction is based on consumption data compared with oil field discoveries, reserves and extraction data. In fact, the R/P ratio gives little information about the long-term fate of a resource. Furthermore, it assumes that production will remain constant over the years, which is highly improbable. So is assumption that the last barrels of oil can be pumped from the ground as easily and quickly as the oil coming out of the wells today. Globally, the demand for oil is expected to grow by some 2% per year over the next decades [5]. Three important parameters must be considered to project the future of oil production: (i) the cumulative production representing how much oil has been produced to date; (ii) the amount of recoverable reserves present in the known oil fields; and (iii) a reasonable estimate of the oil that still can be discovered and extracted. The sum of these represents the ultimate recovery which is the amount of oil that will have been extracted by the time that oil production ceases permanently. The current mean estimate by the United States Geological Survey (USGS) for the ultimate oil recovery is 3 trillion barrels (3000 Gbbl) [24]. This estimate has, however, been questioned as unrealistically high by many geologists, who set the ultimate recovery at more in the region of 2000 Gbbl (Fig. 5.1). Naturally, the yet to discover – and therefore today unknown – new oil fields are the most speculative and controversial part of these estimates. However, the amount of oil is clearly finite and the question is not whether we will run out of readily available oil, but rather when.

In reality, the rate at which a well, an oil field or even a country can produce oil rises to a maximum and then, after more than half of the oil reserve has been produced, declines gradually to unproductive levels. This trend was realized by M. King Hubbert, a noted American geophysicist who worked as a research scientist for USGS and Shell. He found that in any large region, the extraction of a finite resource rises along a bell-shaped curve that peaks when about half of the resource has been exploited. This assumes that no outside regulatory, legislative or other major restraints have been placed on extraction and exploitation. Following this concept, Hubbert predicted in 1956 that oil production in the lower 48 states of United States would reach a maximum sometime between 1965 and 1972 [25] (Fig. 5.2). U.S. oil production indeed peaked in 1970, and has declined ever since. Larger oil fields are generally found and exploited first,

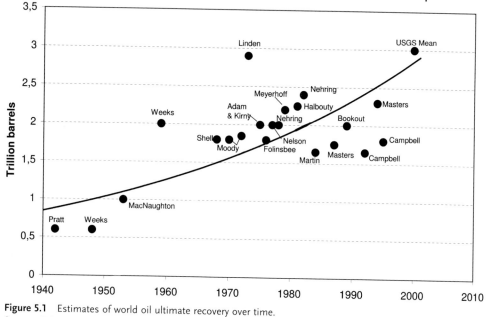

Figure 5.1 Estimates of world oil ultimate recovery over time.
Source: EIA, USGS and Colin Campbell.

giving a discovery peak which, in the lower 48 states of the U.S., occurred during the 1930s. The production peak then follows after a time lag (40 years in the case of the United States) depending on the amount of reserves in place and the rate of exploitation. A similar pattern of peak and decline was observed in many other countries, including the former Soviet Union. Countries such as the United Kingdom and Norway are close to the midpoint, whereas other countries particularly in the Middle-East – namely Saudi Arabia, the United Emirates, Iraq, Iran or Kuwait – are still only at earlier stages towards depletion [26].

Hubbert, as well as several analysts including Campbell, Laherrère [13] and Deffeyes [27], also applied the method to the global world oil production, to determine when the peak for world oil production, the so-called "Hubbert's peak", will occur. For their calculations they estimated ultimate world oil recovery at between 1900 to 2100 Gbbl. Until now, about 900 billion barrels of oil have been produced. Accordingly, we are close to the midpoint of the oil era, corresponding to half of the ultimate recovery. The estimated peak for world oil production will occur between 2005 and 2015 [28,29]. Remarkably, these predictions do not shift significantly even if reserve and production estimates are off by some hundred billions of barrels [27]. After the peak, world oil production will start to decline. What clearly matters, is not so much when our oil will be significantly depleted (not necessarily "gone"), but when the demand will begin to surpass production. Beyond that point, prices will inevitably rise sharply unless demand declines com-

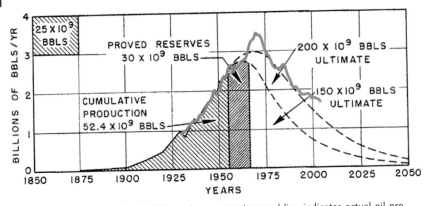

Figure 5.2 Hubbert's original 1956 graph showing crude oil production in the 48 lower U.S. States, based on assumed initial reserves of 150 and 200 billion barrels. The bold, superimposed line indicates actual oil production until 2004 following Hubbert's 200 billion barrels forecast.

mensurably. This leads to the conclusion that a permanent oil crisis is rather close and inevitable.

In the forecasts, it was assumed that around 90% of the oil that can be recovered has already been discovered, putting the reserves at 900 billion barrels and the yet-to-find oil at only 150 billion barrels. This conclusion arises from the fact that today, about three-fourths of the world's oil reserves are located in about 370 giant fields (each containing more than 500 million barrels of oil) that are relatively well studied [4]. As discoveries of these giant fields peaked in the 1960s, large additions to the known oil reserves are unlikely, unless new major oil fields in as-yet unexplored regions of the world are discovered [27]. However, as mentioned, mankind has already increasingly gone to the "end of the world", even under the most hostile climatic conditions in search of oil, and only a few areas, such as Antarctica or the South China Sea, remain to be explored. This leaves little room for the discovery of a new Middle-East. Today, 80% of the produced oil flows from fields discovered before the early 1970s, and a large portion of those fields are already in declining production.

These rather pessimistic – though not unreasonable – forecasts however, usually do not take into account several elements, in particular economics and recovery factors. The oil recovery factor (which is the percentage of oil recoverable in a field) has increased roughly from 10 to 35% over the past few decades due to the introduction of new technological advances such as three-dimensional seismic surveys and directional drilling that are now applied in most oil-producing fields. With continuing progress in so-called enhanced oil recovery (EOR) techniques, recovery factors from 40 to 50% are expected in the future, extending the global oil reserves. Oil prices can also play an important role. Higher prices trigger besides more economical use and savings, the exploration to find new oil fields, the exploitation of fields considered previously non-economical at lower prices, and the development of new extraction technologies. There is much more economic-ally recoverable oil within the $50 to $100 per barrel price range than there was at $20, adding to our reserves.

As discussed, besides conventional oil there are also many non-conventional oil sources, including heavy oils, tar sands and oil shales. These add significantly to petroleum oil sources as their exploitation is becoming profitable with increasing oil prices. These reserves, as discussed, are many. The Orinoco belt in Venezuela has been assessed to contain a whopping 1.2 trillion barrels heavy oil, of which 270 billion barrels are thought to be economically recoverable [5]. The Athabaska and Cold lake tar sands deposits in Canada may contain the equivalent of 300 bil-lion barrels of economically recoverable oil [5]. Non-conventional oil, because of its nature, is more difficult to extract than conventional oil. However, with tech-nological innovations and massive investments, sources that were considered be-fore as too expensive to exploit, are becoming economically viable. In Canada, the operating costs to produce a barrel of oil from tar sands (by *in-situ* recovery) fell from $22 to less than $10 between 1980 and 2003, making this non-conventional oil supply presently competitive with conventional oil [2]. The requirement, how-ever, for very large quantities of natural gas needed for the thermal recovery pro-cess may limit this favorable picture. The amount of oil extracted from the Atha-baska region is expected to grow from 1 millions barrels a day in 2004 to 2 million barrels a day within a decade. While proponents of "Hubbert's peak" theory acknowledge the very large amounts of non-conventional oil resources, they also think that industry would be hard-pressed for the energy, capital and time needed to extract non-conventional oil at a level to make up for the declining con-ventional oil production. In their view, the exploitation of non-conventional oil would have only a limited effect on the timing of the world oil production peak. Production of these substitutes is, as also mentioned earlier, highly en-ergy-intensive. For example, tar sands must be treated thermally, whether *in situ* or in treatment plants, to extract oil, using non-renewable, limited and valu-able natural gas, and generating overall more CO_2 than the production of conven-tional oil. These factors must be seriously considered. Eventually, atomic energy and all other sources of alternative, non-fossil energies could be used to allow the exploitation of these heavy hydrocarbon sources [30–32].

Numerous predictions have been made in the past concerning the peak point of global oil production. Mankind has been said to be running out of oil repeatedly since the beginning of its use on an industrial scale. As early as 1874, the state geologist of Pennsylvania, which was the largest oil producer at the time, esti-mated that there was only enough oil to keep the kerosene lamps of the nation burning for four more years. In 1919, the U.S. Geological Survey, using data based on R/P ratios, predicted that the "end of oil" would come within 10 years. In fact, the 1920s and 1930s saw the discoveries of the largest oil fields known at that time. Since then, many have prophesized the soon to come (gen-erally within a few years or decades) end of oil. Among others, BP predicted in 1979 that the world production peak would occur in 1985 [33], while others fore-cast that the peak would occur between 1996 and 2000. A long history of too-often wrong predictions has led to general skepticism for new and generally pessimistic forecasts for oil's future [34].

While doomsday scenarios are usually given preferential publicity by the media (always ready to report bad news), there are also more optimistic views evolving on the future of oil production [35].

Using an estimated "ultimate" recovery for conventional oil of 3 trillion barrels, similar to the USGS estimate, Odell, for example, predicted in 1984 that global oil production from conventional oil would peak only around 2025 [33]. The steadily increasing exploitation of non-conventional oil sources could push the peak further as far as 2060, but at a cost clearly much beyond the period of "cheap oil" that we currently enjoy.

As we can see, predicting the future is not easy, as quoted by many renowned personalities. What is certain however is that there is only a finite amount of petroleum oil (and natural gas) on the planet Earth. Combined oil production from conventional and unconventional sources will soon – and certainly not later then the middle of the 21st century – reach a maximum and then decline (Fig. 5.3). It is thus imperative to start switching progressively away from oil-based fuels to alternatives. This switch cannot be delayed for too long, so as not to get caught in a real crisis with very high oil prices and all their economic and geopolitical consequences. In that sense, moderately high oil prices are not necessarily detrimental, as they flatten out or decrease excessive demand, encouraging at the same time savings and the development and transition to alternative fuel sources. Oil is

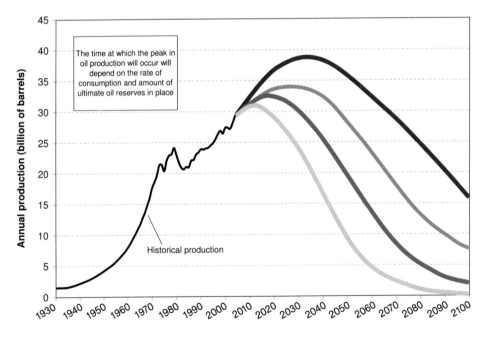

Figure 5.3 Hubbert's peak for world oil production. Source: BP for historical values.

needed for the essential services, fuels and materials it provides us with, including heating, transportation and mobility. If other sources providing the same are found to be competitive or even better, we will switch to them. In the past in England and in most industrialized nations, coal replaced wood when the latter became increasingly scarce and therefore expensive, as well as being inferior to coal in its caloric value and convenience of use. During the 20th century, oil took the place of coal in many uses, not only due to its lower cost but also because it was easier to transport, cleaner, more flexible, and had a higher energy density. Oil was convenient as a transportation fuel, in household use, industrial production, and in electricity generation. Following the oil crises of the 1970s, the utilization of oil in many areas decreased dramatically in favor of natural gas, together with nuclear power for electricity generation as well as a renewed interest for coal. Currently, the largest oil-consuming sector in most industrialized countries is that of transportation, with more than 95% reliance on oil. Therefore, a major reduction in oil consumption will have to come from this sector through more efficient internal combustion engines, the introduction of new technologies such as hybrid propulsion and fuel cells, or the use of alternative fuels. The production of liquid fuels (called syn-fuels) from coal was shown to be technically feasible during the 1930s, and has already been used in some special situations as during World War II by Germany and South Africa during the Apartheid boycott era. Considering its large available coal reserves, similar operations were studied by the United States as a response to the decreasing domestic oil production and increasing oil prices, particularly following the oil crises of the 1970s. These plans, however, were rapidly given up when oil prices stabilized in the mid-1980s. It was generally thought that only when the price of a barrel of oil rose above \$35–40 and remain at that level for a long period of time, would it become economically feasible to consider producing synthetic fuels. Nevertheless, even if the production of liquid fuels from coal or natural gas were to be economically viable on a large scale, it would necessitate vastly increased production of these non-renewable fossil fuels and still be very wasteful from an energy point of view. It would also generate increasing amounts of CO_2, greenhouse gas, SO_2 and other gases, as well as solid waste. Natural gas liquefaction, as discussed earlier, is gaining increasing significance as a means of easier transportation from remote areas, and also as a source for the production of liquid hydrocarbon fuels and products. The Fischer-Tropsch chemistry used first converts natural gas into syn-gas, and then to hydrocarbon fuels. The direct conversion of natural gas (methane) into liquid fuels, primarily through methanol without first producing syn-gas, is therefore of great interest and promise (see Chapter 12). Such processes, which are still under development, could result in the direct commercial production of methanol from methane. Methanol, as a liquid, has the advantage of being much more easily transported than methane. It is also a convenient fuel for internal combustion engines or fuel cells, and can be converted catalytically into ethylene and propylene, and through these into synthetic hydrocarbons and their products.

In a way, natural gas is often seen as a successor to oil because it burns more cleanly, releases less CO_2 per energy unit than any other fossil fuel, and there are still large reserves available. It is, however, more expensive to transport and store than oil, necessitating huge pipeline networks on land and liquefaction to LNG for overseas shipping. Furthermore, due to its lower energy density its use as a transportation fuel has been generally limited to vehicles able to accommodate large pressurized tanks, such as buses. Natural gas is consequently employed mainly in stationary applications such as heating, cooking and electricity generation. Despite the dramatic increase in consumption, having more than doubled since 1970, our proven natural gas reserves are now three times larger than 30 years ago, with a R/P ratio close to 70 years. Regarding the ultimate recoverable amount and the future of natural gas as a fuel, similarly to petroleum oil, there are two major opposing points of view. One side claims that the amounts of natural gas to be ultimately recovered are only equal to or even less than those of petroleum oil. The other side, asserts that there is still enough natural gas left to fill our growing needs for a long time. By adapting Hubbert's concept to predict the future of world gas production, Campbell and Laherrère – the most prominent proponents of this method – forecast that gas field discoveries and production would follow the same pattern as that of oil, and that global supplies of natural gas will decline not long after that of oil. Laherrère estimates the world ultimate natural gas reserves at 350 Tm^3, and conventional sources as representing 280 Tm^3. Unconventional sources such as coalbed methane, tight sand or shale gas are estimated around 70 Tm^3. Based on these estimates, if the consumption rate continues to increase at the current pace, world gas production would peak around the year 2030 [36]. Hubbertians, therefore dismiss gas as a viable long-term alternative to oil to fulfill our future energy needs. However, economic conditions and increasingly higher natural gas prices may also cause the demand to decrease and thus extend its availability. As in the case of oil, there are other estimates that are more optimistic. The USGS assessed the global conventional natural gas resources as over 430 Tm^3, the energetic equivalent of almost 2600 billion barrels or 345 Gt of oil [24]. Other recent estimates for ultimately recoverable conventional natural gas range between 380 and 490 Gt oil equivalent [37]. Consequently, Odell forecasts a conventional natural gas production peak at around 2050 [38, 39]. Taking also into account unconventional natural gas sources, he predicts the extraction peak for combined conventional and unconventional natural gas by 2090. As the recoverable resources for unconventional natural gas are known to an even lesser extent than those for conventional natural gas, this prediction is at best speculative. Some unconventional gas sources such as coalbed methane and tight gas are already exploited on a large scale, principally in the United States. There is a very large potential for the utilization of methane hydrates, which are considered to be present in staggering quantities under the sea floor and arctic permafrost. First, however, these must be assessed more accurately. Further, their practical exploitation must be demonstrated by the development of new and effective technologies. Until then, methane hydrates remain a promising, but as-yet uncertain, energy source.

When considering unconventional sources of natural gas, an interesting but as-yet unproven theory which was developed and vigorously defended by late Thomas Gold (a noted astrophysicist) should also be considered. This involves the possible availability of large natural gas resources of abiological or abiotic origin at greater depths in the Earth's crust (abiological deep methane) [40]. According to this suggestion, extraterrestrial carbon of asteroids could have combined under high temperature and pressure deep under the Earth's surface with hydrogen to form hydrocarbons. The most stable of them, methane, would then migrate to the outer crust and accumulate in geological formations. Following this concept, at least part of natural gas could be of non-biological origin. For the time being however, this suggestion of abiological natural gas remains controversial. Observations that vents deep on the bottom of the oceans discharge methane were later recognized to be due to the discharge of hydrogen sulfide which then converts the CO_2 of sea water to methane under the influence of micro-organisms.

Based on our present knowledge, natural gas must also be considered as a finite resource, the production of which will – not unlike that of oil – reach a peak, most probably during the latter part of the 21st century. As in the case of petroleum oil, new solutions must therefore be found to replace progressively declining natural gas reserves.

Chapter 6
The Continuing Need for Hydrocarbons and their Products

Besides still providing the bulk of our energy needs, fossil fuels also are the sources for our hydrocarbon fuels and derived products. Hydrocarbons are the compounds of carbon and hydrogen. In methane (CH_4), the simplest saturated hydrocarbon (alkane) and the main component of natural gas, a single carbon atom is bonded to four hydrogen atoms. The higher homologues of methane, ethane, propane, butane and so on, have the general formula C_nH_{2n+2}, displaying the tendency of carbon to form chains involving C–C bonds. These can be either straight-chain or branched. Carbon can also form multiple bonds with other carbon atoms, resulting in unsaturated hydrocarbons with double or triple, C=C or C≡C bonds. Carbon atoms are also able to form rings. Cyclic ring compounds of carbons involving both saturated and unsaturated systems are abundant and involve aromatic hydrocarbons, a class of hydrocarbons of which benzene is the parent (Fig. 6.1).

All fossil fuels, natural gas, petroleum and coal, are basically hydrocarbons, but they deviate significantly in their hydrogen to carbon ratio and composition. Natural gas, depending on its origin, contains besides methane (usually in concentrations above 80–90%), some of the higher homologous alkanes (ethane, propane, butane). In "wet" natural gases the amount of C_2–C_6 alkanes is more significant. These so-called natural gas liquids, which were generally only considered for their thermal value, are increasingly perceived as feedstocks for more valuable products such as gasoline. Methane itself, though used mainly as a fuel, is also today's primary source of hydrogen and can be transformed (albeit at a considerable energy cost, via syn-gas) to products otherwise obtained from petroleum. Petroleum or crude oil is a remarkably varied substance, both in its composition and uses. Depending on the source, its color can range from almost transparent clear to amber, brown, black or even green, and it may flow like water or be a semi-solid viscous liquid. Crude oil contains hundreds, if not thousands, of individual hydrocarbons but is predominantly constituted of saturated straight-chain compounds (alkanes) and small amounts of branched alkanes, cycloalkanes and aromatics. Petroleum is the most versatile of our three primary fossil fuels, and can be transformed economically and easily to a vast palette of useful products. Coal, on the other hand, is more hydrogen-deficient and contains large, complex hydrocarbon systems composed mainly of aromatic cycles (Fig. 6.2). Its transformation

Beyond Oil and Gas: The Methanol Economy. G. A. Olah, A. Goeppert, G. K. S. Prakash
Copyright © 2006 WILEY-VCH Verlag GmbH & Co. KGaA, Weinheim
ISBN 3-527-31275-7

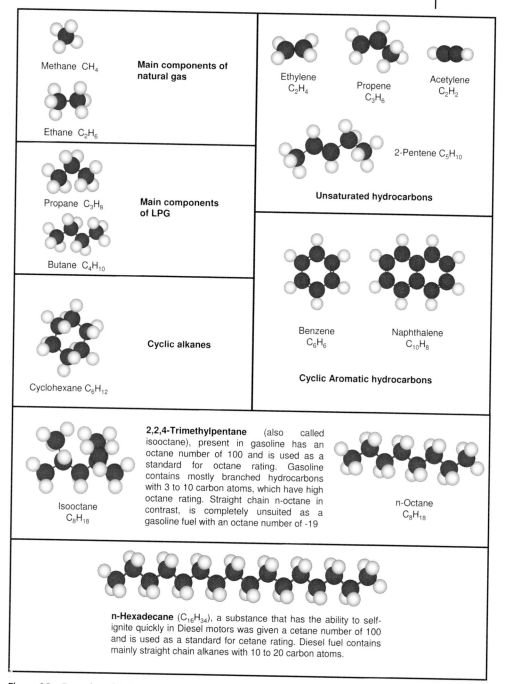

Methane CH₄ — **Main components of natural gas**

Ethane C₂H₆

Ethylene C₂H₄

Propene C₃H₆

Acetylene C₂H₂

2-Pentene C₅H₁₀

Unsaturated hydrocarbons

Propane C₃H₈

Butane C₄H₁₀

Main components of LPG

Cyclohexane C₆H₁₂ — **Cyclic alkanes**

Benzene C₆H₆

Naphthalene C₁₀H₈

Cyclic Aromatic hydrocarbons

Isooctane C₈H₁₈

2,2,4-Trimethylpentane (also called isooctane), present in gasoline has an octane number of 100 and is used as a standard for octane rating. Gasoline contains mostly branched hydrocarbons with 3 to 10 carbon atoms, which have high octane rating. Straight chain n-octane in contrast, is completely unsuited as a gasoline fuel with an octane number of -19

n-Octane C₈H₁₈

n-Hexadecane ($C_{16}H_{34}$), a substance that has the ability to self-ignite quickly in Diesel motors was given a cetane number of 100 and is used as a standard for cetane rating. Diesel fuel contains mainly straight chain alkanes with 10 to 20 carbon atoms.

Figure 6.1 Examples of hydrocarbons.

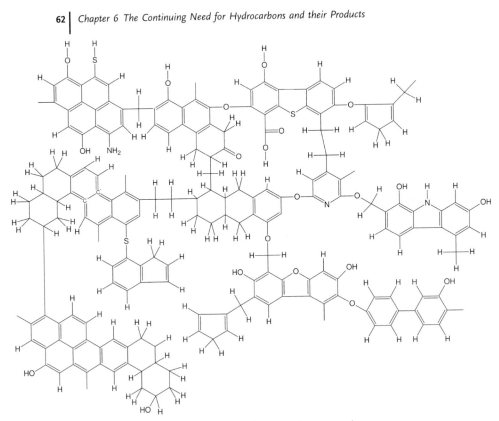

Figure 6.2 Schematic representation of structural groups and connecting bridges in bituminous coal.

to liquid fuels as well as other petrochemicals (coal liquefaction) is feasible and has been demonstrated, involving cleavage and hydrogenation processes, which are very energy-intensive. Coal is thus presently almost exclusively used as a fuel.

In all hydrocarbons, chemical energy is stored in the C–H and C–C bonds. When hydrocarbons react with oxygen, present in the air, in a combustion reaction, they form CO_2 and water. At the same time, energy in the form of heat is released because there is more chemical energy stored in the hydrocarbons and oxygen than in the resulting CO_2 and water. This energy difference is the basis for the utilization of fossil fuels as a source for energy. The burning of hydrocarbons for energy generation is their main use today. A great variety of petrochemicals and chemical products derived from fossil fuels are also made for numerous other applications.

Because of its versatility, our civilization is today especially dependent on oil, and it is so embedded in our daily lives that we hardly think about it. Mankind's use of petroleum is as old as recorded history. As eluded to in earlier chapters, ancient cultures such as the Sumerians and Mesopotamians used asphalt and bitumen from natural pools to seal joints in wooden boats, line water canals or inlay mosaics in walls and floors. At that time, liquid oil was also the fuel of choice for oil lamps. The Egyptians embalmed mummies with asphalt, while the Romans used flaming containers filled with oil as weapons. Native Americans used crude oil for medicinal ointments. As mentioned earlier, the modern petroleum industry however, was only born in the middle of the 19th century in America, with the invention of the kerosene lamp leading to the formation of the first oil companies in Pennsylvania. Commercial production was first aimed at fulfilling the growing demand for kerosene used in lamps. At that time, the lighter gasoline was mainly a wasted byproduct of the distillation of kerosene from crude oil, until the early 1900s when automobiles with gasoline and diesel engines became commonplace. Farm equipment powered by gasoline and diesel fuels soon also became popular, dramatically increasing agricultural productivity. During the 1930s and 1940s, a substantial market for heating oil also developed. Today, our civilization is utterly dependent on oil products that we can find in every area of our lives. The most common are the gasoline used to fuel our automobiles, and the heating oil to warm our homes and offices. Gasoline, diesel and jet fuel provide more than 95% of all the energy consumed in the transportation field by automobiles, trucks, farm and industrial machinery, trains, ships and aircraft. Transportation fuels alone account for about 60% of the petroleum consumed worldwide [14]. Oil is an essential raw material (with natural gas) for the synthesis of fertilizers on which agriculture depends, and it also provides us with the chemicals, dyes, cosmetics, pharmaceuticals, plastics and an innumerable host of other products that are essential for everyday life. As an indicator of our enormous demand for petroleum products, we can take the example of the United States which uses on average more than 20 million barrels of oil per day, representing about 12 L per day for each person in the country.

Oil in its raw state has limited uses, and the processing of crude oil via refining is first necessary to unlock the full potential of this resource. The earliest refineries in the mid-1800s used distillation to produce mainly kerosene, and distillation remains the starting point for oil refining to this day, though many more complex processes such as cracking, reforming and alkylation have been added to convert crude oil into a wide array of desired products.

Fractional Distillation

Atmospheric fractional distillation, the first and core step in the refining process, uses heat to separate the numerous different hydrocarbon components present in oil into several fractions, depending on their boiling point. The first volatile products of the crude oil are gaseous hydrocarbons of one to four carbon atoms that were dissolved in the oil. Liquid petroleum gases (LPG) can be used as fuel or converted to useful chemicals. With increasingly higher boiling range and number of contained carbon atoms, the next fractions are gasoline, naphtha, jet fuel, kerosene, gas oil, and heating and fuel oils. Finally, the heaviest products with boiling temperatures above 600 °C, known as residual oil, can be separated into such individual constituents as coke, asphalt, tars, and waxes. Alternatively, it can be further processed by cracking to produce lighter fractions and maximize the output of the most desirable products such as gasoline and diesel fuel.

After atmospheric distillation, the relative amounts of usable fractions obtained from crude oil do not coincide with the commercial needs, and therefore additional "downstream" processing is needed. In general, these processes are designed to increase the yield of lighter, higher-value products such as gasoline. Downstream operations include vacuum distillation, cracking, reforming, alkylation, isomerization, and oligomerization.

Thermal Cracking

Thermal cracking, the first downstream process that changed the petroleum industry, permitted – by the use of high temperature and pressure – the heavy, low-value feedstock to be broken into lighter, higher-value heating oil, diesel, and gasoline. Thermal cracking was followed by other developments in the 1920s and 1930s. Polymerization (oligomerization) yields high-octane gasoline from olefins produced as byproducts in thermal cracking units. Vacuum distillation takes the residual oil, which is left at the bottom of the column after atmospheric distillation, and allows its further separation. Using thermal action, the visbreaking unit can reduce the viscosity of the residues from the distillation columns, allowing a much easier flow and processing of the product. Coking uses the products left from atmospheric and vacuum distillation and produces, via thermal treatment at high temperature, gasoline, heavy oil, fuel gas and petroleum coke, an almost pure carbon residue.

During World War II, the petroleum industry shifted to products that were essential for the war effort, and especially advanced aviation fuels. This resulted in the development of the alkylation process in which a catalyst (usually sulfuric acid or hydrofluoric acid) is used to combine a branched alkane with an olefin (alkene) to produce high-octane compounds for use as high-quality gasoline components. Nowadays, this process is one of the most important steps in the production of high-octane gasoline for motor cars. Other advances during that period included catalytic cracking which uses a catalyst to accelerate the cracking process, and isomerization converting straight-chain alkanes into branched ones having a much higher octane number. Catalytic reforming produces higher-octane components for gasoline from lower-octane naphtha feedstock recovered in the distillation process.

Over time, these processes have been continuously improved and new schemes such as dehydrogenation (to produce useful alkenes from alkanes) and hydrocracking have been added. Without these it would be impossible to produce economically the large amounts of valuable lighter fractions from the intermediate and heavy compounds that constitute most of the crude oils.

Figure 6.3 BP Grange-mouth refinery, UK (© BP p. l. c.).

Crude oil is a complex mixture of hydrocarbons. Depending on its source, it varies (among other things) in terms of its color, viscosity, and content of sulfur, nitrogen and other impurities. Most commonly, crude oils are generally classified by their density and sulfur content. Less dense, or lighter, crude oils have a higher share of the more valuable light hydrocarbons that can be recovered by simple distillation. Denser or heavier crude oils contain more heavy hydrocarbons of lower value and require additional processing steps to produce the desired range of products. Some crude oils also contain significant amounts of sulfur and heavy metals which are detrimental as they act as contaminants for most refining processes and finished products; they are also pollutants, necessitating additional purification steps. Because the quality of oil varies so widely, refineries differ in their complexity depending on the type of crude oil to be processed as well as the range of products desired. To allow the most flexibility, modern refineries (Fig. 6.3) however are designed to process various blends of different crude oils.

The processing of crude oil opened up the route to petrochemicals, because cracking produces, beside fuels, also unsaturated hydrocarbons containing one or more C=C double bonds, in particular ethylene, propylene, butylene, and butadiene. These compounds are called olefins and, unlike paraffins (the main saturated components of oil), can be readily used and further transformed by chemical reactions. They constitute the basic building blocks for numerous products and are produced in very large quantities. Yearly, some 100 million tons of ethylene and 60 million tons of propylene are manufactured worldwide. They are used mainly for the production of synthetic polymers, and also for many other products. Ethylene, for example, is the starting material for polyethylene, propylene for polypropylene, and synthetic rubber can be produced from butadiene. Besides olefins, aromatic compounds – principally benzene, toluene and xylenes – are also obtained during crude oil refining. These aromatic compounds are important starting materials for synthetic products such as polystyrene, nylon, polyurethane, or polyesters. In total, about 6% of crude oil is used today to produce petrochemicals. Crude oil, together with natural gas, are the sources for some 95% of organic chemicals, yielding products such as lubricants, detergents, solvents, waxes, rubbers, insulation materials, insecticides, herbicides, synthetic fibers for clothing, plastics, fertilizers, and many others. The advance of chemistry in the 20th century has depended – and still depends to a large extent today – on the availability of petrochemical building blocks.

The history of petrochemistry started around the 1900s, at which time the demand for natural rubber collected from *Hevea* trees began to surpass the supply when new applications such as motor car tires were introduced. Replacement materials were needed, and this led to the invention of synthetic rubbers; the process began with the polymerization of butadiene, which turned out to be superior to the natural products.

In 1907, the first fully synthetic plastic – named "bakelite" – was created by the reaction of phenol and formaldehyde. This new liquid resin, when hardened, took the shape of the vessel in which it was formed. Unlike earlier plastics such as celluloid, it could not be remelted, it retained its shape under any circumstances, and it would not readily burn, melt or decompose in common acids or solvents. Bakelite is still used today as an electric insulator. During the next decade, cellophane, the first clear, flexible and waterproof packaging material was developed. The 1920s and 1930s witnessed the introduction of petrochemical solvents and the discoveries of numerous new plastics and polymers including nylon, polyvinyl chloride (PVC), Teflon, polyesters, and polyethylene. The petrochemical industry grew especially rapidly during the 1940s when, during World War II, the demand for synthetic materials to replace costly and often difficult to obtain, less-efficient natural products led the industry to develop into what would become a major factor in today's technological society. During that time, many other synthetic materials such as acrylics, neoprene, styrene-butadiene rubber (SBR) and others went into use, taking the place of dwindling natural material supplies. Among other applications, Nylon was used to make parachutes and to reinforce tires, besides its latter use for synthetic fibers, especially for nylon stockings. Plexiglas was initially introduced during World War II for airplane windows. Lightweight polyethylene insulation made it possible to mount otherwise too-heavy radar units on airplanes. From then on, petrochemical products – and especially polymers – moved into an astonishing variety of areas. Together with oil- and natural gas-based fuels, they touch our daily lives in countless ways. In fact, we are so used to them that we no longer notice their unique nature!

Today, our households are full of products derived from hydrocarbons. In the bathrooms, shampoo and shower gel are composed of synthetic soap formulations and their bottles made out of polyethylene, polypropylene or PVC which have the advantage of being unbreakable. The toothbrushes, hair blow dryers, combs, shower curtains and toilet brushes are all made of plastics. In the kitchen, the refrigerators, coffee machines, toasters, microwaves and other appliances are all composed in part from synthetic polymers. Without proper insulation with polyurethane foam, refrigerators and freezers would consume much more energy. Teflon coating led to non-stick cookware, and plastic packaging of food allows for better conservation and protection against contamination. Plastic bottles and containers for all kinds of beverages, mostly made from polyethylene terephthalate (PET), are safer and lighter than glass bottles. Waste-disposal garbage bags and many other uses of plastics have become essential. In the bedroom, from the "linens" to the alarm clock which wakes us up in the morning, all are made using polymers. A large part of our clothes hanging in the closet or folded in drawers are based on synthetic fibers such as polyesters, polyacrylics, or rayon. To wash our clothes and other fabrics, we use detergents or dry-cleaning solvents, both made from hydrocarbons. In the living rooms, carpeting, furniture and its coverings, televisions, video recorders, home entertainment systems, DVD and CD players together with their remote controls, would not exist without plastics. From the outdated vinyl records to audio and video tapes or the modern CDs and DVDs, all are made using polymers. The electric cables to power all the appliances and equipment would be difficult to run safely throughout the house without plastic insulation. Even the utility lines which bring natural gas or water into our homes are nowadays made out of plastics such as PVC. PVC is also increasingly the material of choice to replace more maintenance-demanding wood for windows, doors and other construction applications. Heating and cooling our homes and buildings uses natural gas, heating oil, and electricity which, for a large part, comes from the combustion of fossil fuels. In our gardens, we relax on weatherproof plastic chairs and let the sprinklers attached to the underlying network of PVC pipes water the lawns treated with synthetic fertilizers to keep them attractive.

Once we step into our motor car, we are literally surrounded by hydrocarbon products. The seats, head and arm rests are made of synthetic fibers, and the upholstery cushioning from urethane foams. The dashboard, steering wheel, door panels, floor mats and almost all the apparent inner parts use different plastics with specific properties. Security features such as bumpers, baby seats and life-saving airbags are also manufactured with polymeric materials. Even structural steel and aluminum frames are increasingly partly replaced by new generations of high-performance composite plastics. The versatility, durability and cost-saving properties of plastics have given them much advantage in the automotive industry. Furthermore, their light weight – especially compared to steel – allows a better fuel efficiency to be achieved. Considering the engine and drive-train, all of the fluids necessary for their proper operation – motor oil, transmission, cooling and steering fluid – are hydrocarbon-based, as are the tires. Our vehicles run on roads which are asphalted, and are powered by gasoline or diesel fuels. Oxygenated and other petrochemical-based additives are added to gasoline in order to improve engine performances and reduce air pollution.

Our furniture, carpeting, computers, printers, telephones, pagers, and mobile phones to ball point pens are also made, at least in part, from plastics. Most of our outdoor activities involve varied hydrocarbon-derived products present in many sport articles, from inline skating, skis and ski boots, snowboards, golf equipment, basket and tennis balls, canoes and boats, bicycle helmets, swimsuits or even the elastic cords for bungee-jumping.

Petrochemicals also contribute to the amazing progress of healthcare and hygiene. For many years, plastic medical products, from flexible intravenous solution and blood bags to disposable syringes to artificial heart valves, have helped doctors and nurses to save numerous lives. Seriously injured or handicapped persons can recover a considerable level of mobility thanks to plastics and resins for artificial joints and limbs. Various plastics are used in medical applications because of their clarity, transparency, flexibility, sterilizability, and ease of processing. Today, an increasing number of tailor-made polymers with very specific properties are being widely developed and used. The crucial factor for plastics placed into the human body for extended periods is their biocompatibility and stability. Petrochemicals are also the basis of synthetic pharmaceutical products. Phenol, for example, has long been used as a starting material for aspirin. Other basic chemicals are the source for medicines aimed at reducing cholesterol, reducing blood pressure, and curing skin diseases. On the prevention side, with the use of condoms made from elastic plastics, the risk of spreading sexually transmittable illnesses such as AIDS can be dramatically reduced. Condoms can also play a major role in population control.

Hydrocarbons also play an essential role in increasing the production of food and other crops. By introducing modern agricultural equipment such as tractors and harvesters fueled by gasoline or diesel fuel, many agricultural operations have been mechanized, reducing dramatically the former need for back-breaking human or animal labor. To increase the yield of crops and to avoid the exhaustion of necessary plant nutriments from the soil, fertilizers are needed. Potassium, phosphorus, calcium, magnesium, iron, and other minerals used as fertilizers can be relatively easily extracted from natural sources, but sources of nitrogen – perhaps the most essential element – are rare in nature. During the 19th century, the main source of nitrogen was "guano", a bird fecal matter collected on islands near the coast of Chile and shipped to Europe or other distant destinations. However, at the start of the 20th century, two chemists – Fritz Haber and Carl Bosch – developed a revolutionary method to produce ammonia (NH_3) from nitrogen contained in the air and hydrogen obtained from methane. Ammonia remains the basis of nitrogenous fertilizers used today, including urea. The use of pesticides synthesized from petrochemical building blocks is also essential to avoid crops losses from diseases and insects, and be able to continue the efficient provision of food for the growing world population.

As we can see, hydrocarbons and their products extracted from fossil resources, together with advances in chemistry, have made our lives better, longer, safer, and more comfortable. They are essential for our society and have transformed or even revolutionized transportation, information, communication, entertainment, medicine, and agriculture, and essentially the way in which we live and work. Further progress will allow future generations not only to enjoy the same quality of life, but most probably also to improve upon it.

In the future, as our non-renewable resources decrease, we will need to replace fossil fuels with different other sources in order to fulfill our energy needs. The demand for derived hydrocarbon products (transportation fuels, plastics, synthetic fabrics, elastomers, rubbers, paints and innumerable other products) is unlikely to regress, however, and it will be necessary to identify sustainable synthetic hydrocarbon sources to produce these commodities and products. In the short term, we can still rely on the available oil and natural gas sources, as well as more extensive coal resources. Using syn-gas from natural gas and coal for Fischer-Tropsch processes allows the production of synthetic hydrocarbons on an industrial scale, as shown during World War II in Germany and in the 1960s in South Africa. At present, Qatar is involved in developing a large-scale Fischer-Tropsch-based industry to produce diesel fuel from abundant natural gas resources. But Fischer-Tropsch chemistry, besides being highly energy-demanding, capital intensive and environmentally polluting, is also based on our diminishing fossil fuel resources. Hence, in the longer term new solutions based on renewable sources from biomass or significantly through methanol obtained by recycling of CO_2 are needed (see Chapters 10–14).

Assuming that mankind can solve its energy needs by using alternative sources and atomic energy, there will still be a need for synthetic hydrocarbons and their products, for convenient transportation fuels, and for various derived materials and products (Figs. 6.4 and 6.5). It has been frequently suggested that agricultural or other bio-sources might be used in this way. Whilst agricultural ethanol, produced by the fermentation of sugar cane, corn or other crops, can be produced in significant quantities and used as a fuel additive, its large-scale use via dehydration to ethylene and subsequently to synthetic hydrocarbons and products, is highly questionable (this was first proposed by Lenin in the 1920s, but soon abandoned). The amount of ethanol required would be staggering, and the production of crops would in turn still require large quantities of hydrocarbon fuels, fertilizers, and pesticides. One reasonable new approach to overcome this problem is based on methanol, which can be produced from a variety of sources and – most importantly – from carbon dioxide and hydrogen (obtained by the electrolysis of water using any form of renewable and atomic energy) to serve as a convenient feedstock for synthetic hydrocarbon and products; this is described in the proposed "Methanol Economy".

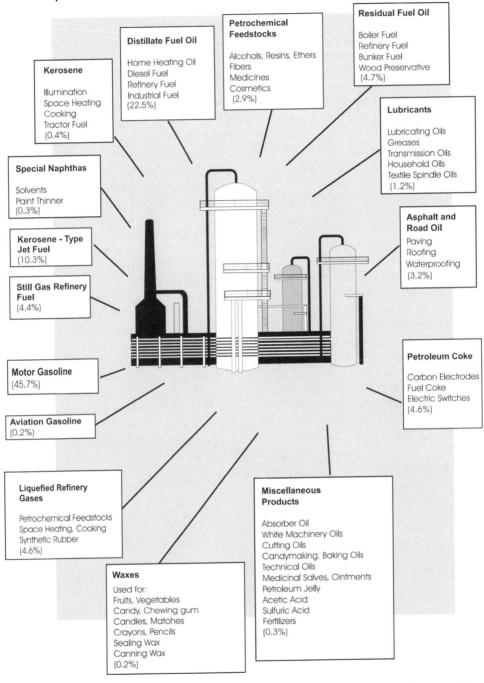

Figure 6.4 Petroleum products and uses in the United States (% refinery yields of finished products in 1997). Source: Petroleum – An Energy Profile 1999, Energy Information Administration (U.S.).

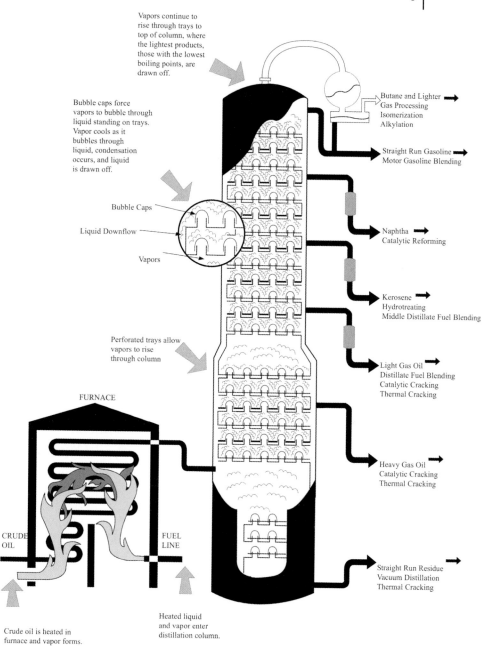

Vapors continue to rise through trays to top of column, where the lightest products, those with the lowest boiling points, are drawn off.

Bubble caps force vapors to bubble through liquid standing on trays. Vapor cools as it bubbles through liquid, condensation occurs, and liquid is drawn off.

Bubble Caps

Liquid Downflow

Vapors

Perforated trays allow vapors to rise through column

FURNACE

CRUDE OIL

FUEL LINE

Crude oil is heated in furnace and vapor forms.

Heated liquid and vapor enter distillation column.

Butane and Lighter
Gas Processing
Isomerization
Alkylation

Straight Run Gasoline
Motor Gasoline Blending

Naphtha
Catalytic Reforming

Kerosene
Hydrotreating
Middle Distillate Fuel Blending

Light Gas Oil
Distillate Fuel Blending
Catalytic Cracking
Thermal Cracking

Heavy Gas Oil
Catalytic Cracking
Thermal Cracking

Straight Run Residue
Vacuum Distillation
Thermal Cracking

Figure 6.5 Crude oil destillation. Source: Petroleum: An Energy Profile 1999, Energy Information Administration (U.S.).

Chapter 7
Fossil Fuels and Climate Change

Climate change, and especially global warming, is receiving much attention and is considered as one of the most pressing and severe global environmental problems facing humanity. Its significance has been widely reported to the general public by the media, always eager to emphasize the possible catastrophic consequences of global warming and link it even to claimed "extreme" weather experienced in the recent years. The Hollywood film *The day after tomorrow*, an environmental catastrophe movie in which a new ice age sets in over the northern hemisphere in a matter of days, gave many people the impression that an increasing greenhouse effect will have practically instant, drastic, apocalyptic consequences on our climate. It adds to the fear that global warming could result in the short-term destruction of ecosystems, cause more powerful and devastating storms and hurricanes as well as melting of the polar ice caps followed by flooding of low-lying coastal areas in Bangladesh, The Maldives, The Netherlands, and Florida. To understand whether such fears have solid foundations we have to look at the facts. The most reliable and internationally accepted and trusted source of information about global climate change is in the International Panel on Climate Change (IPCC) reports. The IPCC was jointly established in 1988 by the World Meteorological Organization (WMO) and the United Nation Environmental Program (UNEP). Its reports are the reference on global climate change for most policymakers, scientists, and experts. The third assessment report published in 2001 shows that the Earth's average surface temperature has increased by about 0.6 ± 0.2 °C from 1861 to 2000 [41] (Figs. 7.1 and 7.2). Closer analysis reveals that most of the warming occurred, without a clear explanation, during two periods, from 1910 to 1945, and from 1976 to 2000. The 1990s has been the warmest decade and 1998 the warmest year since 1861, based on the systematic and global use of thermometers (although measuring the average surface temperature of the Earth is still a complex and difficult task). Temperatures before this date have to be determined indirectly employing so-called proxy data obtained from objects sensible to temperature and still present and measurable today, such as tree rings, ice cores, or corals. The collected data, even if less accurate than direct measurements, indicate that the increase in temperature in the 20th century was probably the largest in the past 1000 years. As a consequence of higher temperatures, there has been a widespread retreat of mountain glaciers in the non-polar regions as

Beyond Oil and Gas: The Methanol Economy. G. A. Olah, A. Goeppert, G. K. S. Prakash
Copyright © 2006 WILEY-VCH Verlag GmbH & Co. KGaA, Weinheim
ISBN 3-527-31275-7

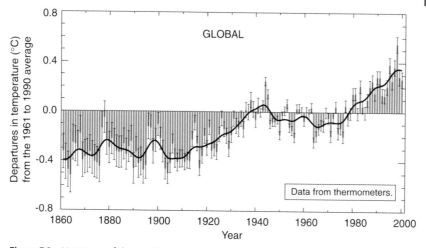

Figure 7.1 Variations of the Earth's surface temperature for the past 140 years. (Source: IPCC, Third Assessment Report, Climate Change 2001: The Scientific Basis.)

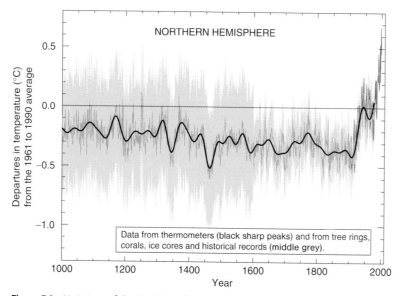

Figure 7.2 Variations of the Earth's surface temperature for the past 1000 years. (Source: IPCC, Third Assessment Report, Climate Change 2001: The Scientific Basis.)

well as a 10–15% decrease in the sea-ice extent during spring and summer in Arctic since the 1950s. The global average sea level rose between 0.1 and 0.2 m during the 20th century, and the global ocean heat content has increased since the 1950s. However, some parts of the globe, mainly in the Southern hemisphere, have not warmed in recent decades and no clear trends in the sea-ice extent of Antarctica are apparent since the end of the 1970s. Furthermore, no significant trends or changes in storm activity, frequency of tornadoes, thunder or hail were noticed over the 20th century. It should also be recognized that, long before human activity on Earth, there were many ice-age periods followed by warming. Thus, human activity caused climate change, although significant, must be considered as superimposed on that caused by natural cycles.

Considering the variations from the past, the question that must be raised is, can we predict climate changes for the future? First, the reasons for past warming periods must be explained. Global warming is now recognized as being based significantly on the greenhouse effect caused by heat-absorbing gases that are present in the atmosphere; these trap some of Earth's reflected infrared radiation of the sun and act like a giant blanket around our planet. These so-called "greenhouse gases" include water vapor, CO_2, methane, nitrous oxide, ozone, and some others. More recently, it was established that man-made chlorofluorocarbons (CFCs) contributed to the depletion of the ozone layer, which protects the Earth from excessive damaging ultraviolet (UV) radiations from the sun. The damage was most pronounced in polar regions, where holes were detected in the Earth's protective ozone layer. Without naturally occurring greenhouse gases such as CO_2, water vapor and methane in the atmosphere, the Earth's average temperature would be much cooler, comparable to the atmosphere on Mars. At the expected −18 °C on average, most of the water would be frozen all year long and the emergence and evolution of life as we know it would have been much more difficult, if possible at all. On the other hand, too much of the greenhouse effect is also detrimental. Such a situation is encountered on Venus, which has an atmosphere rich in CO_2, inducing temperatures above the melting point of metallic lead and making it as hostile to life than the cold Mars. The mankind-caused increased greenhouse effect is real, and of concern. It is important to be concerned about the greenhouse gas concentrations in our atmosphere in order to keep the temperature of the Earth under the control of human effects and to maintain life as we know it. If the concentration of CO_2 or other greenhouse gases in the atmosphere continues to increase substantially, the effect would most certainly lead to further increases in global average temperature. As early as 1895, the Swedish chemist Svante Arrhenius, calculated that a doubling in CO_2 concentration, through human action, would result in an increase of 5–6 °C in the Earth's average surface temperature [42]. He even estimated the amount of coal that would have to be burned, and how long such a process would take. Of course since then we have also burned large amounts of other carbon-based fossil fuels, such as oil and gas. In fact, CO_2 concentrations in the atmosphere have been increasing steadily for more than a century. During several thousand years before the industrial era, the CO_2 concentration was relatively

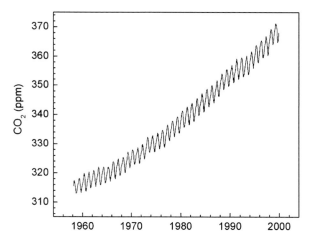

Figure 7.3 Atmospheric CO_2 concentration measured at Mauna Loa, Hawaii. (Source: CDIAC, Carbon Dioxide Information Analysis Center.)

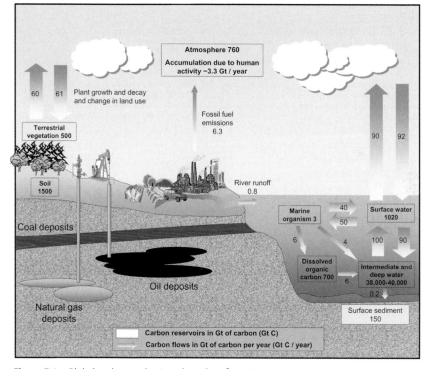

Figure 7.4 Global carbon cycle. Based on data from IPCC.

stable around 280 ppm, but since 1750 the atmospheric concentration of CO_2 has increased by 31% to reach some 370 ppm today (Figs. 7.3 and 7.4). The present CO_2 concentration has not been exceeded in the past 420 000 years, and the current rate of increase is unprecedented in at least the past 20 000 years. It is now widely accepted that the observed increase in CO_2 concentration is significantly due to anthropogenic source – that is, due to human activities. The combustion of fossil fuels is by far the largest contributor to man-made anthropogenic CO_2 emissions, the remainder being mainly a result of land-use change, especially deforestation (Figs. 7.5 and 7.6). About half of the anthropogenic CO_2 emissions is absorbed again by the oceans and vegetation of land areas, whereas the remainder is added to the atmosphere, increasing CO_2 concentration. It should be remembered that the overall concentration of CO_2 in the air is only about 0.037% (370 ppm), but this plays an essential role in maintaining life on Earth.

So far, most of the concerns about greenhouse gases have been focused on CO_2, which represents about 60% of the human-caused greenhouse gases present in the atmosphere [21]. Less attention has been given to other greenhouse gases, the concentrations of which have also been increasing, especially methane, nitrous oxide and, more recently, chlorofluorocarbons. Atmospheric methane concentrations have risen by 150% since 1750, and continue to increase, being now at the highest in the past 420 000 years. Methane has both natural and human-related origins. Decaying plant matter in wetlands and rice paddies ac-

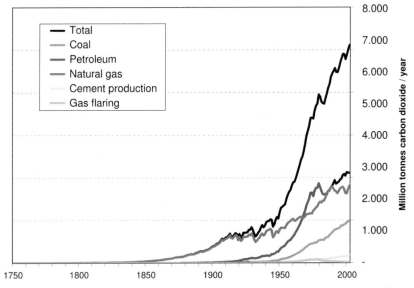

Figure 7.5 Global CO_2 emissions from fossil fuel burning, cement production, and gas flaring for the period 1750 until 2002. Source: Marland, G., T.A. Boden, and R. J. Andres. 2005. Global, Regional, and National CO_2 Emissions. In Trends: A Compendium of Data on Global Change. Carbon Dioxide Information Analysis Center, Oak Ridge National Laboratory, U.S. Department of Energy, Oak Ridge, Tenn., USA.

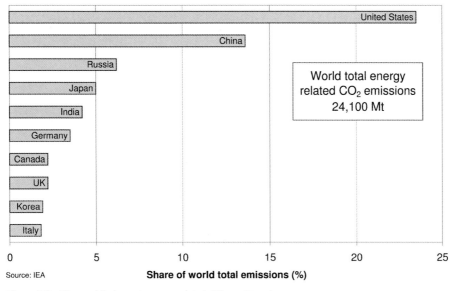

Source: IEA

Figure 7.6 The world's largest energy-related CO_2 emitters in 2002. Source: IEA.

counts for some three-fourths of natural methane emissions. Other sources include oceans, gas hydrates releasing methane due to changes in temperature or pressure, and termites which produce methane as a part of their normal digestive process. Similarly, in the digestive system of domesticated animals such as cattle, buffalo, sheep, goats or camels, cellulose from plant material is broken down by microbial fermentation in their digestion process, producing also significant amounts of methane as a byproduct. This constitutes a major source of human-related methane emissions besides wet rice agriculture, oil and gas production, and landfills. Methane generated in landfills, as waste decomposes under anaerobic (oxygen-lacking) conditions, is increasingly used as an energy source and is not allowed to escape into the atmosphere. In total, about half of the current methane emissions are anthropogenic. Even with a concentration increase of only 1060 ppb in the atmosphere (much less than that of CO_2), methane contributes to 20% of the increased greenhouse effect [21]. Methane is a more effective greenhouse gas because it has a higher Global Warming Potential (GWP). The GWP is a measure of the relative heat absorption capability of a given substance compared to CO_2 and integrated over a chosen time. Over a 100-year period methane, has a 23-fold higher GWP than CO_2.

The atmospheric concentration of nitrous oxide (so-called "laughing gas"; N_2O) has also increased by about 16% during the industrial era to levels not seen in the past thousand years. Because of its high GWP factor of 296, N_2O contributes to 6% of the increased greenhouse effect [21].

Chlorofluorocarbons (CFCs) destroy the Earth's ozone layer and also absorb infrared radiations. The latter also contributes to global warming. Due to their ozone-depleting properties, CFCs have been phased out by the Montreal Protocol and their concentrations in the atmosphere are now diminishing, or at least increasing more slowly. Substitutes for CFCs may be less harmful for the ozone layer, but some compounds such as hydrochlorofluorocarbons (HCFC) and hydrofluorocarbons (HFC) still have GWP factors reaching up to 12 000. The concentration of HCFC and HFC in the atmosphere is increasing. All halocarbons combined represent today some 14% of the increased greenhouse effect [21].

Atmospheric aerosols are small airborne particles and droplets produced by a variety of processes that can be natural (such as volcanic eruptions or sand dusts), or anthropogenic (mainly through fossil fuel and biomass burning). Sulfate aerosols from fossil fuel burning and other aerosols from volcanoes and biomass burning reflect and scatter solar energy before it can reach the Earth's surface, and thus have a cooling effect on the atmosphere. Aerosols can also enhance the condensation of water droplets and consequently favor cloud formation, increasing the reflection of incoming sunlight back into space. Whereas aerosols are cooling the atmosphere, the magnitude of this effect does not compensate for the heating induced by greenhouse gases. Furthermore, aerosols have a much shorter lifetime than greenhouse gases in the atmosphere.

Table 7.1 Global warming potentials (GWPs) of some greenhouse gases.

Gas	Formula	Global warming potential[a]	Atmospheric lifetime (years)
Carbon dioxide	CO_2	1	
Methane	CH_4	23	12
Nitrous oxide	N_2O	296	114
Hydrofluorocarbons (HFC)		12–12 000	0.3–260
Examples:			
HFC-23	CHF_3	12 000	260
HFC-32	CH_2F_2	550	5
HFC-134a	CH_2FCF_3	1300	14
Fully fluorinated species		5700–22 200	2600–50 000
Examples:			
Perfluoromethane	CF_4	5700	50 000
Perfluoroethane	C_2F_6	11 900	10 000
Sulfur hexafluoride	SF_6	22 200	3200

[a] Over a 100-year time horizon.
Based on data from IPCC, Third Assessment Report, 2001.

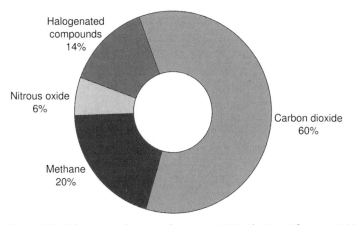

Figure 7.7 Relative contributions of greenhouse gases to the increased greenhouse effect induced by human activity. Source: IPCC Third Assessment Report. Climate Change 2001: The Scientific Basis, Table 6.1. Total forcing change between 1750 and 1998 is 2.48 W/m².

When considering the question of global temperature change, it must be remembered that the greenhouse gas of overwhelming concentration is the moisture in the air, and that we have only extremely limited control of this parameter. Nature, therefore, has its own major controlling effect on the climate, including the cycle of celestial alignment of the Earth relative to the Sun [43]. Although temperature and the CO_2 content of the air are clearly related, some point out that increasing temperature would cause increasing CO_2 levels, but not necessarily vice versa [44].

By using the vast amount of data collected about greenhouse gases and other elements able to influence the Earth's atmosphere, computer simulations have been created to investigate the causes of climate changes and, most importantly, the observed warming trend of our time. These models have become more sophisticated over the years, but some aspects can still not be effectively simulated due to the extreme complexity of climate systems. Nevertheless, they now provide a reasonably accurate model of climate variations that have occurred during the past 150 years. Over the past 50 years, the rate and increase of warming seems to be directly correlated to the increase in greenhouse gas concentration in the atmosphere. This led the IPCC to state in its third assessment report that "…there is new and stronger evidence that most of the warming observed over the last 50 years is attributable to human activities". Whereas there is no question that the human activities of a growing population on Earth will continue to affect our climate, the degree to which the human effect is superimposed on Nature's own cycles remains to be questioned. By continuing to emit greenhouse gases, humanity will continue to influence the atmospheric composition well into the 21st century. CO_2 emissions from fossil fuel combustion especially are expected to remain a major factor for coming trends in global warming. However, as our fossil fuels reserves are being steadily depleted, this trend will not go on indefinitely as fossil fuel will not last significantly for more than a century or two. Based on mathematical models similar to the ones used to describe the past, projections for future climate changes have been made. Of course they are highly dependent upon assumptions such as future greenhouse gas emissions, and therefore a large number of potential scenarios can be envisioned. Nevertheless, they all project a global average temperature increase, from 1.4 to 5.8 °C over the remainder of this century, with atmospheric CO_2 concentrations ranging from 540 to 970 ppm. Realistically however, a rise of 3 °C or less is more likely than a 6 °C increase. Nonetheless, such a temperature change would have widespread repercussions on the Earth. Initially, it would affect the sea level, which is expected to rise up to 0.88 m during the next century, primarily due to thermal expansion and water added from melting glaciers and ice caps. Even if the overall greenhouse gas concentrations were to be stabilized, the local warming of Greenland for example, if sustained, could lead to a further meltdown of the ice sheet with a resulting increase of sea level. This would certainly be very bad news for low-lying islands and countries such as The Netherlands, Bangladesh, and The Maldives. Globally, due to warming of the oceans and increased evaporation more intense and heavy precipitations can be expected with increased risks of flooding. At the same time, more cloud formation and precipitation could mitigate atmospheric warming.

As a result of higher temperatures, not only sensitive ecosystems but also agricultural productivity in many regions would be drastically affected. The consequences will differ greatly between the industrialized world and developing countries. Globally, industrialized countries in temperate climates will gain longer growing seasons, while the higher CO_2 concentrations will act as a fertilizer, increasing crop growth and yield [45]. Higher temperatures could also extend the growing range of some crops such as wheat further north in Canada or parts of Siberia. At the same time, crop yields in other areas may be reduced by more frequent droughts. Improved irrigation, changes in farming methods or selection of more adapted crops would be necessary.

Overall, it is likely that the adverse impact of global warming will be the greatest for lower-income populations, particularly in tropical countries where fewer resources are available to adapt to the negative effects of climate change. But with the advantages that can also be expected from a moderate temperature increase, the overall balance of the global change would be difficult to predict. It is clear that man's contribution to climate change should be of great concern to all of us, and it is important that steps be taken as soon as possible to mitigate greenhouse gas emissions and their consequences. It is also necessary to point out that, in the longer time scale, the man-made causes of climate change (primarily considered as global warming) will in many instances be only of a temporary nature. As the world's fossil fuel reserves are finite and not renewable, and there is a leveling of the world's population, then excess CO_2 release will by necessity decrease. It is not only Nature's own technology but also effective CO_2

chemical recycling technologies (see Chapters 10 to 14 on the methanol economy) that will tend to keep CO_2 levels balanced.

Mitigation

Several options are available in the short range to limit or reduce greenhouse gas emissions. The efficiency of current processes which consume fossil fuels can be improved to increase the amount of useful energy per unit of CO_2 emitted. This can, for example, be achieved by the construction of more efficient electric power plants using improved technologies, by driving cars with higher fuel efficiencies, and by employing appliances that consume less energy. Conservation, perhaps through better insulation of commercial buildings as well as private homes, can also result in tremendous cuts in heating or air-conditioning energy consumption. These measures not only reduce the CO_2 emissions and other environmental impact, but they are also economically advantageous as they reduce the amounts of fuel necessary, and therefore the costs. At the same time they can reduce the dependency of energy-importing countries, such as Europe, the United States and Japan, from oil and gas suppliers. Such savings alone, however, can only extend the availability of needed and accessible fossil fuels in the relatively short term.

A switch to fuels that emit less or no CO_2 per unit of energy produced will be necessary. Natural gas will continue to play, as long as it is readily available, an important role in emission reduction by partially replacing coal in electricity generation. Non-fossil fuel energy sources will, however, need to play an increasingly important role to provide for our future energy needs. Among them, hydropower is already widely used and well developed for the generation of electricity, but suitable hydropower resources (rivers, waterfalls, etc.) are limited by their nature. Wind, solar and geothermal energy and energy from the combustion of biomass represent an increasing – but still small – fraction of our energy needs. One of the main obstacles to a wider application of these renewable energy sources is their cost, as well as technological limitations. All this makes the use and extension of nuclear fission power, which is a well-established and reliable source of energy that does not emit CO_2, inevitable on a massive scale for the future. Of course, nuclear power should be made even safer, and problems of the storage and disposal of radioactive waste must be solved. There is also a need to develop new generations of nuclear reactors, including breeder reactors and eventually controlled fusion.

As long as we use fossil fuels, the CO_2 produced during their combustion should be captured from industrial exhausts, particularly flue gases from coal- or gas-fired power plants or cement factories, in order to avoid its emission to the atmosphere. The removal and capture of CO_2 from exhausts can be achieved using a variety of techniques, including chemical absorption, adsorption onto solids, and permeation through membranes. Until now, none of these has been applied on the scale of large commercial power-plants, but the system is feasible and has been tested on a limited scale at a few locations. The removed CO_2 can then be compressed and, according to present plans, sequestrated underground in geological formations, depleted oil or gas reservoirs, or even at the bottom of the sea [46]. For example, since 1996, Statoil in Norway has been re-injecting CO_2 contained in the natural gas produced at its Sleipner platform back into a deep saline aquifer beneath the North Sea [47]. CO_2 can also be used in existing oil fields for enhanced oil recovery. But CO_2 sequestering provides only a temporary solution, and in the long run the "Methanol Economy" (see Chapters 10 to 14) will allow recovered CO_2 to be chemically recycled to methanol used for energy storage, as fuel, or to be converted to synthetic hydrocarbons and their products. This will also provide economic value for the re-use of CO_2, thus lowering the cost of CO_2 removal.

Among the greenhouse gases other than CO_2, methane and nitrous oxide production from agricultural origins (livestock, animal waste, rice culture, nitrogen fertilizer) could be reduced by making adequate changes in farming procedures. Methane eventually also can be separated and recycled. Methane from landfills is already used as a source of energy and for other uses. N_2O emissions from industrial productions (e.g., of adipic acid) are being progressively eliminated.

Fluorinated gases are only produced industrially, and their presence in the atmosphere is purely of human origin. Due to their very high global warming potential, these gases should be monitored and controlled rigorously. Emissions can

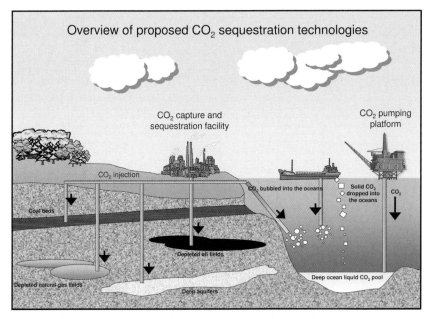

Figure 7.8 An overview of proposed CO_2 sequestration technologies.

be minimized through process changes, improved recovery, recycling or containment, or avoided by the use of alternative compounds and processes.

Two decades ago, with the discovery of a hole in the Earth's ozone layer due to the emission of CFCs, it was clear that we already faced an environmental challenge of global dimension. This problem threatened to have long-term consequences and could only be resolved by a concerted worldwide response. For the first time, international action was taken through the ratification in 1987 of the Montreal Protocol, progressively phasing out the manufacture and use of CFCs, which had been linked to ozone depletion.

Today, scientific evidence for humanity's responsibility for at least part of the observed climate change is becoming clear. However, this problem is infinitely more complex than that of CFCs, which could be relatively easily banned and replaced, necessitating truly major and difficult international action. This was the intention of the Kyoto Protocol, which was ratified by some 140 nations, but not by the United States, Australia, and some others. The Protocol rightly stated the problem, and suggested regulatory limits for greenhouse gas emissions but, besides the trading of carbon quotas, it offered no technical solutions. Today, new technological answers to the problem are essential, and these may be offered by the suggested "Methanol Economy", which is based on the recycling of CO_2 to produce new fuels and materials. Moreover, such an approach would provide a renewable, inexhaustible carbon source, whilst mitigating any human-caused global climate change and liberating mankind from its dependence on diminishing fossil fuel reserves.

Chapter 8
Renewable Energy Sources and Atomic Energy

As discussed in the earlier chapters, we rely today for a significant part of our energy needs and related hydrocarbon fuels and products primarily on non-renewable fossil fuel sources. In pre-industrial times, as is still the case today in some developing countries, energy supply was based primarily on renewable sources. The power of watermills and windmills was used to grind grain, press oil or to pump water, while wind energy at sea moved ships, and biomass energy sources such as wood and dung warmed us and cooked our food. With industrialization, however, the role of renewable energy in the global energy supply was gradually taken over by fossil fuels – first coal, and later oil and natural gas. During the past two centuries – which is a relatively short time in human history – our energy needs have relied predominantly on fossil fuels. While the reserves of these fossil fuels are still significant, they are nevertheless limited and diminishing, and they cannot sustain our life-style and development in a permanent manner over the centuries to come. Therefore, in order to satisfy our future energy requirements, the use of energy alternatives to fossil fuels – including renewable sources and nuclear energy – must be increasingly relied upon and developed. But this increasing need to produce synthetic hydrocarbons for fuels and materials will also necessitate large amounts of energy, which must be obtained from non-fossil fuel sources. Therefore, it is necessary not only to discuss the availability and feasibility of alternative energy sources and atomic energy upon which we will need to rely in the future, but also to identify new, efficient ways in which to store, transport, dispense and use energy, while highlighting the advantages of the proposed "Methanol Economy"

Unlike fossil fuels, renewable energy is derived from sources that are not subject to depletion, and simply replenish themselves. These include heat and light from the Sun and wind, organic matter, hydropower, tides, waves and geothermal heat from the Earth's crust. Unfortunately, "renewable" does not necessarily mean non-polluting or "green". Renewable energy gained significance, especially in the United States, after the oil crises of the 1970s. The sudden oil shortages, which led to sharp increases in gasoline and electricity prices, shocked those nations which were used to cheap energy and prompted governments to find solutions to become more energy-independent. Steps were taken to restructure the energy industry and to reduce dependence on imported oil, in part by encouraging the

Beyond Oil and Gas: The Methanol Economy. G. A. Olah, A. Goeppert, G. K. S. Prakash
Copyright © 2006 WILEY-VCH Verlag GmbH & Co. KGaA, Weinheim
ISBN 3-527-31275-7

development of renewable and alternative energy sources through government incentives and tax credits. In the early 1980s, many of these measures were in place,
but by the mid-1980s fossil fuel prices had fallen substantially and ideas of renewable energy sources had faded as the core of attention of energy policies, because
of their higher costs. Recently however, European countries with few or no fossil
fuel resources, as well as Japan, driven by the desire of energy independence and
adopting the Kyoto Protocol that mandates greenhouse gas emission reductions,
have begun to invest again intensively in renewable energy sources. The increased
use of renewable energy technologies can contribute to meeting, in parts, both
environmental and energy security goals. However, as the overall energy needs
of mankind are enormous, this cannot alone be considered as the solution, at
least for the foreseeable future. Since few of the alternate energy sources depend
on combustion to generate heat or electricity, they also offer substantial environmental benefits compared to fossil fuel technologies (*vide infra*). Renewable energies are typically based on indigenous sources whose supply is not easily dis-
rupted, therefore their development and use also enhances energy security.

There is hope that renewable sources of energy could become a major factor in
developing a secure energy supply for the future. In principle, renewables having
the advantage of not being based on limited natural resources, have enormous potential. About ten thousand times more energy from the Sun reaches the surface
of the Earth than is generated by all of the fossil fuels consumed. Although the
large amount of exploitable primary energy is often scattered, it can be converted
in many ways to usable heat and power. However, widespread use is faced with
numerous challenges, and many forms of renewable energy are not currently economically viable; moreover, there are also technical problems associated with the
integration of renewables into existing systems. These energy sources are also
often less environmentally friendly than is suggested. Despite popular belief,
no form of energy – whether renewable or subject to depletion – is pollution-
free. Some adverse effects occur either during the construction, operation or disposal of generating facilities and fuels:

- Geothermal energy, for example, is not strictly a renewable
  resource as underground reservoirs become depleted over longer
  periods, and solid waste and poisonous vapors can also be generated.
- The construction of large hydroelectric plants requires the
  draining of rivers and submerging of vast areas of land.
- The use of solar energy necessitates the production of photovoltaic solar cells, and the large amounts of energy needed for this
  may be derived from fossil fuels. This process also often involves
  the use of hazardous materials such as cadmium and arsenic.
  The extensive use of solar energy also would involve large areas of
  land to be covered by light-absorbing panels.

The main obstacle against the massive and widespread use of renewable energy,
together with other renewable sources such as wind, solar or geothermal energy,

is that of cost when compared to conventional fossil fuels. These renewable processes are relatively capital intensive, and require major investment to capture diffuse energy sources, making them unattractive for the short term. However, in the long term, when the initial investment has been made, the economics of renewables improves, since operation and maintenance costs are relatively low in comparison with conventional energy sources using fossil fuels, particularly if the latter are subject to unavoidably significant price increases. In recent years, the progressive deregulation of electricity markets in the United States and Europe, resulting in rising competition between energy suppliers, was directed more towards the short term, a cost-minimizing strategy. This made renewable energies appear at a disadvantage. Governmental support, in the form of research funds, incentives, subsidies and tax credits, is therefore essential to develop sustainable and meaningful energy policies. The government must play an essential role in promoting the development and use of renewable energies, which still account today for only for a relatively small part of the overall global energy use.

In 2002, according to the International Energy Agency, renewables accounted for only 13.5% of the total primary energy supply (TPES) of the world (Fig. 8.1). More traditional sources, such as combustible biomass and waste comprised the bulk at 10.8%, with another 2.2% contributed by hydropower. New renewable energy sources for electricity or heat generation, including geothermal, wind and solar, accounted for less than 0.5% of the TPES.

Today, renewable energies are at various stages of development. Some are well established, such as the use of hydropower, steam from geothermal wells, and burning of biomass and waste. Others are newer but actively developed, such as wind power, photovoltaic cells (which convert sunlight directly into electricity), and the conversion of biomass into gaseous or liquid fuels. Still others are only emerging concepts in the research and development stage, as in the case of power production from ocean tides, waves, currents and temperature gradients. Some of the advantages, limitations and possible future of these energy sources are discussed in the subsequent sections.

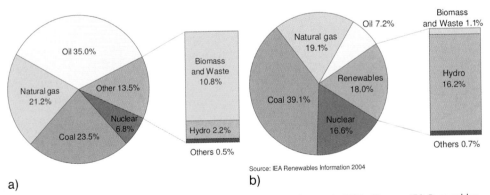

Source: IEA Renewables Information 2004

a)

b)

Figure 8.1 (a) Share of renewables in world total primary energy supply (TPES) in 2002. (b) Share of renewables in world electricity production in 2002. (Source: IEA Renewables Information 2004.)

Hydropower

Hydropower is the world's largest used renewable energy source for the production of electricity. The use of water power dates back to antiquity, with the use of water wheels turned either by the flow of a river or the weight of water falling from a dam or waterfall. Following the invention of the electric generator and hydraulic turbine, hydropower has been an important part of electricity generation since the 19th century. Indeed, today it supplies 17% of the world's electricity needs. For a number of countries, hydropower is the major electricity source (Fig. 8.2); for example, it generates 50% of electricity in Canada and Sweden, over 80% in Brazil, and nearly 100% in Norway. Hydroelectric power, although still capable of being improved, is a mature technology. New large-scale projects in developed countries, where most suitable sites have already been developed, are however by necessity limited. Current world hydroelectricity production exploits only 18% of the overall potential [5], and in developing countries and transition economies, the unexploited resources are still vast. Latin America for example, has tapped only 20% of its potential, Asia 11%, and Africa about 4%. Therefore, 90% of future hydroelectric plants are expected to be constructed on these

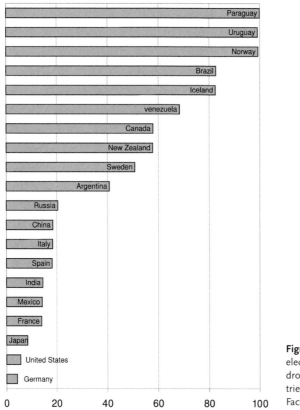

Figure 8.2 Percentage of electricity produced from hydropower in different countries. (Source: CIA World Factbook, December 2003.)

continents. A rational utilization of hydroelectric resources could help these developing regions to modernize their economies and raise their standard of living. Even if hydropower has lower operation costs and longer life expectancy than most other modes of electricity generation, resulting in generally lower energy prices for the consumer, it is also highly capital intensive. The large initial investment necessary for the construction of significant hydropower plants is an important issue, and many developing nations may find it difficult to finance such large new ventures.

Between the 1930s and the 1970s, the construction of large dams was, in the eyes of many, synonymous with development and economic progress. These dams not only produced electric power, but also provided water for irrigation and helped in flood control. In the United States, the Hoover dam on the Colorado River (Fig. 8.3), constructed during the height of the great depression and completed in 1936, was the largest dam of its time and viewed as a symbol of modernization and man's ability to harness nature. It opened the way to the development of the western United States by powering cities as far as Los Angeles and nearby Las Vegas, which was at that time hardly more than a water refilling stop on the Union Pacific railroad for the steam engine locomotives. In the former Soviet Union, major hydroelectric projects were essential to the industrialization of selected areas. The construction of dams accelerated dramatically after World War II, to peak during the 1970s when numerous large dams were commissioned around the world, not only for hydroelectric power generation but also for flood control and irrigation purposes. Today, there is an estimated 45 000 large dams on our planet, of which 22 000 are located in China alone!

Figure 8.3 The Hoover dam on the Colorado River in the United States.

[48]. Large dams are some of the biggest structures built by mankind. The Itaipú hydroelectric plant on the Paraná River in South America, the largest in the world, has an output of 12 600 MW, equivalent to about 12 large commercial nuclear reactors, and provides 25% of the energy supply in Brazil and 78% in Paraguay. In recent times however, large hydro projects have raised many criticisms on environmental and social grounds. They inevitably require large dams and reservoirs which inundate the affected land areas and can disturb local ecosystems, reduce biodiversity, modify water quality, as well as cause major socio-economic damage through the displacement and relocation of local populations. Land inundations are large, some running into thousands of square kilometers. The world's two largest dam reservoirs, Ghana's Akosombo on the Volta River with 8730 km^2 and Russia's Kuybyshev on the Volga River with 6 500 km^2, approach the size of small countries such as Lebanon or Cyprus. In tropical regions, these waters can also create feeding grounds for malaria-bearing mosquitoes and other water-borne diseases. Until the 1960s, most hydroelectric megaprojects did not involve the massive relocation of population, but as large dams began to be constructed in more densely populated areas, especially in Asia, the population to be resettled grew in size. It is estimated that during the 20th century, between 40 and 80 million people have been displaced because of large dam construction. The Three Gorge Dam (Fig. 8.4), which is under construction on the Yangtze River in China, will alone cause the relocation of more than one million people. When completed in 2009, this 2 km-long and 200 m-high dam will be the world's largest hydroelectric plant, with a production capacity of 18 000 MW, which is much needed by China's booming economy, but with a price tag in excess of $20 billion. To make way for this enormous project, the more than a 500 km-long artificial lake behind the dam will submerge more than 4500 towns and villages, ancient temples, burial grounds, and spectacular scenic canyons that have inspired poets and painters for centuries and attracted tourists from all around the world. Environmentalists argue that the dam will doom migratory fish and wipe out a number of rare species, including the Yangtze River dolphin. There

Figure 8.4 The Three Gorge dam, under construction in China.

are also concerns that the lake will trap millions of tonnes of pollutants spewing from Chongqing, one of China's largest industrial cities. In Egypt, the High Aswan dam was completed in 1970, and captures the Nile River in the world's third largest reservoir, Lake Nasser. Before the dam was built, the Nile overflowed its banks once a year, depositing millions of tonnes of nutrient-rich silt on the valley floor, and making the otherwise dry land fertile and productive. In some years however, when the river did not rise, it caused drought and famine. By constructing the dam the floods could be controlled, as could the drought by water release. At the same time, huge amounts of electricity were generated. Unfortunately, the rich silt that normally fertilized the dry desert land during annual floods is now stuck at the bottom of Lake Nasser, forcing the use of about one million tonnes of artificial fertilizer to substitute for natural nutrients that once fertilized the arid floodplains [49]. This exemplifies one of the major problems associated with the construction of dams: the progressive silting of reservoirs over time, especially in erosion-prone regions such as the high mountain ranges of the Himalayas and China's Loess plateau. Possible technical solutions are to de-silt reservoirs by flushing or adapted dredging or heightening of the dam walls.

A significant potential for expanding hydropower lies in small systems that have relatively modest and localized effects on the environment and populations. They have generating capacities in the range of 1 to 30 MW, or even less for mini- and micro-hydro installations, and are especially attractive to supply electricity to remote areas far from any electrical grid. These units are typically "run-of-river" plants which transform the kinetic energy of rivers and streams into electricity, and have little or no storage capacity. This lowers their environmental impact compared to large hydroelectric installations, but it also means that they are vulnerable to seasonal fluctuations and will not be able to produce electricity in dry seasons. Today, "small-hydro" represents only about 5% of the worldwide hydro-electricity generation.

Most concerns associated with the development of hydroelectric power have been addressed, and many were successfully mitigated. The key to future projects will be prior planning not only of the economical feasibility, but also of environmental and social factors. Potential benefits as well as the downsides of each project should be carefully weighed. Adverse publicity and negative media comments on dams should not overshadow the many benefits associated with their construction. Hydropower installations, beside providing electricity at the lowest cost compared to any other energy source, have also provided other important benefits such as flood control, irrigation, drinking water, and improved navigation. The human displacement of thousands of people associated with the construction of a new dam should be compared against the benefit to millions provided with electricity. The further expansion of hydropower will continue to be a key step in the modernization of many developing countries as a reliable and affordable source of energy. The construction of hydroelectric instead of fossil fuel-burning plants also reduces the emission of significant air pollutants, mainly SO_2 and NO_x as well as CO_2, the main greenhouse gas.

Geothermal Energy

The temperature of the Earth's gradually increases with depth, reaching 4500 °C at its core. Some of this heat is a relic of the formation of our planet about 4.5 billion years ago, but most is a result of the decay of naturally occurring radioactive elements. As heat flows from warmer to cooler regions, the Earth's heat gradually flows from the core to the surface, where an estimated 42 million thermal MW are continually radiated away, averaging merely 0.087 W m^{-2}. The majority of this immense heat cannot be practically captured. However, in some locations heat reaches the surface more rapidly, generally on the margins of the tectonic plates. There, the concentrated energy can be released by natural vents: volcanoes, hot steam or springs. It is in these locations that geothermal plants can preferably be inserted and are the most efficient. While hot springs and pools for bathing were used long ago by the Romans, the utilization of geothermal resources for other uses is relatively recent. In 1812, steam from geothermal fields in Larderello, Italy, was used for the manufacture of boric acid. The same geothermal sources were first exploited to produce electricity in 1904, but significant development in other parts of the world did not start before the second half of the 20th century. Despite its short history, geothermal energy is now relatively common in many parts of the globe (Figs. 8.5 and 8.6). In New Zealand, two power plants were built in the early 1960s to produce 170 MW, while in the United States, California's Geysers (Fig. 8.7), the world's largest geothermal power development, came online in 1960. Today, the United States is the leading producer of geothermal electricity, followed by the Philippines, Mexico and Italy. The total generating

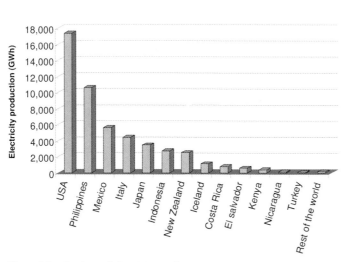

Percentage of geothermy in the country's total electricity generation	
Philippines	25.6
Iceland	15.8
El salvador	15.8
Costa Rica	13
Kenya	8.6
New Zealand	6.6
Nicaragua	4.7
Indonesia	3.2
Mexico	2.9
Italy	1.7
USA	0.4
Japan	0.3
Turkey	0.1
World	**0.3**

Figure 8.5 Geothermal electricity production, 1999. (Source: IEA, World energy outlook 2001 Insights.)

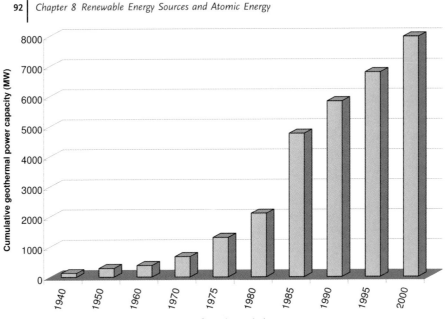

Figure 8.6 Worldwide development of geothermal electric power.

Figure 8.7 The Geysers, California. (Source: NREL.)

capacity is however, only close to 10 000 MW, and production stands around 50 TWh, which represents 0.3% of the global electricity production. Nevertheless, in some developing countries geothermal power plays a key role accounting in 1999 for 26% of the electricity generated in the Philippines, 16% in El Salvador, and 13% in Costa Rica [5].

The conventional type of natural geothermal reservoir used for electricity generation is hydrothermal, consisting of the accumulation of hot water or steam trapped in fractured porous rock. The most profitable and valuable – but also the rarest – are vapor reservoirs which yield mainly high-temperature, super-heated steam above 220 °C, also called dry steam. This is produced from wells up to 4 km deep and is able to power directly gas turbines to generate electricity. Notable examples are Larderello in Italy, Geysers in northern California, and Matsukawa in Japan. More common are systems based on hot water at temperatures in the range 150 to 300 °C. In this case the hot water, when brought to the surface, depressurizes and boils explosively, forming large quantities of steam which is fed to turbines to produce electricity in so called flash-steam power plants. In both cases, residual water as well as the condensed steam after utilization are re-injected into the reservoir to maintain pressure and prolong productivity. Lower-temperature geothermal systems between 100 and 150 °C do not allow efficient flash-steam power production. There, the extracted hot water can be used to vaporize another fluid with a lower boiling point (e.g., isopentane) that will drive the power-producing turbine/generator units in so-called binary-cycle plants.

Even if some geothermally active areas have extensive hydrothermal systems, most of them consist only of hot dry rocks with no water or steam. Collecting the heat present in these formations is much more challenging. A pair of wells must be drilled at depths typically more than 4 km, and the rock artificially fractured to allow water to circulate into the injection well. Steam or hot water are then returned to the surface via the production well. The potential for this resource is enormous, and technical feasibility studies are under investigation in Japan, United States, and Europe. The most advanced research on this topic at this time is being conducted on a hot dry rock unit in Alsace, France [50].

Low- to moderate-temperature (from 35 to 150 °C) geothermal resources can also make direct use of the thermal energy to heat buildings, offices and greenhouses, and even for fish-farming in colder areas. For some time now, most of Iceland, which sits right on the geologically very active Mid-Atlantic Ridge, has been heated by geothermal energy. Geothermal heat can also be produced by heat pumps that use the relatively constant soil temperature to provide heating for buildings in winter and cooling in summer. Today, geothermal heat is used by more than 55 countries with an installed total capacity over 17 000 MW of thermal power. The largest installed capacity in the world is presently in the United States, followed by China, Iceland, and Japan.

Geothermal energy is, in the strictest sense, not a renewable resource. The energy that is taken out of the Earth will not be replaced in the future. At a given location, a resource can be depleted relatively quickly if too much hot water or steam is extracted and not re-injected after heat extraction, as happened for the Geysers in California. However, geothermal resources around the globe are so widespread and immense that there is little possibility of their exhaustion on a human timescale. Today, for electricity generation, only the highest grade geothermal resources can be utilized economically, averaging a production cost of $0.03 to $0.10 per kWh [5], the lower end of this scale being quite competitive with fossil fuels. Geothermal resources have the advantage of being available all the time, with little or no fluctuation in power output, allowing reliable and predictable electricity production. Their environmental impact is low because the water used is generally re-injected in the reservoir, while gas emissions are limited and addressed by strict regulations. Small amounts of solid byproduct materials such as salts or heavy metals require disposal, while others such as silica, sulfur, or zinc can be extracted for sale. The principal barriers to faster worldwide geothermal development are mostly technological with high costs of exploration, drilling and plant construction, but there are also limited numbers of sites that are economically exploitable with present technologies. Most near-future development is expected in already known fields, principally along the tectonically active margins of the Pacific region, notably in East Asia. As mentioned, geothermal energy represents today only 0.3% of the global electricity generation. Despite forecasted regular growth, this share is only expected to increase slightly during the next decades. Therefore, from a global aspect, geothermal energy will remain a marginal source of electricity, despite playing an important role on a local scale in some countries. The successful development of hot dry rock technology might alter this picture, however, by allowing the installation of geothermal units in a larger number of formerly unsuitable locations.

Wind Energy

Wind, which actually is also a form of solar energy, is generated by the uneven heating of the Earth and its atmosphere by the sun. As air is heated, it expands and rises because its density decreases. To replace the rising hot air, cooler air from elsewhere flows into the region, producing wind. Topographic features such as hills, mountains, valleys or large open spaces can also influence the speed, density, and direction of the wind.

The utilization of wind power, not unlike water power, dates back thousands of years. The Egyptians built and used sailing boats in 2000 BC, while later sailing ships made world exploration and travel possible, allowing Columbus to discover America and Ferdinand Magellan to complete the first voyage around the world. It was not until the end of the 19th century that wind, as the prime energy source for maritime transportation, declined rapidly with the development of steam-powered vessels. On land, windmills provided energy for milling and pumping of water in early Mediterranean and Eastern civilizations, as well as in medieval Europe. Windmills continued to play an important role well into the 20th century in the rural United States or Australia, providing energy for isolated farms, mostly to pump water for irrigation and livestock. In the late 19th century, windmills also began to be used to produce electricity, making it possible for farmers and ranchers far from electricity distribution lines to generate their own electricity and to operate small electrical devices such as a radio. Their use was however progressively eliminated before World War II by the availability of cheaper, more reliable, fossil fuel- or water-based electricity from central generating plants delivered through steadily extending electrical transmission grids. Until the early 1970s, the era of cheap oil and gas induced very little interest in electricity generated by wind, and it was only with the oil price shocks of the 1970s and early 1980s that efforts were made to develop and design new wind machines capable of producing electrical power efficiently at a competitive price. In the early 1980s, state subsidies, incentives and tax credits began to revitalize the wind energy industry. Almost all growth was concentrated in the United States, as well as Denmark. In 1985, the United States alone accounted for more than 90% of the installed generating capacity, with 1 GW, and had the world's largest windmill facility at the Altamont Pass, near San Francisco. With the expiration of tax incentives in the United States in 1985 and declining oil prices however, wind power development came to a standstill. Subsequently, thanks to improvements in turbine designs as well as the active promotion of wind energy, development was restarted in the 1990s and was this time led by European countries such as Germany, Denmark, and Spain. Over the past decade, the installed capacity has increased from 2500 MW in 1992 to over 47 000 MW at the end of 2004, at an annual growth rate of almost 30% [51]. Today, wind power is the fastest-growing renewable energy resource, and indeed the fastest growing of any energy source. In Denmark, wind produces 20% of the country's electricity needs. The spectacular increase in wind power capacity in recent years has placed Germany, with an installed capacity close to 17 000 MW, meeting 7% of the national electricity needs, as the world leader in power production from wind, followed at a distance by Spain with an installed capacity of over 8000 MW [51]. Almost three-fourths of the world's wind power capacity is now concentrated in the European Union, where wind met 2.4% of the electricity demand in 2003. Nine of the ten largest wind turbine manufacturers are based in Europe, and European companies control some 90% of the global wind energy market. Outside Europe, the United States and India are leading markets. Substantial development of wind energy is also expected in numerous countries, including Brazil, Canada, Australia, and China (Fig. 8.8).

When considering the installation of a wind-power generating facility, the single most important factor is the wind speed. As the power increases in relation to the cube of the speed, a doubling of the average wind speed induces a wind power increase by a factor of 8. Consequently, even small changes in wind speed can produce large changes in energy output. Therefore, the installation of new production capacity must be preceded by careful studies of wind flow conditions. Relatively constant winds, even with medium speeds, during extended periods of time are more suitable for reliable power generation than faster but more intermittent or unstable winds. Once the wind resource has been explored, wind turbines have to be constructed to transform the wind energy into useful electrical power. Modern wind turbines generally consist of a three-bladed rotor connected to an electric generator, able to operate over a large range of wind speeds and fixed on the top of a hub. In most cases clusters containing up to several hundreds of wind turbines are concentrated in so-called wind farms (Fig. 8.9). Over the past 20 years, the power of these turbines has increased dramatically as the rotor diameter increased from around 20 m and generating capacities of 20–60 kW, to more than 100 m and generating units over 2 MW in commercial applications, as well as prototypes close to 5 MW [52]. Larger and taller turbines can also take better advantage of the fact that wind speed increases with increasing height above the ground level. These increases in size and technological know-how, coupled with the economy of scale from a fast-growing production, have greatly

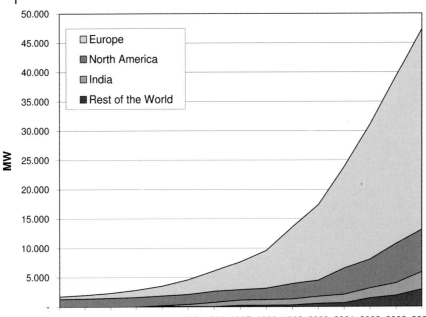

Figure 8.8 World wind power installed capacity. (Source: Global Wind Energy Council, European Wind Energy Association, IEA.)

Figure 8.9 Offshore wind farm, Denmark. (Courtesy: Elsam A/S, Denmark.)

reduced the cost of wind power to the point that some of the most productive on-shore wind farms are approaching price competitiveness with energy from fossil fuels. The exploitable onshore wind resource for the European Union is conservatively estimated at 600 TWh and the offshore wind resource up to 3000 TWh [52]. Taken together, this may represent more than the present total electricity consumption of the European Union. Onshore wind potential is being developed first because of obvious lower installation costs compared to offshore facilities.

To date, only Europe has installed wind capacity offshore, mainly in the North Sea, in relatively shallow waters and using large turbines to minimize the cost per unit of energy produced. However, because of the considerable potential of offshore facilities and often higher average wind speeds and low turbulence at those sites, projects are under way all around the World.

One major problem with wind-generated electricity is that the resource is highly variable, changing on a daily, seasonal, and annual basis. Therefore, power output prediction can be very challenging and additional generating capacity, as well as sufficient backup power, may be necessary in case of low wind speeds. Wind turbines being constructed in exposed, highly visible areas, also raise public resistance because of the visual impact of wind farms as well as noise levels, which should also be taken into account by building them far from populated areas and sensitive landscapes. Although only limited land is necessary for the installation of each turbine, they must be placed at sufficient distances in order not to interfere with each other, and this can result in extensive land requirement for their establishment. However, as the footprint of the wind tower is small, much of the land comprised in wind farms can be used for other purposes such as agriculture or grazing livestock. On the other hand, wind farms also represent other environmental hazards by killing birds and thus interfering with ecosystems. Nonetheless, the environmental impacts can be greatly reduced or eliminated by the construction of offshore wind farms.

Based on the large and widespread resources, costs, maturity of the technology and relatively limited environmental impacts, wind power has a promising future in supplying an increasing share of electricity in countries willing to invest in it. The absence of greenhouse gas and pollutant emissions is also a strong advocate for electricity generation from wind.

Solar Energy: Photovoltaic and Thermal

The Sun is an enormous, effective, and far-away (some 150 million km from Earth) nuclear fusion reactor that can supply the Earth with energy now, and for several billions of years to come. Sunlight, or solar energy, emitted from the Sun is the most abundant energy source. At any given time, sunshine delivers to Earth as light and heat about 10 000 times more energy than the entire world is consuming. Solar radiation, before entering the Earth's atmosphere, has a power density of 1370 W m^{-2} [4]. Of all the sunlight that passes through the atmosphere annually, only about half reaches the Earth's surface; the other half is scattered or reflected back to space by clouds and the atmosphere, or absorbed by atmospheric gases such as CO_2 and water vapor, the atmosphere, and clouds. As 71% of our planet is covered with water, most of that energy that makes it to the surface is absorbed by the oceans. On a clear day, the radiation received on the Earth's surface around noon is about 1000 W m^{-2} [5]. The amount of energy received on the surface (called insolation) is usually measured in kWh m^{-2}. Annual average insolation varies from maxima in hot desert areas to minima in polar regions. Seasonal variations are more important in regions far from the equator, as differences between winter and summer day lengths are more pronounced. In the extreme case, at the north pole, insolation is near zero during the six months of polar winter when the sun never rises. In the United States, the highest insolations with a daily average in excess of 6 kWh m^{-2}, can be found in the dry and most of the time cloudless South-West. This includes states such as Arizona, New Mexico and Nevada, as well as Southern California and especially the Mojave Desert, where major solar energy technologies are tested. Using the current solar technology, an area of 160 × 160 km in this region could generate as much energy as the entire United States currently consumes [53]. The enormous flux of inexhaustible (at least on the human time-scale) solar energy holds a tremendous potential for providing humanity with clean and sustainable energy. As with the wind, real interest in utilizing the Sun's energy arose in the aftermath of the oil crises of the 1970s. Today, the various ways in which solar energy can be used to generate electricity, provide hot water, and heat or cool our buildings are at different stages of technological development.

Electricity from Photovoltaic Conversion

The conversion of daylight into electricity, called the photovoltaic effect, was first discovered by the French scientist Edmond Becquerel in 1839. The explanation for this effect was later provided by Albert Einstein who received for this work (and not the theory of relativity) the Nobel Prize in physics. However, the development of the first practical photovoltaic cell occurred only in 1954 at the Bell Telephone Laboratories, with the production of a silicon-based cell with 6% efficiency in converting light into electricity. The technology that was originally developed for space applications in the 1960s to power military and later commercial satellites, has since been given much attention as a potential energy source for civilian uses.

Photovoltaic (PV) systems that convert the energy of photons from sunlight directly into electricity using semiconductor devices are commonly known as solar cells. When photons enter the cell, electrons in the semiconducting material are freed, allowing them to flow and generate electricity. Solar cells are typically combined into modules in sizes from less than 1 W to 300 W to produce higher voltage and currents. If larger electricity production is required, these modules can be connected in series or parallel to form photovoltaic arrays. The modular nature of solar cells means that they can be used in applications ranging from a fraction of a Watt, such as a solar wristwatch, to large multi-MW power plants containing millions of solar cells.

Solar cells are most commonly made from mono- or polycrystalline silicon. Their efficiency in converting sunlight into electricity is constantly improving, from around 15% in the mid-1970s to more than 25% today under laboratory conditions [54]. Other materials such as cadmium telluride, gallium arsenide, copper-indium diselenide or amorphous silica, usually in the form of deposited thin films, have also been used to produce solar cells, but their large-scale utilization has met limited success because of a shorter life-time, low efficiency or toxicity issues in production and disposal. Photocells based on organic materials are now also being designed and tested.

During the last decade, the cumulative PV capacity in the world has increased at a rate of more than 30% per year, reaching by the end of 2004 more than 2.5 GW compared to less than 0.1 GW in 1990 [54] (Fig. 8.10). This growth is impressive, but still equals only the production capacity of two nuclear reactors. Over three-fourths of the world capacity is installed in only three major countries: Japan, Germany, and the United States. Not surprisingly, 54% of the solar cells produced in the world in 2004 came from Japan, followed by Germany as the second largest producer, and the United States.

PV technology has a wide range of applications. It can provide electricity to isolated systems far from power lines, such as telecommunication towers or water pumps in developing countries. A major future increase in solar power is, however, expected to come from the PV installed to generate all or part of the electricity needed in grid-connected buildings.

Solar cells have many advantages: they are silent, produce no emissions, have no moving parts, are easy to maintain, can be installed almost anywhere, are easily adapted to the customer's specifications, and use no fuel.

The downside of solar cells is that energy generation is limited to daylight. The Sun does not shine during the night, and even during the day cloudy conditions also reduce the energy output to only 5–20% of that of full sunlight output. The

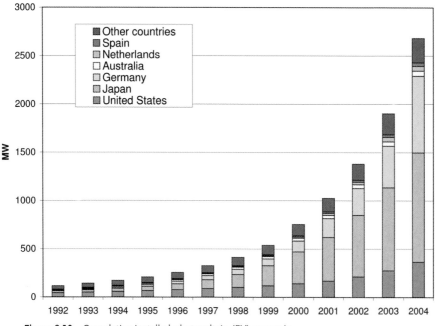

Figure 8.10 Cumulative installed photovoltaic (PV) power in reporting IEA countries. (Source: IEA-Photovoltaic power systems programme.)

intermittent nature of PV systems is an issue that must be addressed by the installation of supplemental conventional generating capacity or effective energy storage for cloudy or low-sunlight conditions. The main present challenge, however, is the cost of generating electricity from solar cells; this is in a range from $0.3 to $0.8 per kWh, depending on the location [55], and is an order of magnitude higher than the electricity production from natural gas at some $0.03 per kWh. As demand increases and larger quantities of PV are produced, the costs will decrease. However, unless a major breakthrough in solar cell production occurs that will dramatically decrease costs, its contribution to global electricity generation in the foreseeable future is likely to remain relatively limited and strongly dependent upon governmental incentives.

Solar Thermal Power for Electricity Production

Solar thermal power systems use mirrors and optical devices to redirect, focus, and concentrate the Sun's rays onto a receiver, where heat is generated. The first application of this technique was recorded in 212 BC, when Archimedes is said to have used mirrors to burn Roman ships attacking Syracuse. Today, the thermal energy can than be used to produce steam-driven generators for electricity production, or to initiate chemical reactions. The concentration of sunlight may be achieved with different technologies, including parabolic troughs, solar power towers and parabolic dishes differing in the way in which they collect solar radiation. Parabolic trough collector systems are commercially available and are the least expensive solar thermal technology. They use single-axis sun-tracking parabolic mirrors to focus sunlight onto a glass tube which runs in the trough and contains a heat-transfer fluid that reaches temperatures up to 400 °C and, via an heat exchanger, generates steam to drive an electric generator. Nine plants, ranging from 10 to 80 MW have been installed in the Southern California Mojave Desert since 1991, providing a total capacity of 354 MW of peak electricity to the grid [56]. Other projects are under way in several countries such as Spain, which has tested at the Plataforma Solar de Almerìa, the possibility of using superheated steam directly as the heat-transfer fluid (Fig. 8.11).

One commercially less mature technology than parabolic trough is that of solar towers, which use many large mirrors called heliostats that track the path of the Sun throughout the day and focus the rays on the solar receiver used to heat a

Figure 8.11 Two EuroDish-Dish/Stirling systems in operation at the Plataforma Solar de Almeria, Spain. (Source: Schlaich Bergermann und Partner, World-wide Information System for Renewable Energy, WIRE.)

working fluid, typically molten salt, to generate steam and produce electricity. Because solar energy is concentrated on a point rather than a line in the case of parabolic trough, the temperature generated is higher; in a range from 500 to 1500 °C. The Solar One and Solar Two projects, which were also operated in the Mojave Desert until 1999, demonstrated the technology at the 10-MW pilot-scale.

Smaller solar towers were also field tested in different countries, including Italy, France, Spain, and Japan, not only for electricity generation but also other applications requiring high temperatures (such as methane production and material testing) [56]. Solar dishes, which resemble satellite dishes, use parabolic mirrors that track the Sun in two axes, concentrate the sunlight at their focal point, and generate temperatures similar to those of solar towers. In the most common approach, the concentrated solar energy is used to heat the hydrogen or helium working fluid of a Stirling engine [56]. The generated mechanical energy is then converted to electrical energy. On average, the dishes are between 8 and 10 m in diameter, though some can be much larger, like the "Big Dish" in Australia with a surface area of 400 m^2. Existing units are small, producing between 10 and 25 kW. Like PV set-ups, they can be installed in large groups to serve utility needs, or in smaller numbers for decentralized energy generation. Stirling Energy System recently announced its plan to construct the largest solar installation in the world, based on solar dishes, in the Mojave Desert in California. When completed in 2010, the facility will be composed of 20 000 25-kW solar dishes with a production capacity of 500 MW, comparable to a typical fossil fuel power plant [57, 58]. In addition, a 300-MW plant containing 12 000 solar dishes will also be constructed to provide electricity to the city of San Diego, California. Of all solar technologies, solar dish/engine systems have demonstrated the highest solar to electric conversion efficiency at 29%, compared to around 20% for other solar thermal technologies [56]. With receiver temperatures in excess of 1000 °C for solar tower or solar dish technology, processes to generate hydrogen thermochemically could also be exploited and are under investigation. The production cost of electricity with a parabolic trough, the most mature technology, is still high at $0.10–0.15 per kWh for the best power plants [53]. In order to bring solar thermal power generation to the market, the key cost reduction issue is currently being studied through research and technological development in numerous projects around the world. Solar thermal power plants are currently the most likely to become the first large scale solar electricity producers.

Electric Power from Saline Solar Ponds

A solar saline pond is a few meters in depth and artificially maintained so that the degree of its salinity, and consequently density, is higher at the bottom than at the surface. The difference in salinity is created by dissolving large amounts of salt at the bottom of the pond and keeping the surface supplied with low-saline water, to maintain the necessary salinity gradient. Because of the difference in salinity there is minimal mixing between the layers, and convection is prevented. Absorption of solar energy by the bottom of the pond heats the lower depths of water, which are prevented from rising by higher density relative to upper part of the pond. Under these conditions, water at the bottom of the pond can attain temperatures close to 90 °C. Research on saline solar ponds began in the 1950s in Israel, and resulted in the early 1980s in the construction of a 25-hectare demonstration plant in Beit Ha'aravah near the Dead Sea [59]. A 5-MW low-temperature turbine using a low-boiling working fluid was used to transform the temperature difference between the pond's water layers into electricity. The overall efficiency of such a relatively low-temperature system was only about 1%. On a continuous basis, the pond provided approximately 800 kW. The unique feature of the saline pond, however, is that it can store the energy of the Sun in the form of heat and provide electricity even at night and during cloudy days. The Israeli solar saline pond power plant was operated until 1990. Since then, only one other plant using this technology has been built in Texas to provide a food cannery with both electricity and heat. Due to the relatively low efficiency of the system and the large area of land required, the potential of energy production from solar saline ponds is, however, limited.

Solar Thermal Energy for Heating

Solar thermal systems are widely used to provide heating and hot water for residential, commercial or industrial applications. Since their use took off in the 1970s as a result of the high oil prices, several million units for the production of hot water have been installed worldwide. In Tokyo alone there are 1.5 million buildings equipped with solar water heaters [5]. Some small countries such as Israel are also widely using solar thermal energy for homes and other installations [59]. They are the most widespread and easiest way to utilize solar power, as only moderate temperatures are needed and the technology is simple. Solar heating systems are composed of a collector in which a liquid, generally water, is heated, and a pump to transfer the heat to living spaces or a storage tank for later use. In the thermal collector, mounted on the roof facing the sun, a fluid running through polymer or copper tubing is heated by sunlight. For some applications, solar water heating systems have been especially popular: as an example, in the Western United States, heating of swimming pools accounted for more than 90% of the demand in 1999 [5]. Although governmental incentives and further cost reductions are still desirable, the technology is mature and already considered to be competitive in an increasing number of countries. On the other hand, solar heating and cooling of buildings is not yet competitive with conventional energy sources, and therefore generally still rare.

Economic Limitations of Solar Energy

Today, the major problem facing solar energy – especially for the production of electricity – is its cost, which is still high when compared not only to fossil fuels but also to renewables such as wind or hydropower. Despite the free and inexhaustible solar power source, the up-front cost for the equipment to collect and store solar energy is high. Likewise, due to the diffuse nature of sunshine, the collecting area necessary to produce large amounts of solar energy is by necessity large. The absence of sunlight during night-time and reduced insolation in cloudy conditions also necessitates back-up generators or expensive and limited-energy storage batteries. With continued research and technological improvements however, solar energy will in the long run certainly become an important part of our energy-mix.

Biomass Energy

Biomass energy or bioenergy refers to the use of a wide range of organic materials as fuels. These are produced by biological processes, and include forest products, agricultural residues, herbaceous and aquatic plants, and also municipal wastes. In principle, biomass is inexhaustible and renewable, provided that new plant life is grown to replace the ones harvested for energy. Biomass can either be burned to produce heat or electricity, or transformed into liquid fuels such as ethanol, methanol, or biodiesel. Biomass has been our predominant source of fuel well into the 19th century. Whereas dried plants, plant oil, animal fat or dried dung were used for lighting and cooking needs, the most common biomass source was wood. Its dominance was progressively replaced by fossil fuels, first coal and then oil and gas during the 19th and 20th centuries. Today, biomass supplies about 11% of the world's primary energy consumption. Indeed, in many poor countries it continues to be the most important source of energy for heating and cooking purposes, whereas in developed nations only a small fraction of the energy needs are covered by biomass.

Electricity from Biomass

The cheapest, most used, and simplest way of using biomass to generate energy is to burn it. On a commercial scale, this is done in a process similar to the one burning coal to produce electricity or heat. In these applications, wood, wood waste and municipal solid waste are the most utilized fuels. The average plant has generally a small size (around 20 MW) and an efficiency ranging from 15 to 30% for conversion to electricity [5]. With co-generation of electricity and heat, the total efficiency can reach 60%. Methane-rich biogas, if captured and collected from landfills, can also be used to generate sizeable amounts of energy.

New technologies that are commercially available for converting biomass to electricity include co-firing and gasification. Co-firing power plants use biomass as a supplementary energy source with a conventional fuel, typically coal. Gasification converts solid biomass through partial oxidation at high temperature into a combustible gas, containing mainly carbon monoxide and hydrogen. The gas produced can then by burned in a gas turbine or internal combustion engine to generate electricity. It is worthwhile noting that during the great depression and World War II, small gasifiers were used for cars to convert wood and charcoal

Table 8.1 Production of electricity from biomass and waste in 2003.

Country	Production (TWh)	Percentage of world electricty production from biomass	Percentage in the country's total electricty production
United States	64.6	32.4	1.7
Japan	25.9	13	2.5
Germany	12.3	6.2	2.2
Finland	10.6	5.3	12.6
Brazil	10.4	5.3	3.2
Canada	8.3	4.1	1.6
United Kingdom	6.2	3.1	1.7
Spain	5.5	2.8	2.4
Rest of the world	55.4	27.8	0.7
World	199.1	100	1.4

Data source: EDF and IEA key statistics.

to gas that was fed to the engine. These vehicles were not very efficient and needed extensive maintenance, but nevertheless functioned quite well.

The cost of biomass energy varies widely depending on the fuel, its quality, and the technology used. Electricity-generating costs are however generally higher than those for fossil-fueled plants because of lower efficiencies, higher capital, and fuel costs. Most estimates for the fuel cost are in a range of $150 to $250 per tonne, but this can be much lower in cases where the fuel is a byproduct from some other process [5].

Among the OECD countries (Organization for Economic Co-operation and Development, including most developed nations), electricity from bioenergy represented 1.6% of the total electricity generation in 1999 [5]. More than half was produced from solid products such as wood and agricultural residues, while waste accounted for 35% of electricity produced by biomass. In 2003, in forest-rich Finland, electricity from bioenergy represented 12.6% of the total electricity production, but only 1.7% in the United States (which has overall the largest electricity generation capacity from biomass, but also the highest electricity consumption).

Liquid Biofuels

Biofuels are liquid fuels produced from biomass feedstock through different chemical or biological processes. Today, biomass is the only available renewable source for producing high-value liquid biofuels such as ethanol or biodiesel. These fuels can offer renewable alternatives to transportation fuels that presently are obtained almost exclusively from oil. Ethanol, the most common biofuel, is produced by fermentation of annually grown crops (sugar cane, corn, grapes, etc.). In this process, starch or carbohydrates (sugars) are decomposed by microorganisms to produce ethanol. Ethanol can be produced from a wide variety of sugar or starch crops, including sugar beet and sugar cane and their byproducts, potatoes and corn surplus. In Russia after the Bolshevik revolution, Lenin proposed the use of agricultural alcohol to produce industrial fuels and products. This diversion amounted to the use of Russian people's beloved source of vodka, however, the plan was soon abandoned. During World War II in Europe, blends of ethanol with gasoline were used, but only anhydrous ethanol is miscible with gasoline, phase separation otherwise causing stalling of the engines. Ethanol has been promoted and used more recently extensively in Brazil and the United States as a response to the OPEC oil embargoes and rising gasoline prices (but also to subsidize farmers). Beginning in that period, Brazil – one of the largest sugar cane producers in the world – gave farmers financial incentives to switch from sugar to ethanol production. This plan was implemented, and by the mid-1990s about 4.5 million vehicles were running on pure ethanol, and the remainder of the fleet on a blend of gasoline containing up to 24% ethanol by volume. With falling oil prices and the end of price subsidies for ethanol at the end of the 1990s however, the sales of pure ethanol cars fell almost to zero [60, 61] (Fig. 8.12). Increasing oil prices, experienced in the recent past, have now revived interest in ethanol. The production of ethanol from sugar cane in Brazil reached some 15 million m^3 per year in 2004 [62]. The extraordinarily high productivity of

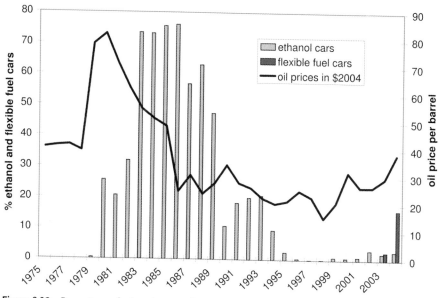

Figure 8.12 Percentage of ethanol-powered automobile produced in Brazil compared to the country's total automobile production.

Source: Associação Nacional dos Fabricantes de Veículos Automotores ANFAVEA, Brazilian Automotive Industry Yearbook 2005.

sugar cane (up to 80 t cane per hectare, compared to 10–20 t for most plants cultivated under temperate climates), associated with low wages, has contributed to a great extent to the competitiveness of ethanol in Brazil. So-called flexible fuel vehicles (FFV), able to run on any mixture of gasoline and ethanol, have also been introduced recently. However, it must be pointed out that the amount of ethanol produced annually in Brazil represents only the equivalent of some 7–8 million tonnes of petroleum oil, less than the quantity consumed by the world in a single day. Besides alcohol, sugar cane byproducts – principally bagasse (sugarcane husk) – are burned to supply Brazil's grid with about 600 MW of electricity [63]. This again, is less than the output of a single large-scale fossil fuel or atomic power plant.

In the United States, ethanol produced from corn is used in gasohol, a blend of 10% ethanol and 90% gasoline, as well as an oxygenated additive in gasoline since the early 1980s (Fig. 8.13). However, ethanol is only economically competitive because of a significant tax subsidy (currently $0.54 per gallon). Growing corn for ethanol production is also very energy-intensive because of the need and cost of fertilizing, harvesting, and transporting the corn, as well as subsequent fermentation and distillation which requires large amounts of energy that are generally provided from oil and natural gas. It should be emphasized that, in fact, ethanol produced from corn produces at most only 25–35% more energy than was consumed in its production [64, 65]. Indeed, some claims state that the process is a net energy user [66–68]. In any case, corn used for ethanol production is far from an ideal feedstock, though dedicated energy crops and new genetically engineered crops could increase the energy efficiency. Utilization of the cellulose content of plants, which is resistant to fermentation, to produce ethanol has in the past only been possible by prior hydrolysis with sulfuric acid. Currently, the

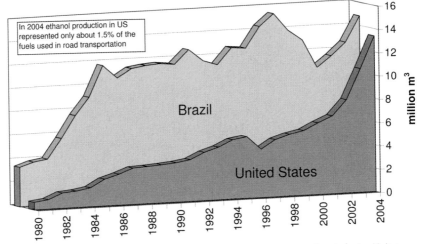

Figure 8.13 Historic production of ethanol in the United States and Brazil. Based on data from Renewable Fuel Association and Sao Paulo Sugarcane Agroindustry Union (UNICA).

development of new strains of microorganisms capable of digesting cellulose directly is being explored, and this may allow the use of other types of vegetation with lower production costs to be processed to ethanol, making the overall process cheaper and more efficient.

Biodiesel, processed from seed crops such as rape, sunflower and soy, is currently mainly produced in Europe and the United States on a limited scale (Fig. 8.14). Market penetration is small and the production costs relatively high, although interest is growing. In 2004, biodiesel represented less than 1% of the 270 Mt fuel (gasoline and diesel fuel) consumed by road transport in Europe [69, 70]. The direct use of plant oils in diesel engines is not recommended as it will considerably reduce the engine's lifetime. This drawback has been mitigated by reacting these oils with methanol or ethanol in a so-called transesterification process to yield commercial biodiesel. Biodiesel can be blended without major problems with regular diesel oil in any proportion. The production of biodiesel is also less energy-intensive than ethanol from corn because no fermentation and distillation is necessary. However, biodiesel from oil seed crops requires up to five times more land per unit of energy produced than ethanol.

Methanol was once produced from wood, and is therefore sometimes still referred to as wood alcohol. This process (called pyrolysis) was still being used at the beginning of the 20th century, and involves the heating of wood in the absence of air to yield a mixture of solid, liquid, and gaseous products from which methanol could be extracted in low yields. Today, methanol is predomi-

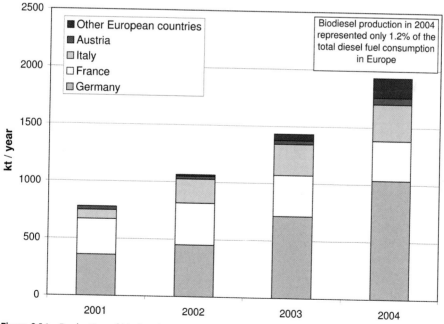

Figure 8.14 Production of biodiesel in Europe.
Source: European Biodiesel Board and Eurobserver.

nantly produced by steam reforming of methane (natural gas) and subsequent re-action of the produced syn-gas to methanol. However, any source of carbonaceous material could be converted to syn-gas and thus methanol. This includes also bio-mass which could therefore become a source of methanol in the future.

Biomass as an energy source has many advantages. It is renewable, provides a convenient way of storing energy (e.g., in the form of wood), which is not the case for wind or solar energy, and it can be found in different forms all over the world. It can reduce the energy-dependence on foreign countries. Biomass is also versa-tile as it includes solid fuels such as wood or crop residues, liquid biofuels such as ethanol, and biodiesel, as well as gaseous fuels in the form of biogas or syn-gas. From an environmental point of view, biomass can help mitigate climate change, as the CO_2 released by use of bioenergy is captured from the atmosphere by the growing plants and should therefore induce no net CO_2 emissions (it is carbon neutral). On the other hand, the conversion of solar energy to biomass is only achieved at an efficiency of around 1%, which is very low even compared to the inefficient conversion of solar energy to electricity with solar cells (in excess of 10%). In order to generate bioenergy on a large scale, vast areas of land are ne-cessary. Also, great care must be taken in choosing crops for energy production, as these should have a high photosynthetic efficiency, grow rapidly, use minimal amounts of fertilizers, herbicides and insecticides, and have limited water needs to minimize energy input into their cultivation. These "energy crops" should also preferably be grown on land not dedicated to food crops in order to avoid compe-tition with food production. The vast expanses of the seas can also be used to grow algae, which can be utilized to produce bioenergy, and experimental facil-ities in the United States and Japan have explored this possibility. With the pres-ent technology, a large part of the world's agricultural land would have to be de-voted to energy crops if they were to supply a substantial amount of our energy needs. Even if the use of bioenergy could be cost-effective in certain cases for the production of heat and electricity, it has generally higher costs than conven-tional energy sources. An economical and sustainable large-scale use of this re-source will therefore require technological advances or breakthroughs, especially in the bioengineering field to design suitable high-yield energy crops. In the transportation sector, the production of cellulose-based ethanol and other liquid fuels, particularly methanol through biomass gasification, will allow a higher yield per unit of land [71]. Algae grown in the sea might also eventually extend the scope of bioenergy.

Ocean Energy: Thermal, Tidal, and Wave Power

Energy contained in the oceans, which cover more than 70% of the Earth's sur-face, exists in two forms: (i) mechanical energy in the waves, tides and marine currents; and (ii) thermal energy from the Sun's heat. Because ocean energy is abundant, all of these sources have been considered for energy production and are at variable stages of development.

Tidal Energy

Tidal energy exploits the rise and fall of tides caused by interaction of the gravitational fields of the Moon and Sun. Tidal movements are both periodic and predictable, and were already used before 1100 AD in tide mills in the United Kingdom and France to grind grain. Their modern version, the tidal power plant, which takes advantage of the difference in water levels between low and high tide, operates on the same principle as an ordinary hydroelectric plant. Behind a dam constructed across a bay or estuary, water flowing through sluices is collected during high tide and later released at low tide through turbines to generate electricity. In order to be exploitable for electricity generation however, differences between high and low tides must be significant, and only a few locations worldwide, generally in river estuaries, are suitable. Today, the only large-sized tidal power station with 240-MW generating capacity is situated at the estuary of the Rance River, near Saint Malo, France. It was built in the 1960s and has now completed more than 30 years of successful operation. Since then, the only other somewhat significant tidal power plants constructed have been a 18-MW project in Annapolis, Bay of Fundy, Canada, a 3.2-MW device in China, and a very small 0.4-MW unit near Murmansk, Russia [72]. In the European community, the technically exploitable resource is estimated at 105.4 TWh per year, with only around 50 TWh per year economically viable, and the majority of sites (90%) being located in France and the United Kingdom [73]. Beyond the EU, Canada, the states of the Former Soviet Union, Argentina, Western Australia and Korea have potential sites that have been investigated. Taking advantage of this resource, South Korea has recently announced its plans to construct the world's largest tidal plant (260 MW capacity). Tidal energy per se is a mature technology, but it requires high capital expenditure, long construction times, and has low load factors, leading to long payback periods and high electricity generating costs. Thus, governments are likely to remain the only entities to undertake large-sized projects in this field. One significant factor for the limited use of tidal power raised by ecologists has been the potential environmental impact of such projects. Studies conducted in the United Kingdom, however, concluded that this technology does not necessarily cause major environmental changes; rather, economic considerations are the real barrier to further development of tidal power. Nevertheless, even if all the potentially technically exploitable sites for tidal energy were to be developed, this alternative source of energy would still only represent a very small fraction of our energy needs.

Tides also drive *marine currents*. Although energy in marine currents is generally diffuse, it is concentrated in some locations near the coast where sea flows are channeled through constrained topographies such as straits and islands. Tidal turbines which resemble wind turbines but are placed under water can be used to generate electricity from these currents. Since the density of water is about 1000 times that of air, the power density of these currents are appreciably higher than that of wind, and therefore much smaller turbines are required. In contrast to wind, marine currents are also highly predictable, and they would also have considerably less environmental impact than the construction of a tidal dam.

Tidal turbines, however, are still only in the research and development phase and very few prototypes are currently in operation in the United Kingdom and Norway [74]. Tidal turbines with a total generating capacity of 200 kW were also recently installed in New York City [75].

Waves

Waves are generated by wind blowing across the sea surface, and wave energy is thus a concentrated form of solar energy. The amount of energy in waves depends on the wind speed and power. Deep ocean waves with large amplitude in particular, contain considerable amounts of energy. The strong winds blowing across the Atlantic Ocean, and creating large waves, make the western coasts of Europe ideally suited for wave energy. Other wave-rich areas in the world include the coasts of Canada, Northern United States, Southern Africa and Australia. Wave energy is a relatively new technology and is still in the research & development and demonstration stages, which was pursued vigorously in the 1970s and 1980s. A wide variety of designs have been proposed to harness wave energy with many different extraction methods. However, only a fraction of them have actually been deployed and tested as demonstration units. Wave energy can be converted into electricity in both onshore and offshore systems. A number of onshore systems have been built in Scotland, India, Norway, Japan and other countries using mainly three different designs: (i) systems that funnel waves into reservoirs; (ii) pendulor systems driving hydraulic pumps; and (iii) oscillating water column systems that use waves to compress air within a container [76]. The mechanical power created with these devices is then used directly or via a working fluid, water or air, to drive an electric turbine/generator unit. Shoreline systems have the advantage of being easier to maintain, but the energy potential is higher in offshore locations where several systems have also been tested. Recently, construction of the first commercial wave-farm was started in Portugal [77], with electricity being generated by devices called Pelamis, named after a giant sea snake in Greek mythology. The name comes from their shape, a series of connected cylindrical modules with a total length of 150 m and a diameter of only 3.5 m. In the initial phase of the project, three of these machines, with each a capacity of 750 kW, will be located 5 km off the coast. The global wave power resource is estimated in excess of 2 TW, with the potential of generating more than 2000 TWh annually (about 10% of the world's present electricity consumption). However, like tidal power, wave power – despite continuing cost reduction – is unlikely to be economically competitive with other power sources in the near future except in special cases such as isolated coastal communities far from any electric grid.

Ocean Thermal Energy

Ocean thermal energy is also a potential source of energy. Covering more than 70% of Earth's surface, the oceans are the world's largest solar collectors, absorbing enormous amounts of Sun energy in the form of thermal energy. A process called Ocean Thermal Energy Conversion (OTEC) uses this heat stored in the oceans to generate electricity. It exploits the difference in temperature between the sea's upper layer, which is warmed by the sun, and the colder deep water. The warmer water at the surface is used to vaporize a working fluid, or is transformed to steam under vacuum, to run a turbine/generator system producing electricity. Cold water pumped typically from water depths of around 1000 m is then employed to recondense the vapor and close the cycle. To produce significant amounts of power, the temperature between the upper and lower water layer should differ by at least 20 °C; these conditions are encountered in tropical seas. Tapping the thermal energy of the oceans was initially proposed in 1881 by the French physicist Jacques Arsene d'Arsonval. It was, however, his student Georges Claude who built the first OTEC land-based system in Cuba in 1930, followed by a floating model off the coast of Brazil. Due to poor location selection and technical difficulties these projects were abandoned before they actually produced more power than was necessary to run the systems. In the late 1970s, interest in OTEC rose again and a few experimental units were constructed in Hawaii, Japan and the island republic of Nauru in the Pacific Ocean. Besides electricity production, the spent cold deep seawater rich in nutrients from an OTEC plant could also be used for the culture of both marine and plant life near the shore, or on land. In case water is used as the working fluid, fresh water obtained by vacuum distillation could be produced from seawater. High construction costs are still an obstacle for this technology, however. The fact is that OTEC is trying to produce electricity from a temperature difference which would be typically considered unusable for power generation. To compensate for this, massive amounts of water from the ocean's surface and their depths have therefore to be pumped through the plant to generate reasonable quantities of energy. The challenge is to achieve this at a reasonable cost. Thus, even if the potential resources are very large, OTEC is still in the research phase and is likely to remain there in the foreseeable future. If developed and economically viable, the first market for this technology would most likely be in tropical island nations for the combined production of energy and desalinated water.

Nuclear Energy

In 2002, nuclear power – which is used almost exclusively to produce electricity in commercial applications – generated some 2700 TWh of electric power, representing about 17% of the world's electricity consumption (Fig. 8.15). Its share of the total primary energy supply amounted to 7% [14]. Today, some 440 commercial nuclear reactors are operating in 30 countries with over 360 000 MW of total production capacity. Currently, the United States has 104 commercial nuclear power plants accounting for 20% of the electricity generation. In Western Europe, nuclear energy generates around 35% of the electricity; more than from any other source. France and Belgium, for example, produce respectively, 78% and 55% of their electricity through nuclear power. Other industrialized countries with lim-

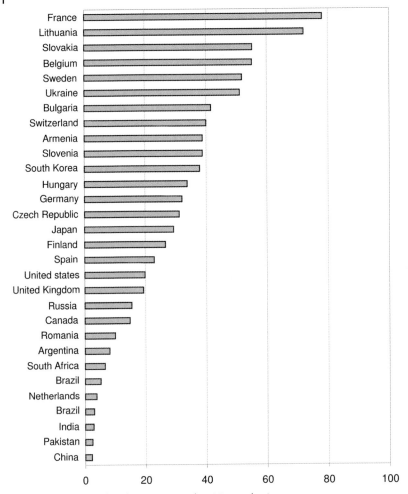

Figure 8.15 Share of nuclear energy in electricity production in 2004. Based on data from IAEA.

ited or no fossil fuel resources, such as Japan and South Korea, also rely heavily on nuclear energy for their electricity supply.

All commercial nuclear plants presently use uranium as fuel. Uranium is a slightly radioactive metal that occurs naturally throughout the Earth's crust. It is about 500 times more abundant than gold, and as common as tin, tungsten, or molybdenum. Uranium is originally formed in stars which, at the end of their life exploded, with some of their shattered dust aggregating together to form our planet. Uranium is present in most rocks in concentrations of 2 to 4 ppm. In phosphate rocks used as fertilizers, its concentration can be as high as 400 ppm, while some coal deposits have uranium concentrations in excess of

100 ppm. Uranium is also dissolved in sea water at a concentration of 3–4 ppm. There are, however, a number of areas where the concentration of uranium is much higher and economically exploitable. These deposits are particularly important in Australia and Canada, which are therefore currently the largest uranium producers in the world. Some Canadian deposits contain more than 100 kg of uranium per ton of raw ore. Like coal, uranium has to be mined in underground or surface mines depending on the depth at which the deposit is located. It is then sent to a mill where the ore is crushed to powder and leached with a strongly acidic or alkaline solution to extract the uranium from the rock. By precipitation from this solution, uranium oxide (U_2O_3) powder referred to as "yellow cake" (because of its color) is obtained. Natural uranium consists of a mixture of two isotopes: uranium 235 (^{235}U) and uranium 238 (^{238}U). Only the isotope ^{235}U, which represents merely 0.7% of the natural uranium, is capable of undergoing fission, the process by which energy is generated in nuclear reactors. Even if some reactors are able to use natural uranium to produce energy, the vast majority of them require a higher concentration in ^{235}U. Thus, it is necessary to enrich uranium from its original concentration in ^{235}U of 0.7% to typically 3–5%. For this enrichment process to occur, the uranium must be in a gaseous form, and this is carried out by conversion into uranium hexafluoride (UF_6), which is a gas at relatively modest temperature (solid UF_6 sublimes at 56 °C). UF_6 is the feedstock for the two enrichment processes used today on a large commercial scale: gaseous diffusion and gas centrifugation. Both of these take advantage of the difference in mass between ^{238}U and ^{235}U, to separate the two isotopes. However, because the masses of the two isotopes are very close, their separation requires advanced technology. Once enriched to the desired level, UF_6 is converted to enriched uranium dioxide (UO_2, with a melting point of 2800 °C) powder which is pressed into small pellets inserted into long thin tubes, made from a zirconium alloy, to form fuel rods. The rods are then sealed and assembled into clusters to form fuel assemblies ready for use in nuclear reactors.

Energy from Nuclear Fission Reactions

Inside the nuclear reactor, the released energy comes from the fission of ^{235}U atoms (Fig. 8.16). When a ^{235}U nucleus is struck by a slow neutron, it splits, releasing two smaller atoms and two or three neutrons in a process called "fission". The two new atoms may themselves undergo further radioactive decay, releasing beta or gamma radiation to achieve stability. The energy released by the fission is a result from the fact that the fission products and emitted neutrons, weigh together less than the original ^{235}U atom. Following Einstein's famous equation, $e = mc^2$, the difference in weight is converted into energy. In fact, a tremendous amount of energy, as 1 kg of pure ^{235}U can generate over 2 million times more energy than 1 kg of coal! (Table 8.2). Because each fission event liberates two or three neutrons able to split other ^{235}U atoms, themselves releasing two or three neutrons and so on, a rapidly multiplying sequence of fission events, known as a nuclear chain reaction, can occur and emit increasing quantities of

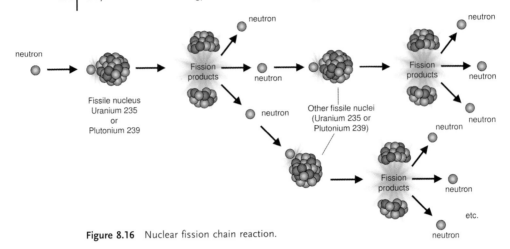

Figure 8.16 Nuclear fission chain reaction.

energy. If controlled, this chain reaction can be used to produce a large and sustained amounts of energy. This possibility was demonstrated in 1942, following Leo Szilard's original suggestion, by an outstanding team of scientists led by Enrico Fermi, who constructed the first nuclear reactor in a squash court under the stands of the University of Chicago's football field. At that time, the project funded by the U.S. government was directed to investigate the possibility of making an atomic bomb based on nuclear fission. This first nuclear reactor, called "atomic pile" consisted of highly purified graphite, uranium, uranium oxide and cadmium control rods and was about the size of a two-car garage. To control the chain reaction, it was necessary to control the neutron flow. When a ^{235}U nucleus is broken by fission, the neutrons emitted have high kinetic energy, meaning that they have high speeds in the order of 20 000 km s^{-1}. These fast-moving neutrons will most probably be captured by ^{238}U atoms which still constitute the major part of the uranium, but do not undergo fission. Slower-moving neutrons have a much lower energy content. They are still able to split ^{235}U nuclei and sustain the chain reaction, but are less likely to undergo reactions with other kinds of atoms. To reduce the neutron's speed, so-called moderators are used. The best moderators are low-mass atoms such as deuterium, helium or carbon which de-

Table 8.2 Energy content of various fuels.

Fuel	Average energy content in 1 g [kcal]
Wood	3.5
Coal	7
Oil	10
LNG	11
Uranium (LWR, once through)	150 000

crease efficiently, by successive collisions, the neutron's energy, keeping neutron losses at a minimum and do not undergo fission themselves. Many early atomic reactors used graphite as a moderator because it is inexpensive and easy to handle. Water and heavy water (water in which the hydrogen atoms have been replaced by deuterium, D_2O), however, are more efficient moderators and are the most widely used for nuclear reactors today. To control the rate of the chain reaction, control rods able to absorb neutrons without re-emitting them, are also necessary. These generally contain cadmium or boron, and can be moved in and out the reactor to regulate the flux of neutrons and thus the amount of energy produced. Fully inserted, the control rods will stop the chain reaction.

Using a combination of uranium, uranium oxide, graphite moderator and cadmium control rods, Fermi's first nuclear reactor when tested successfully in December 1942, produced only about a few watts of energy, but proved that a controlled use of atomic energy was possible.

Most importantly at the time, it also showed that the construction of an atomic bomb was achievable, triggering the Manhattan project, which led to the explosion of the first atomic bomb in 1945 and put an end to World War II. In the early years, military applications dominated the use of nuclear energy. In 1954 for example, the first nuclear submarine, the *U.S.S. Nautilus* was launched; this was able to stay underwater without refueling for long periods of time – an unimaginable situation before the advent of this new energy source.

Interestingly, Fermi's nuclear reactor was not really the first one on Earth. As we now recognize, about 2 billion years ago, natural nuclear reactors operated in a rich deposit of uranium near what is now Oklo, Gabon. At the time, the concentration of ^{235}U in all natural uranium was above 3% instead of today's value of 0.7%, due to radioactive decay. A natural chain reaction started spontaneously in the presence of water acting as moderator and continued for about 2 million years before stopping. During that time, fission products as well as plutonium and other transuranic element were naturally formed [78].

The world's first commercial-scale nuclear power plant opened in 1956 in the United Kingdom. It was equipped with a Magnox reactor using graphite as moderator and CO_2 gas as a coolant. It used, like Fermi's reactor, non-enriched natural uranium containing only 0.7% ^{235}U. In the United States, the first commercial nuclear power began operation in 1957 in Shippingport, Pennsylvania. It was a so-called pressurized water reactor (PWR), which is still the technology used in 60% of the plants currently in operation worldwide. The heat generated by the fission reaction is used to heat water in which the fuel rods are immersed. Reaching temperatures of around 300 °C, the water, however, does not boil because it is kept under high pressure. The pressurized water serves both as a moderator and coolant. Via a heat exchanger it is used to boil water in a secondary loop, producing steam to propel a turbine which in turn spins a generator to produce electric power.

The second most common type of nuclear reactor is the so-called boiling water reactor (BWR), with more than 90 units operating worldwide. The design of the BWR has many similarities with the PWR, except that the water cooling the core is allowed to boil and the steam generated is used directly to drive turbines. After condensation, the water is returned to the reactor to close the cycle. This system has a simpler design than the PWR, but as the water around the core is contaminated with traces of radioactive material, although generally with a short half-life, the turbine must be shielded in order to avoid the escape of radiation. For safety reasons, PWRs are thus the preferred reactors in the Western World. In France, for example, all 58 nuclear reactors in operation are of the PWR type.

Both the PWR and BWR, representing together more than 80% of the commercial nuclear reactors worldwide, are light-water (H_2O) moderated and use uranium enriched at 3–5% in fissile ^{235}U isotope. Because light-water not only slows neutrons but can also absorb them, it is not as selective as a moderator than heavy-water (D_2O) or graphite. Therefore, the CANDU (Canada deuterium uranium) pressurized water reactors developed in Canada, using natural uranium (0.7% ^{235}U) are moderated with heavy-water (D_2O). The cost of uranium enrichment is avoided, but extensive amounts of expensive D_2O have to be employed. About 40 CANDU reactors are presently operating in seven different countries, including Canada, India, South Korea, and China. In the United Kingdom, advanced gas-cooled reactors (AGR) derived from the earlier Magnox reactors using graphite as moderator, CO_2 as coolant, and uranium enriched at 2.5–3.5% in ^{235}U are in operation.

The commercial reactors used today (PWR, BWR, CANDU, AGR, etc.) constructed between the years 1970s and 2000 are considered as second-generation reactors (Fig. 8.17). Currently, the transition to a third generation of reactors is under way. Two units have already been completed in Japan, and several others are under construction or planned in countries such as Taiwan, France, or Korea. They are an evolution from the second-generation reactors, but feature enhanced safety systems and are less expensive to build, maintain and operate. At the same time, revolutionary designs known as generation IV systems, which have new and innovative reactor or fuel cycle systems are well under development [79].

Most of the commercial reactors currently in operation use enriched uranium in a once-through cycle (Fig. 8.18); this means that the uranium is used only once and must then be disposed of. This cycle is the most uranium reserve-intensive, as only ^{235}U contributes by fission to the production of energy. ^{238}U, which constitutes up to 97% of the fuel, and 99.7% of natural uranium is left almost untouched. The solution to limit this waste of resources is to use a different fuel cycle. In the typical reactor, fast neutrons are slowed down by moderators to increase the probability of collision between these slow neutrons and the fissile ^{235}U nucleus, and thus increase the amount of energy generated by fission. Fast neutrons, however, have the ability to convert the ^{238}U isotope, which does not directly undergo fission, to plutonium 239 (^{239}Pu), a fissile material which can thus produce energy. Therefore, ^{238}U is referred to as a "fertile" isotope. ^{239}Pu was formed during the universe creation, but due to its half-life of 24 110 years it has disappeared a long time ago in Nature. In the existing fuel cycle, the enriched uranium, once used, still contains 1% ^{235}U but also 4% fission products, 0.1% of minor actinides, and 1% ^{239}Pu. The spent fuel can be reprocessed and the useful ^{235}U sent back to the enrichment plant. ^{239}Pu also can be separated and mixed with uranium in the form of oxides to form a mixed oxides fuel commonly known as MOx. MOx is currently in regular use to generate electricity in a number of countries, including Germany, Belgium, and France. In the United States, the reprocessing of nuclear fuel was banned in 1977 by president

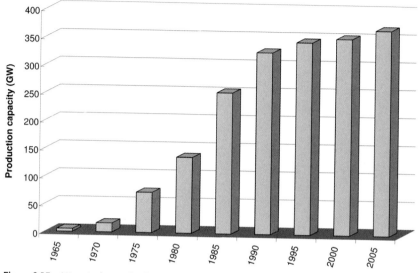

Figure 8.17 Historical growth of nuclear energy. (Source: IAEA.)

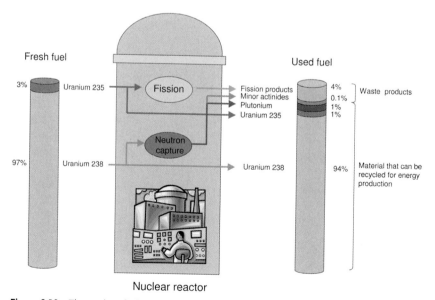

Figure 8.18 The nuclear fuel cycle in a once-through reactor.

Jimmy Carter, and despite a lift of the ban in 1981, until now no reprocessing of spent nuclear fuel has been carried out. The use of ^{239}Pu in nuclear reactors to produce energy allows the destruction of highly radioactive ^{239}Pu which otherwise would have to be stockpiled or disposed of, and also induces significant savings in ^{235}U.

Breeder Reactors

Even more intriguing is the possibility of actually producing more fuel than is consumed, using the so-called "breeder reactor". In this reactor, which is based on fast neutrons, more fissile material is produced through the conversion of ^{238}U into ^{239}Pu than is consumed through fission (Fig. 8.19). As fast neutrons are desired, no moderator is necessary in this type of reactor. Because it does not appreciably slow down neutrons and is an excellent heat-transfer material, liquid sodium is typically used as the coolant. Several fast breeder reactors have been constructed over the years in a number of countries. The best known are France's "Phénix" and "Superphénix" reactors. Phénix is an experimental reactor with a power capacity of 250 MW. Its successor, Superphénix, was a commercial-scale reactor that was connected to the grid and had a capacity of 1300 MW, but it was closed down in 1997, after some technical problems, but mainly for administrative and political concerns [80]. In Russia, Japan and India, breeder reactors are well under development with several units under construction. The United States, which pioneered this field with the construction of the first breeder reactor in 1951 at the Argonne National Laboratory [81], is now reconsidering this technology for future power plants. In fact, the United States initiated in 2000 the Generation IV International Forum (GIF) with nine other countries aimed at determining and developing the most promising generation IV reactor designs that can provide future worldwide needs for electricity, and also produce hydrogen and other products. Among the six selected systems, three are fast-neutron reactors which are all operated in a close fuel cycle, meaning that the fuel is recycled [79].

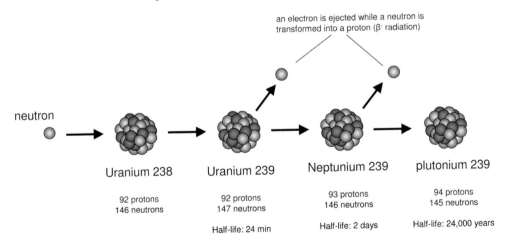

Figure 8.19 Uranium 238 fertilization by neutron capture.

About 4.6 million tonnes of proven uranium reserves are estimated to be exploitable at costs below $130 kg^{-1}. At a current worldwide demand of 60 000 t per year, this represents about 75 years of consumption. Based on geological surveys, potential resources exploitable at a cost of $130 kg^{-1} uranium are estimated at an equivalent of 280 years of consumption [82]. With present low prices of around $20–30 kg^{-1} and still considerable reserves, there is nevertheless little incentive to open new mines or to start explorations in the quest of uranium. However, if nuclear power were to supply a much larger part of our energy demand, in the long term the transition to breeder reactors is necessary. All the uranium could then be used as fuel, and not only the 0.7% of ^{235}U contained in natural uranium, multiplying the reserves by a factor of roughly 100 and fulfilling our energy needs for a least thousand years. Eventually, the 4 billion tonnes of uranium contained in the oceans in the form of carbonates could also be exploited. Although very dilute (3–4 mg m^{-3}) and with an estimated extraction price of up to $1000 kg^{-1}, it could become a viable source because the cost of uranium is only a minor component of the price of electricity generated by breeder reactors. This would leave us with an almost unlimited source of energy [83].

Besides ^{238}U, another "fertile" isotope which is susceptible to be transformed into fissionable atoms exists in nature, namely thorium 232 (^{232}Th). Almost 30 isotopes of thorium are known, but only ^{232}Th (which has a half-life of 14 billion years) is present naturally. When bombarded with neutrons, ^{232}Th becomes ^{233}Th, which eventually decays to ^{233}U. ^{233}U is a fissionable material with similar properties to ^{235}U, and can be used as nuclear fuel. Thorium is considered to be about three times more abundant than uranium, and so the potential energy available from its exploitation is therefore tremendous. The use of thorium for electricity generation has been studied and successfully demonstrated in several reactor prototypes, including the high-temperature gas-cooled reactor (HTGR) and the molten salt reactor (MSR). The MSR was first considered by the United States in the early 1950s for aircraft propulsion. Using a ^{232}Th-^{233}U fuel cycle, MSR is now one of the six systems selected by the Generation IV International Forum on future nuclear reactors. Countries such as India, with limited deposits of uranium but large thorium resources, are particularly interested in such a technology. The use of thorium fuels, compared to uranium fuels, also produce much less plutonium and other actinides, so that induced radiotoxicity is considerably reduced. This is due to the fact that ^{232}Th has two protons and four neutrons less than ^{238}U, which makes it less probable for ^{232}Th through a succession of neutron captures to be transformed into long-lived toxic transuranic actinides such as neptunium, americium, curium, or plutonium. From the viewpoint of decreased nuclear waste production, the thorium-based fuel cycle is thus more desirable.

The Need for Nuclear Power

Since the construction of the first nuclear reactor (pile) by Fermi and his team more than 60 years ago, the history of nuclear energy has been a mixed one. In the early years, after World War II, the concept of nuclear power holding the promise of cheap and abundant energy was overwhelmingly supported by public opinion. On the other hand, since nuclear energy was first used in the construction of the atomic bomb, many still associate atomic energy with destruction and killing. Over the years, after the Three Mile Island incident in 1979 and, more importantly, after the Chernobyl accident in 1986, serious public concerns emerged and public opinion increasingly turned against nuclear power, forcing many governments to drastically reconsider their policies for electricity generation from atomic energy. In the United States, after the Three Mile Island incident – which, incidentally did not lead to any casualties – anti-nuclear activism successfully slowed down and eventually stopped the construction of any new nuclear power plant ordered after 1973. By filing innumerable lawsuits against utility companies, construction was halted at the cost of millions, if not billions, of dollars to power companies. It effectively scared any corporation which planned to build a new nuclear power plant. The actual cost of the Shoreham nuclear power plant in New York for example, was first estimated at $240 million but finally rose to $4 billion at its completion. Seabrook, in New Hampshire, which was supposed to cost around $1 billion, was abandoned before completion after years of delays in construction and with $6 billion already been spent. In Europe too, countries such as Sweden, Spain and Germany, fearing possible accidents involving nuclear reactors, have put a hold on their construction. Other nations with no or limited fossil fuel resources such as France, Japan and South Korea, however are actively developing and expanding their nuclear power sector. France's decision to launch a large nuclear program dates back to the 1970s, and was aimed at minimizing the country's dependence on imported oil. Following the popular saying: "no oil, no gas, no coal, no choice" the French authorities of the time quickly came to the conclusion that nuclear power was not a choice

Table 8.3 Nuclear power reactors under construction.

Country	Units	Total capacity [MW]
India	8	3602
Ukraine	2	1900
Russian Federation	4	3775
Japan	2	1933
Iran	1	915
China	2	2000
Taiwan	2	2600
Argentina	1	692
Romania	1	655
Finland	1	1600
Total	**24**	**19 672**

Source: IAEA (as of august 2005)

but a necessity if their country was to keep its energy independence. Both benefits as well as risks were efficiently explained to the population, which understood that life would be very difficult without nuclear energy. The fact that Japan, the only country which ever experienced a nuclear bomb attack, also strongly favors the use of atomic power, should tell us much about the safe and peaceful use of this technology. In recent years, a less passionate, more rational, facts-based view of nuclear energy, coupled with an increasing overall need for energy started to emerge, especially in Asia where numerous nuclear power-plants are under construction. The United States is currently also again slowly considering the construction of new nuclear power plants.

Economics

Surprisingly, in some way anti-nuclear activism had also a positive long-range influence on nuclear energy. It forced energy companies and governments under much higher public scrutiny to make nuclear power plants even safer and more productive and reliable than almost any other industry. At the same time, in the United States, the average capacity factor of nuclear power plants increased from 58% in 1980 to 70% in 1990, and close to 90% in 2003. The increased capacity factor resulted in lower generation costs and an increased electricity production using nuclear power. The increase in electricity production from 1990 to 2003 was the equivalent of adding more than 20 new nuclear reactors to the U.S. capacity. The cost of producing electricity at U.S. nuclear power plants, including fuel, operation and maintenance, has been declining over the past decade, from more than \$0.03 kWh^{-1} in 1990 to less than \$0.02 kWh^{-1} in 2004 (Fig. 8.20). This is comparable to the cost of electricity production from coal, but much lower than that from oil and gas. Even including construction capital as well as the decommission cost of old power plants and the treatment and storage of nuclear waste, the cost of electricity from nuclear origin is estimated in France to be less than €0.03 kWh^{-1} (Fig. 8.21). This low generation cost, due in large part to the homogeneous and standardized nature of the French reactors, with all 59 reactors based on a PWR design, allows the country to be the largest electricity exporter in the World [14]. Often-cited arguments against nuclear energy based on high costs have thus become baseless.

Safety

Nuclear power plants are closely regulated by national and international agencies which provide rigorous oversight of the operation and maintenance of these plants. The safety record of nuclear power plants has been exemplary over the years. This was achieved through improved plant designs, high-quality construction, regular staff training, safe operation, and careful emergency planning. Diverse and redundant systems prevent accidents from occurring, and multiple safety barriers are placed to mitigate the effect of accidents in the highly unlikely event they occur. The safety of the nuclear industry has been significantly im-

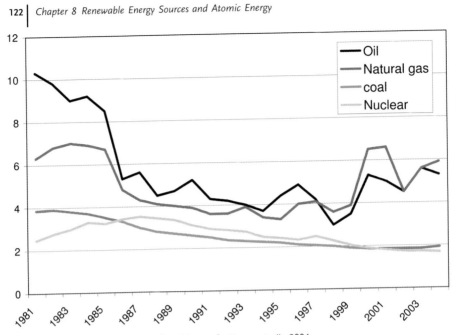

Figure 8.20   United States electricity production costs (in 2004 cents kWh^{-1}). Costs include fuel, operation and maintenance. Based on data from NEI.

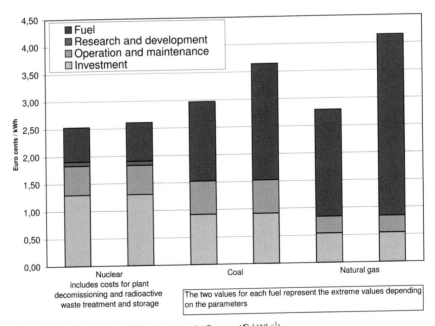

Figure 8.21   Electricity production costs in France (€ kWh^{-1}). Based on data from CEA (1997).

proved since the Three Mile Island incident in 1979 which, although it was the most serious nuclear accident in the Western World, did not lead to any casualties.

The Chernobyl disaster was on a completely different scale. It was a consequence of human error, lack of safety measures, poor construction and design. The power plant had no reactor containment building preventing radioactivity from escaping to the atmosphere, and would never have been licensed and allowed to operate in any Western country. The explosion that occurred in Chernobyl however was not nuclear, but chemical. When the reactor went out of control, huge amounts of heat generated by nuclear fission evaporated virtually instantaneously the water used under normal conditions to cool the fuel rods. The steam produced reacted with extremely hot zirconium to produce hydrogen, and with the graphite used as moderator to form carbon monoxide and hydrogen. The pressure increased dramatically and blew off the 2200-ton concrete lid of the reactor pressure vessel. The remaining graphite then ignited, with flames carrying, in the absence of any containment, highly radioactive material into the atmosphere. Thirty-five persons who tried to extinguish the fire died shortly after the accident from the direct effects of radiation. According to the latest and most comprehensive study from the United Nations, 20 years after the accident, fewer than 50 deaths have been directly attributed to radiation from the disaster [84]. Over time, however, a total of up to 4000 people could eventually die from radiation exposure from the Chernobyl accident. Poverty and "lifestyle" diseases, rampant in the former Soviet Union, pose far greater threat to local communities than the exposure to radiations resulting from the accident. Although this was the most serious nuclear accident, it should also be compared with other energy-related losses of life such as coal mining accidents which draw much less public attention but nevertheless cause thousands – if not tens of thousands – of deaths every year. The Chernobyl accident is considered as the archetype of the worst conceivable civilian nuclear disaster. The probability for a disaster of this magnitude to occur in the Western World has been estimated by nuclear safety experts in the order of one millionth per reactor per year of operation in currently used reactors, and even less in the next generation of reactors with advanced safety features. Other potential accidents such as a dam rupture or the explosion of a LNG tanker have a much larger probability to cause large casualties. It is also important to point out that, contrary to widespread belief, it is impossible for a civilian nuclear reactor to undergo a nuclear explosion of the kind generated by nuclear bombs. Nuclear fuel used in commercial units contains at most 5% ^{235}U, whereas for nuclear bombs the uranium used must contain at least 90% ^{235}U and be contained in specifically designed devices before a nuclear explosion can occur.

Radiation Hazards

Unstable atomic nuclei can split to form other particles, ejecting at the same time different types of radiation (alpha, beta, gamma, and X-ray) in a process called radioactivity, discovered by Henri Becquerel in 1896. These radiations are able to penetrate matter and can disrupt biological systems and thus essential processes in human body cells. The degree of penetration will depend on the energy of the radiation, with gamma and X-rays being the most energetic. Today, radioactivity is a part of our daily life, and is present everywhere from various natural sources: cosmic rays, uranium and thorium contained in the Earth's crust, granite used as a construction material, radon gas produced by the natural decay of uranium, potassium in fertilizers and food, etc. The average natural irradiation to which a human is exposed in a year is around 2.4 millisieverts (mSv, the unit which quantifies the biological effect of radiation in our body).

Besides natural radioactivity, populations in developed countries are also exposed to artificial sources of radiation amounting less than 1 mSv, essentially for medical purposes (X-rays) but also daily activities such as watching television (0.015 mSv) or taking an airplane trip (0.05 mSv for a Paris to New York round trip) (Table 8.4; Fig. 8.22).

Nuclear power plants are responsible for emissions of around 0.0002 mSv per year, similar to the commonly used smoke detector containing americium, but 10 000 times less than naturally occurring radiation. Thus, they present no risk in terms of radiation in normal operation. Nuclear power plants *de facto* are even emitting less radiation than coal-burning power plants! A part of the radioactive contaminants in coal, uranium and thorium pass up the chimney stack and

Table 8.4 Radiation exposure (mSv year) from different activities.

Natural background radiation	2.4
Working at a nuclear power plant	1.15
One diagnostic X-ray	0.2
Living in a stone, brick or concrete building	0.07
One round-trip flight Paris-New York	0.05
Living at the gate of a nuclear power plant	0.03
Watching television	0.015
Luminous wrist watch	0.0006
Coal-fired power plant, average within 80 km	0.0003
Average radiation from nuclear power production	0.0002
Smoke detector	0.00008

Source: UNSCEAR, NEI and the environmental case for nuclear power, R. Morris 2000.

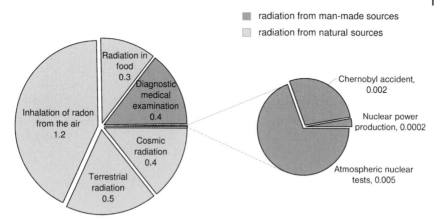

radiation from man-made sources
radiation from natural sources

Figure 8.22 Average radiation dose to the public (in mSv per year). Based on data from UNSCEAR, Sources and effects of ionizing radiation (New York, 2000).

are released to the atmosphere, while another part stays in the ashes. In fact, it has been calculated that that about 13 000 t thorium and 5000 t uranium are presently released yearly into the environment as a result of coal combustion [85].

Nuclear Byproducts and Waste

Nuclear wastes are classified according to their activity and lifetime. Some radioactive elements have a half-life of only a few seconds, while others have half-lives of millions or even billions of years, the danger of nuclear waste being inversely proportional to its lifetime (Table 8.5). The longer the half-life of a nucleus, the lower the number of disintegration events per unit of time.

With regard to nuclear waste management, most attention has been focused on high-activity spent nuclear fuels. Nuclear fuel rods are generally present in a commercial reactor for three to four years, during which time they are progressively used up; subsequently they have to be replaced by new rods in order to keep the electricity output constant. Used nuclear fuel is not a waste, however, as it still contains very large amounts of compounds with high energy value: 95% uranium (of which about 1% is ^{235}U) and 1% plutonium. It also contains about 0.1% of actinides (neptunium, americium, curium, etc.) which could eventually be used to produce energy, and 4% of fission products which have no energy potential and are thus considered as waste which must be disposed. Treatment of the used fuel in special facilities in Europe and Japan enables the recycling of unused ^{235}U, and plutonium. It also reduces by the same time the amount of highly radioactive waste produced and allows a better use of natural resources. For political reasons based on fears of nuclear proliferation, the reprocessing of nuclear fuel was banned in the United States in 1977. Despite a lifting of the ban some years later, however, reprocessing was only reconsidered in recent years. Origin-

Table 8.5 Half-lives (years) of some radioactive elements.

Uranium 238	4 470 000 000
Uranium 235	704 000 000
Neptunium 237	210 000
Plutonium 239	24 000
Americum 243	7400
Carbon 14	5730
Radium 226	1600
Cesium 137	30
Strontium 90	29
Cobalt 60	5.3
Phosphorus 30	2.55 min

ally forbidden by fear of nuclear proliferation, reprocessing is now, ironically, the preferred option to destroy stocks of military-grade plutonium by transforming it to usable MOX fuel for nuclear reactors. This enables the generation of energy while simultaneously reducing the amount and toxicity of spent fuels. It was also found to be cheaper than considering plutonium as waste and trying to dispose of it in depositories, such as under the Yucca Mountain in Utah. This facility was designed and built to store nuclear waste safely for at least 10 000 years, but following pressure from groups opposing the disposal of nuclear waste, standards were recently increased by the Environmental Protection Agency (EPA) to a 1 000 000 years guarantee to protect the next 25 000 generations of residents living near the site [86]. However, the underground storage of used nuclear fuel without reprocessing is clearly a waste of energy potential and resources, by placing in the ground toxic materials that would be perfectly useful to generate energy. It is thus not the best approach for nuclear waste management. Having said that, the disposal of nuclear material in specifically chosen underground facilities is feasible, safe for ten thousands of years, and does not involve insurmountable technical difficulties. Furthermore, the small quantities of used nuclear fuel generated each year from a commercial power plant are in the order of only a few tonnes and thus should be easily manageable. Technical solutions also exist for the treatment of all other nuclear wastes. The disposal of nuclear waste has never been an "unsolvable problem", as some antinuclear activists claim and would like us to believe. Rather than technical, the problem is more an ethical and political one. If we were able to build the atomic bomb, we certainly should be able to solve the problems of radioactive byproducts and waste.

In order to reduce the radiotoxicity of used nuclear fuels and make much better use of the uranium resources, reprocessing and the construction of breeder reactors to convert not only ^{235}U but also most of ^{238}U to energy is a preferable solution. The fast neutrons in breeder reactors are also able to brake the highly active minor actinides such as neptunium, americium and curium produced during fission reactions, which contribute (besides plutonium) to most of the radioactivity of the spent nuclear fuel.

Emissions

Of all the large-scale energy sources used today, nuclear has probably the lowest impact on the environment. In contrast to fossil fuel-powered plants, nuclear power plants do not emit any greenhouse gases or air pollutants such as SO_2, NO_x, and particulates. Thus, the use of nuclear energy helps to keep the air clean, to mitigate Earth's climate changes, avoid ground-level ozone formation, and prevent acid rains. Although not emphasized to the public, in 2003, U.S. nuclear power plants saved us from the emission of 3.4 million tons of SO_2, 1.3 million tons of NO_x, and from 680 million tons of CO_2 entering the Earth's atmosphere [87]. Air pollution, especially from SO_2 emissions and sulfate aerosols produced in large quantities in coal-burning power plants, is estimated to be the cause of approximately 40 000 deaths per year in the U.S. alone [88] – a number which must be compared with that of no casualties due to the operation of commercial nuclear reactors in the U.S. over a period of almost 50 years. Coal also contains mercury, cadmium, arsenic and selenium and, as mentioned above, even radioactive elements such as uranium and thorium, all of which are released into the atmosphere or are present in coal ashes. Because there is virtually no regulation regarding the millions of tonnes of coal ashes produced every year, these can be dumped almost anywhere without much concern for public health. To generate electricity, coal is clearly much more dangerous to use than nuclear power. Even natural gas – the cleanest fossil fuel – still generates large quantities of CO_2 and NO_x. In the context of increased concern about the greenhouse effect and global climate changes due in large part to CO_2 emissions, "clean" nuclear energy will therefore have an increasing role to play.

Nuclear Power: An Energy Source for the Future

Nuclear power generates a significant part of the world's electricity and has much to offer for the future, not only for electric power generation but also for other applications such as hydrogen production (see Chapter 9) and water desalination. Over the years, the safety record in western-built reactors has been remarkable, and advanced reactor design promises to be even safer. With near-zero emissions, nuclear power has a clear advantage over fossil fuels with regard to air pollution and growing concerns about global climate change. Moreover, with most resources of uranium and thorium being located in stable areas of the world, stability in prices and production would also be greatly improved. The progressive introduction of breeder reactors should allow a better usage of the these resources and secure the world's energy supply far into the future, until better energy-generating technologies – perhaps even nuclear fusion – emerge.

Nuclear Fusion

The Sun, like innumerable other stars, is a giant thermonuclear reactor which obtains its energy from nuclear fusion reactions of small nuclei, primarily hydrogen, to produce larger ones, mostly the transformation of hydrogen into helium. The mass of the fused reaction products being less than that of the initial particles, the difference is converted into energy in accordance with Einstein's equation, $e = mc^2$. This represents a tremendous amount of energy – about ten times greater than a typical fission reaction for the same mass of nuclear fuel. Since the positive electric charges of the nuclei provide strong repulsive forces, the energy of the particles, and consequently the temperature, must be very high for the fusion reaction to occur at a sufficient and sustained rate. In the center of the stars this is achieved by intense gravitational forces in a high temperature plasma (15 million °C). Plasma is (beside solid, liquid and gas) the fourth state of matter where the atoms nuclei and electrons are separated and move independently from each other. On Earth however, natural gravitational confinement is impossible. Thus, different technologies must be used to contain very energetic particles and prevent them from escaping before reacting. The two techniques presently considered to achieve this goal are: (i) magnetic confinement, in which a very high-temperature plasma is contained by a strong magnetic field for suitable extended periods of time; and (ii) inertial confinement, where fusion is realized in a small concentrated volume of plasma heated and compressed extremely rapidly with high-energy lasers. For electricity production, magnetic confinement is presently the most advanced and favored option. Experimental studies have been conducted in several countries since the 1950s, but the real breakthrough came in 1968 when the Russians obtained for the first time a 10 million °C plasma in a so-called Tokamak ("Toroidalnaya Kamera c Magnitinymi Katushkami" or toroidal vacuum chamber and magnetic reel in English). Since then, Tokamaks became the reference, leading to the construction of large-scale fusion experiments in the 1980s: the Japanese JT-60, American TFTR, or European JET. A Tokamak is basically a donut-shaped cylinder in which a strong magnetic field is created. The plasma, being a combination of almost independent electrically charged electrons and ions, is trapped in this magnetic field and circulates in a circle in the tokamak. The temperature necessary to maintain a fusion reaction in such systems is in the order of 100 million °C. To achieve such staggering temperatures, the plasma is heated by the injection of highly energetic neutral particles which will also serve as fuel, and with electromagnetic waves at specific frequencies which transfer their energy to the plasma through antennas placed in the confinement chamber. The magnetic field contains the hot plasma far enough from the inner tokamak's wall to avoid its cooling and allow fusion to occur. The heat generated by the reaction is removed through heat exchangers placed in the reactor's wall. This heat is then used to produce steam and finally electricity via a turbine/generator system, much like in today's fossil fuels and nuclear power plants.

To be viable from an energetic standpoint and to reach the so-called break-even point, the energy produced by the fusion reaction must be at least equal to the amount of energy furnished to heat the system. Beyond break-even, the important task will be to create much more energy than is injected into the system and to establish the feasibility of nuclear fusion as an economically viable source of energy.

Currently, the best ratio of energy produced over energy injected into the system, named gain and represented by the symbol Q, has been obtained at the Joint European Torus (JET) with a Q = 0.65, close to break-even. ITER (International Thermonuclear Experimental Reactor), an international collaboration between Europe, Japan, Russia, China, India, United States and South Korea on fusion reaction, is expected to yield a gain between 5 and 10 (Fig. 8.23). With fusion reactions being more efficient the larger the installation, this next generation Tokamak, which will be built in Cadarache, France, will be much bigger than its predecessor. It constitutes the preliminary step to the construction of commercial reactors of even larger size with Q superior to 30.

On Earth, the most accessible and practical nuclear fusion reaction on which all efforts have been concentrated, is the one between deuterium (D) and tritium (T), the two heavier isotopes of hydrogen which offers by far the highest gain (Fig. 8.24). Deuterium is a stable element, abundant in water (33 g m^{-3}) from which it can be extracted. Existing deuterium resources represent more than 10 billion years of annual world energy consumption! Tritium however, is a radioactive element which does not exist in nature and has to be prepared from lithium by neutron bombardment. Lithium resources are estimated at 2000 years, but can be extended to several million years if it is extracted from sea water. In existing experimental reactors, tritium is generated outside and subsequently injected into the plasma. In future systems however, tritium will be produced directly inside the thermonuclear reactor. The inner walls will be made of lithium-containing materials which, under bombardment of neutrons from the fusion reaction, will be transformed to tritium.

Fusion reactors are intrinsically safe because at any time only small amounts of fuel are present in the plasma chamber, and any uncontrolled perturbation will

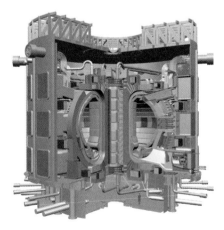

Figure 8.23 The ITER Fusion reactor
(Source: ITER, www.iter.org).

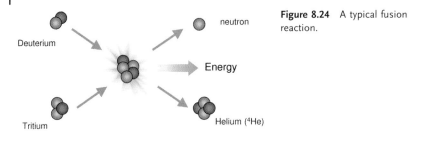

Figure 8.24 A typical fusion reaction.

immediately induce a temperature drop, leading to a cessation of fusion reactions. The risk for potential runaways is therefore eliminated. Like fission, fusion does not produce any air pollutant contributing to acid rains or greenhouse gases responsible for global climate change, and could thus mitigate the environmental risks associates with fossil fuel burning. None of the basic fuels for fusion, deuterium and lithium, nor the reaction product helium, is radioactive or toxic. Tritium, which will be entirely produced on site, decomposes to helium by emission of a low-energy beta ray and has a relatively short half-life (12.3 years). Its radioactivity is low, but reactors will have to take into account the permeation of this gas through matter. As in every system under intense high-energy particles flow, the materials constituting the reactor will be activated. However, a judicious choice of materials with rapidly decreasing activity will allow a minimization of the amounts of radioactive waste to short-lived, low-activity materials. The deuterium-deuterium fusion reaction involves lower radioactivity, but is less practical as it requires about an order of magnitude higher plasma pressure to produce the same power than the deuterium-tritium reaction. The reaction of ^{3}He with deuterium generates even less radioactivity. ^{3}He, however, is not available on Earth and must be prepared. Eventually, with increased knowledge and advanced technology both of these fusion reactions could be used in fusion reactors.

Nuclear fusion is still an energy source of the future. Much progress has already been accomplished, but extensive research and development will still be needed. With mega-projects such as ITER and its followers, however, there is little doubt that this technology will be made practical during the 21st century. Fusion, when compared to present sources of energy, offers numerous advantages: the fuels, deuterium and lithium, are widespread and virtually inexhaustible, obliterating energy security problems and resource-based conflicts. Fusion also has large-scale power-generating capacity with minimal environmental impacts, and is inherently safe. Water used as working fluid in heat exchangers to cool the reactor and produce electricity could also be replaced by liquid metals or helium to achieve higher operating temperatures (1000 °C) and allow, besides electric power generation, the production of hydrogen by thermochemical splitting of water, as has already been proposed with nuclear fission (see Chapter 9). Because of the high cost and complex technology of eventual fusion energy plants, there will most likely initially be only relatively few such large installations serving major power needs, with smaller and more dispersed atomic fission plants providing more decentralized electricity production. Advances in superconductive power transmission lines could, however, substantially help and alleviate present electric transmission limitations.

Future Outlook

Alternative energy sources to fossil fuels are numerous, but have their drawbacks and limitations. Hydropower has been used on a large scale for over a hundred years, but the installation of new capacity is becoming more limited because the best sites are already developed and environmental and socio-economic considerations must increasingly be taken into account before flooding large areas for reservoirs. Energy from geothermal wells can play an important role on a local scale for some countries, but resources are limited on the global scale. Solar energy, photovoltaic and thermal is still too expensive and intermittent. Suitable energy storage or supplementary other energy sources are needed as, there is only limited solar energy production under cloudy conditions, and none during the night. Wind power is also reliant on the intermittent power and speed of winds, but is much cheaper than present solar energy and is therefore of great potential. Biomass can provide a significant but nevertheless limited amount of energy that is inadequate to sustain our modern society's needs. Ocean power in the form of tides, waves and thermal energy is unlikely to represent a significant share of the global energy production in the foreseeable future. Taken together, these renewable energy sources must – and will – certainly play an increasing role in our future global energy mix. However, they will be unable to replace by themselves the energy obtained from non-renewable fossil fuels. With decreasing petroleum and gas reserves, we could rely for some time more heavily on the larger resources of coal. This, nevertheless is only a temporary solution until coal production eventually also starts to decline. It would also imply much larger air pollutant and greenhouse gases emissions, with severe health consequences and contributions to global climate change. By considering environmental, energy security and long-term stability viewpoints, it is clear that nuclear power – albeit made even safer and with problems of radioactive byproduct reprocessing and storage solved – is the major energy source of choice based on current knowledge. Indeed, for the foreseeable future it may produce the vast and increasing amounts of energy needed by humanity. Advanced breeder reactors and new fusion technologies could provide our energy needs for centuries or millennia to come. Eventually, it may be possible to find more efficient, as-yet unknown, ways of using the Sun's energy, but this is for future generations.

Nuclear reactions should not be viewed as purely human inventions. Rather, they are naturally occurring events throughout the Universe, and humankind has only relatively recently succeeded in controlling and harnessing the energy of the atom. Without thermonuclear reactions in the Sun and nuclear decay inside our own Earth, there would be no renewable energies such as solar, wind, biomass, geothermal or hydro energy, or even fossil fuels on our planet. Once we produce energy, it still must be stored, transported and provided in suitable form for subsequent use. This is a major unresolved challenge. We must also identify new solutions to provide convenient hydrocarbon-based fuels for transportation and household needs, as well as the variety of products and materials derived from renewable and sustainable sources. These points are discussed in the following chapters.

Chapter 9
The Hydrogen Economy and its Limitations

Inexhaustible and non-polluting, hydrogen is described by many as the fuel to our future energy needs. Being involved in some way in the so-called "Hydrogen Economy" seems these days to be quite obligatory for governments and any large energy-related company, automobile manufacturers and other industries. The idea sounds rather simple: take hydrogen, one of the most plentiful elements on Earth and in the cosmos, and use it as a clean-burning fuel or in fuel cells to power cars, heat houses and offices, generate electricity, etc. It produces only water as a byproduct and none of the CO_2 and other pollutants obtained by burning fossil fuels as in the current carbon (fossil fuel)-based economy. As some 60% of our oil consumption is used in transportation, numerous programs have been launched around the globe for the development of hydrogen-based fuel cell-powered cars. In 2003, President George W. Bush announced a five-year, $1.2 billion budget for hydrogen research to commercialize hydrogen-powered cars by 2020. The European Union launched a €2.8 billion public-private partnership over a 10-year period to develop hydrogen fuel cells. In the past year, the Japanese governmental budget for fuel cell research was almost doubled to $270 million, and other nations such as China and Canada are also increasing their efforts in this field. Most automobile manufacturers have already invested large amounts of money in the development of hydrogen fuel cell-powered cars, and major energy and oil companies are testing ways to provide and refuel these new vehicles with hydrogen. Despite all these efforts, however, the challenges that lie in the way to the hydrogen economy are enormous. Fundamental problems will have to be solved if hydrogen gas is ever to become a practical, everyday fuel that can be filled into the tanks of our motor cars or delivered to our homes as easily and safely as gasoline or natural gas are today.

The Discovery and Properties of Hydrogen

Hydrogen is the lightest element of the Periodic Table. Hydrogen was first formed as the universe began to cool down after the Big Bang, and still represents 90% of the atoms present in the cosmos, the rest of it being mostly helium. By fusion reactions in the stars, hydrogen subsequently formed the heavier elements, and

Beyond Oil and Gas: The Methanol Economy. G. A. Olah, A. Goeppert, G. K. S. Prakash
Copyright © 2006 WILEY-VCH Verlag GmbH & Co. KGaA, Weinheim
ISBN 3-527-31275-7

can thus be considered as their common ancestor. Hydrogen is the fuel of the stars. Every second, 600 million tonnes of hydrogen are converted into helium in our Sun alone by nuclear fusion, releasing enormous amounts of energy, providing also the light and heat which makes life on our Earth possible.

Hydrogen is also one of the most widespread and plentiful elements on Earth. Due to its high reactivity however, hydrogen combines with other elements. As our atmosphere contains 20% oxygen, molecular hydrogen (H_2) is not present in it except in small amounts in the upper atmosphere. In nature, hydrogen is nearly always found combined with other elements. In every water molecule (H_2O), covering 70% of the Earth's surface, two hydrogen atoms are attached to an oxygen atom. Hydrogen can also be found in hydrocarbons as well as in every living organism, plants or vegetation.

Figure 9.1 The Sun converts 600 million tonnes of hydrogen to helium every second (photo source: NOAA).

Table 9.1 Properties of hydrogen.

Chemical formula	H_2
Molecular weight	2.0159
Appearance	colorless and odorless gas
Melting point	$-259.1\,°C$
Boiling point	$-252.9\,°C$
Density at $0\,°C$	0.09 kg/m^{-3}
Density as a liquid at $-253\,°C$	70.8 kg/m^{-3}
Energy content	$28\,670$ kcal kg^{-1} 57.7 kcal mol^{-1}
Octane number	130+
Autoignition temperature	$520\,°C$
Flammability limits in air	4–74%
Explosive limits in air	15–59%
Ignition energy	0.005 milli calorie

Hydrogen as a distinct element was first identified, and some of its properties described, in 1766 by the English scientist Henry Cavendish who called it "inflammable air" (see Table 9.1). By applying a spark to hydrogen, water was produced. This later led the French chemist Antoine Lavoisier to name the gas hydrogen from the Greek "*hydro*" and "*genes*" meaning "water" and "born of". Shortly after the French revolution (during which Lavoisier literally lost his head on the guillotine), the first practical use for hydrogen was found in the military in the form of reconnaissance balloons filled with hydrogen gas able to fly high above enemy lines. The large quantities of hydrogen gas needed were initially produced by passing steam at high temperature over iron filings. The possibility to generate hydrogen and oxygen gases by water electrolysis was discovered in the early 1800s by two Englishmen, William Nicholson and Anthony Carlisle. William Grove, in 1839, found how to reverse the electrolysis process and generate electricity by combining hydrogen and oxygen in what would be later called a fuel cell.

As mentioned, unlike wood, coal, oil or natural gas, hydrogen is not found in its free form on Earth and thus cannot be collected for combustive energy production. A significant amount of energy must first be expended to produce hydrogen, which is bound to other elements such as in water or hydrocarbons, in order to be able to use it as a fuel. Hydrogen is thus not a primary energy source but only an energy carrier. Some of its physical characteristics, however, are not well-suited for this purpose, especially as a transportation fuel. Paradoxically there is presently great interest for it for such use. The lightness of hydrogen (indeed, it is the lightest of all elements) represents a handicap for its storage, transmission, and use in its gaseous form. The small H_2 molecule can also diffuse through most materials and make for example steel brittle, especially at high pressure or/and temperature. Being a volatile gas, it can only be condensed to a liquid at a very low temperature of $-253\,°C$, only $20\,°C$ over absolute zero. Hydrogen can also ignite or explode in contact with air and should thus be handled with substantial care.

The Development of Hydrogen Energy

From the early 19th century on, hydrogen obtained from coal and combined with carbon monoxide in a mixture called "town gas" was widely used to heat and light homes, apartments, businesses and to provide street lighting. However, with the advent of electricity, and the development of naturally occurring oil and natural gas that could be used directly without former processing, the importance of hydrogen as a fuel rapidly declined. Today, the use of hydrogen as a fuel is limited to niche markets, principally as a rocket propellant and to potential development as a transportation fuel. Nevertheless, starting in the 19th century, the unique properties of hydrogen fascinated generations of scientists, futurists and even science fiction writers. As early as 1874, Jules Verne in one of his visionary books, *The Mysterious Island*, describes in a discussion between his characters what would happen to America's commerce and industry when the world runs out of coal several centuries later. Cyrus Harding, the engineer of the group explains that one will then turn to another fuel, proposing to the astonishment of his companions that water would be the fuel of the future. Or more precisely, "…water decomposed into its primitive elements and decomposed doubtless, by electricity which will then have become a powerful and manageable force".

> *"Yes my friends, I believe that water will one day be employed as fuel, that hydrogen and oxygen which constitute it, used singly or together, will furnish an inexhaustible source of heat and light, of an intensity of which coal is not capable. Some day the coalrooms of steamers and the tenders of locomotives will, instead of coal, be stored with these two condensed gases, which will burn in the furnaces with enormous calorific power…. I believe, then, that when the deposits of coal are exhausted we shall heat and warm ourselves with water. Water will be the coal of the future."*

To our knowledge this is probably the earliest reference to a "hydrogen economy". Verne, however, never mentioned where the primary energy necessary to produce the needed hydrogen from water electrolysis would come from.

In the 1920s, Canada's Electrolyser Corporation Ltd. opened the way to commercial-scale hydrogen production through water electrolysis. This technology allowed hydroelectric power plants to utilize their excess capacity to produce hydrogen and oxygen. The generated gases were used mainly for non-fuel-related applications such as steel cutting and synthesis of fertilizers. At about the same time, German engineers, especially Rudolf Erren, experimented with hydrogen as a fuel for trucks, automobiles, trains, buses and other internal combustion driven devices [89]. In aviation, hydrogen was first exploited in the German Zeppelins which offered regular transatlantic flights, some 20 years before airplanes did (Fig. 9.2). During these trips, large amounts of liquid fuels were consumed, gradually reducing the weight of the airship. To maintain the buoyancy, part of the hydrogen that kept the vessel afloat in the air was used as extra fuel instead of being simply blown-off. The catastrophic fire of the airship *Hindenburg* in 1937, however, ended the era of the hydrogen-filled Zeppelins. During World War II, hydrogen fuel attracted some interest for submarines and trackless torpedoes. After the war however, and during the era of cheap oil and gas, the potential

Figure 9.2 Zeppelin LZ-129 *Hindenburg* flying over New York.

use of hydrogen as a fuel (except for space and military applications; Fig. 9.3) was widely ignored. It only resurfaced with the oil crises of the 1970s and the necessity to find alternatives to petroleum oil. Also, helped by a growing public awareness of pollution problems, suggestions involving hydrogen fuel flourished. This was also the time when the term "hydrogen economy" was introduced and the International Association for Hydrogen Energy was created. Interest by governments and private companies however, lasted only as long as the cost of oil remained high. With sharply declining oil prices in the 1980s, funding for the development of hydrogen energy and alternative energy sources was significantly reduced. For example, the budget for renewable energy in the United States was cut by almost 80% in the early 1980s. Regardless, in the former Soviet Union, a Tupolev 155 experimental airplane tested the use of liquid hydrogen and natural gas as alternatives to jet fuel in 1988 [90]. The use of cryo-fuels however was found impractical, and technically too challenging for regular operation. The large insulated spherical tanks needed to keep the gases liquid were too voluminous and had only enough capacity for relatively short flights. Liquid hydrogen was also deemed too expensive compared to kerosene fuel. Interest in hydrogen fuel began to rise again in the 1990s, based on concerns about decreasing petroleum and gas reserves, and reports on increasing CO_2 emissions that were considered to be a major cause of global climate change. At the same time, considerable advances in the development of fuel cells, and especially Proton Exchange Membrane (PEM) fuel cells, have made a commercial hydrogen fuel cell-powered motor car potentially feasible. This resulted in the transportation sector making by far the most of hydrogen-related investments. Most major carmakers, including Daimler-Chrysler, Honda, Toyota, General Motors and Ford, have built prototype fuel-cell cars, buses or trucks, and consequently the term "Hydrogen Economy"

Figure 9.3 A space shuttle launch at Cape Canaveral, Florida.

became very popular and attracted much public attention. Today, hydrogen-powered vehicles are attracting funding and wide media coverage, and numerous organizations have been created to promote hydrogen fuel via publications, television, meetings and exhibitions. As earlier indicated, governments in the industrialized world themselves have also pledged and provided significant funding for the development of the hydrogen economy.

Hydrogen as a fuel, undoubtedly has many advantages. Its oxidative conversion to produce electricity or heat is clean, producing only water and generating no pollutants. But the crucial question is how to generate, economically, the large quantities of hydrogen needed. If a significant part of this hydrogen is to be produced by reforming of fossil fuels, as is the case today, it would only displace and possibly even increase – but not solve – the pollution and greenhouse gas emissions problem. However, if hydrogen could be generated by the electrolysis of inexhaustible water sources using nuclear and renewable energy, it could become a low-emission or emission-free source of fuel and energy storage medium. On the other hand, because of its unfavorable physical properties and high reactivity, hydrogen storage, transportation and use present major challenges.

The Production and Uses of Hydrogen

Today, hydrogen is used for the most part on a large scale as a feedstock in the chemical and petrochemical industry, to produce principally ammonia, refined petroleum products and a wide variety of chemicals. Its utilizations also include the metallurgic, electronic, and pharmaceutical industries (Fig. 9.4). Except as a propellant for rockets and space shuttles, hydrogen is still rarely used today as a fuel. Industrial facilities often build their own hydrogen production unit to ensure secure supply and safety, avoiding also transportation difficulties. Consequently, the hydrogen market is presently mainly a "captive" market. To cover the present needs, about 50 million tonnes of hydrogen are produced yearly worldwide, representing some 140 toe, or less than 2% of the world's primary energy demand. Using hydrogen as the main energy source would thus imply enormous investments to increase the production capacity and to establish the needed infrastructure for storage and distribution.

As mentioned earlier, hydrogen is not a primary energy source, but rather an energy carrier. It must first be manufactured before it can be used as a fuel. Currently, almost 96% of the world's hydrogen needs are produced from fossil fuels [91], with almost half being generated by the steam reforming of methane (Fig. 9.5). Although relatively inexpensive, this process relies on diminishing natural gas (or oil) resources and emits large amounts of CO_2. At present, water electrolysis is much costlier, and represents only 4% of the production. It is preferentially used when high-purity hydrogen is needed. At the same time, as mentioned, it uses inexhaustible water resources and the energy required can come from any source, including in the future atomic and alternative energy sources, and not fossil fuel-based energy.

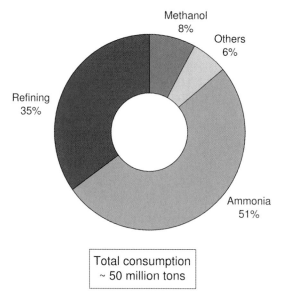

Figure 9.4 The main hydrogen-consuming sectors in the world.

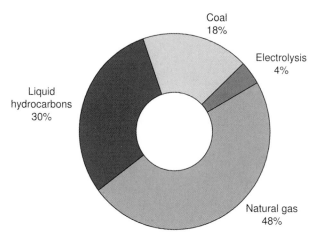

Figure 9.5 Sources for hydrogen production currently used in the world.

Hydrogen from Fossil Fuels

Hydrogen can be obtained from hydrocarbons by reforming or partial oxidation. Compared with other fossil fuels, natural gas is the most suitable feedstock for hydrogen production because of its wide availability, ease of handling and it has the highest hydrogen-carbon ratio, which minimizes the amount of CO_2 produced as a byproduct. Methane can be converted to hydrogen by steam reforming or partial oxidation with oxygen, or by both in sequence (autothermal reforming). Steam reforming is presently the preferred method, accounting for 50% of the hydrogen produced worldwide, and for more than 90% in the United States. In this process, natural gas reacts with steam over a metal catalyst in a reactor at high temperatures and pressures to form a mixture of carbon monoxide (CO) and hydrogen. In a second step, the reaction of CO with steam (water gas shift reaction) produces additional hydrogen and CO_2. After purification, hydrogen is recovered, while the CO_2 byproduct is generally vented into the atmosphere. In the future, however, it could be captured and sequestered, if stricter measures to mitigate climate changes are imposed. Methane steam reforming can also be performed at a smaller scale using various converters. Hydrogen could thus be produced directly locally, such as in filling stations. These decentralized units would have higher hydrogen production costs and lower efficiency than larger industrial ones, but could avoid the cumbersome (and dangerous) transportation of hydrogen from distant centralized production centers. In this case however, the cost of CO_2 capture would be prohibitively high. Partial oxidation and autothermal reforming are more efficient than simple reforming, but require oxygen, the separation of which from air at low cost is still technically difficult. The concept of producing hydrogen from petroleum, albeit established, is not attractive for the long run because it would not solve our energy dependence on diminishing oil reserves.

With the largest reserves compared to all other fossil fuels, coal could supply significant amounts of hydrogen well into the next century. The current technology to achieve this goal is the so-called integrated gasification combined cycle (IGCC). This "clean coal" technology would allow the co-generation of hydrogen and electricity and therefore significantly improve the overall energy efficiency compared to current commercial plants. In this process, much as in methane reforming, coal is gasified by partial oxidation with oxygen and steam at high temperature and pressure. The created synthesis gas, a mixture containing mostly CO and H_2 (but also CO_2), can be further treated with steam to use CO to increase the H_2 yield by the water gas shift reaction. The gas can then be cleaned to recover hydrogen. However, because coal has a low hydrogen/carbon ratio, it also releases much more CO_2 per unit of hydrogen or electricity produced than methane or even petroleum. In current projects, this issue is considered to be addressed by capturing and sequestering the CO_2 emitted into geologic formations or depleted oil and gas reservoirs. Using this technology, Vision 21 and the billion-dollar FutureGen programs funded by the United States government through the Department of Energy (DOE) have the goal to co-produce hydrogen and electricity in zero-emission coal-fueled facilities [92]. To validate the concept, a 275-MW prototype of such a power plant is now in the planning stage to be built and tested. This is justified by DOE citing the large amounts of coal reserves still available. Coal gasification is, however, a less mature technology than other hydrogen-generation processes, although the cost of hydrogen production using this technology is among the lowest available. In large centralized plants, the current cost of producing hydrogen is estimated to be just above $1 per kg [91], with substantial potential for improvement and further price reduction. However, the planned sequestration of CO_2 produced in large amounts from coal and other fossil fuel-burning power plants will be technologically and economically very challenging. None of the existing CO_2 separation and capture technologies has yet been adapted for a large-scale power plant, and costs are uncertain. CO_2 sequestration has so far only been tested on a relatively small scale, and the environmental impacts of large-scale sequestration must be carefully assessed before we start pumping billions of tons of CO_2 into subterranean cavities and under the seas. The energy needed for the capture and sequestration processes will also reduce the overall efficiency of the power plants by as much as 14% [93]. Although coal, as long as it is readily available, is a viable alternative to generate hydrogen in large centralized plants, it is unsuited to decentralized hydrogen production.

The use of fossil fuels to produce hydrogen necessitates that the CO_2 emitted as a byproduct is, according to present plans, captured and sequestered in order to reduce greenhouse gas emissions and to mitigate global climate change. Such storage, although possible, is not without dangers, as Earth movements or volcanic eruptions could lead to the catastrophic release of large amounts of CO_2. Regardless, the use of fossil fuels would only be a temporary solution as oil, natural gas and eventually coal will all be depleted. Thus, more sustainable methods are needed, such as to generate hydrogen from biomass or from water by electrolysis using not only all renewable energy sources such as hydro, wind, solar and biomass, but also atomic energy.

Hydrogen from Biomass

Biomass could potentially become an important source of hydrogen [94]. Biomass includes a large variety of materials such as agricultural residues from farming and wood processing, dedicated bioenergy crops such as switchgrass, and even algae in the sea. As mentioned in Chapter 6, biomass can be used for the production of such liquid fuels as ethanol, biodiesel and methanol. Like fossil hydrocarbons, biomass can also be converted into hydrogen by gasification or pyrolysis coupled with steam reforming. This approach can greatly benefit from the extensive knowledge accumulated over the years in the field of fossil fuel transformation and refining. Gasification plants designed for biomass are generally limited to midsize-scale operations, due to the high cost of gathering and transporting the usually dispersed and relatively limited amounts of available biomass. Currently, such plants operate only at some 26% efficiency with estimated hydrogen production costs over $7 kg^{-1} H_2$ [91]. Although with technology improvements and increased efficiencies, lower hydrogen prices can be expected, the cost is expected to remain at or above $3 kg^{-1}. An interesting alternative, which has been demonstrated commercially [95], is to gasify biomass together with coal, with mixtures containing up to 25% biomass. The construction of biomass-specific gasification units would be unnecessary, and in case the biomass feedstock is seasonal, the plant could operate on coal only.

With regard to greenhouse gas emissions, biomass combustion releases CO_2 that was previously captured from the atmosphere, so that in this recycling the net CO_2 emission is near zero. However, the cultivation of crops requires fertilizers (that need hydrogen in the form of ammonia) and water, as well as energy for the production, harvesting and transportation. All of these factors, together with their environmental impacts on soil, water supply and biodiversity, must be taken into account and the possible consequences of a large-scale intensive energy-related crop farming carefully assessed. Dedicated high-yield energy crops such as switchgrass, which can be grown with minimal energy input, would be preferable, although biomass for energy would still have to compete for huge areas of land with other agricultural products. Algae grown in the vast expanses of the sea could, however, change this picture in the future. In any case, biomass could only be expected to supply a part of the large quantities of hydrogen required.

Photobiological Water Cleavage

Besides biomass transformation processes, a technology aimed at producing hydrogen by the direct cleavage of water with microorganisms without first producing biomass is emerging. This direct photobiological process could potentially be several-fold more efficient than existing biomass gasification, but it is still in the early research stages and significant efforts and breakthroughs will be needed if it is to become a potential hydrogen source.

Water Electrolysis

Electrolysis, the process of cleaving water into hydrogen and oxygen using electricity, is an energy-intensive but well-proven method of producing hydrogen. It is presently about three to four times more expensive than the production of hydrogen from natural gas reforming, which explains its present small share in global hydrogen production. However, it is potentially the cleanest method of producing hydrogen with respect to greenhouse gas emissions, as long as the electricity needed comes from renewable or nuclear energy sources and not from fossil fuels. One should always bear in mind that hydrogen energy is only as clean and environmentally friendly as the process used to produce it. Commercial electrolysis is a mature technology that has been around for over a century to produce high-purity hydrogen. It has been used, however, to a significant extent only in locations where cheap electricity sources exist, such as hydropower in Canada and Norway. A typical commercial electrolyzer has an efficiency of about 70–80%, but a higher efficiency can be obtained with more elevated temperature water or steam electrolysis. Because the efficiency of the electrolysis reaction is independent of the size of the cell or cell-stack, electrolyzers allow both centralized and also decentralized hydrogen production, such as in local service stations. The absence of moving parts requires low maintenance, and electrolyzers are well-suited for use with intermittent and variable power sources, such as wind or solar. Furthermore, any excess electricity generated during off-peak periods could be stored in the form of hydrogen, which then could be used to produce additional power during peak demand.

Basically, any energy source that produces electricity can be used to produce hydrogen by electrolysis. Today, more than 60% of the electricity in the world is still produced by fossil fuel-burning power plants. It would however, be unreasonable to use fossil fuels to generate electricity and then use the electricity to generate hydrogen. As each transformation involves energy loss, the overall efficiency would be lowered and much more CO_2 would be emitted than had fossil fuels been used directly or transformed to hydrogen by reforming. In order to be sustainable and environmentally friendly in the long term, electricity for water electrolysis must be derived from renewable or nuclear energy sources which do not emit CO_2 and air pollutants, such as SO_2 and NO_x.

Hydropower, which is by far the largest renewable electricity source today, is clearly well-suited to produce hydrogen, although its availability, as discussed in Chapter 6, is limited.

Considering its enormous potential, *wind power*, compared to all the other renewable energy sources, has probably the greatest impact for the production of pollution-free hydrogen at a reasonable cost in the foreseeable future. Electricity from wind is already competitive with power from fossil fuels in some areas, and constant developments and improvements in turbine technologies are expected to further reduce significantly its costs. On the other hand, the intermittent nature of wind energy, with capacity factors of only about 30%, is a serious drawback, resulting in coupled electrolyzers for hydrogen production operating at full capacity only for limited periods of time, and requiring considerable hydrogen storage capacity to offset lack of production when the wind is not blowing. With significant optimization of wind-coupled electrolysis and hydrogen storage systems, however, costs for hydrogen generation could fall from current estimations of $6–7 kg^{-1} to less than $3 kg^{-1} [91].

Another source for hydrogen, which could potentially meet all our energy needs into the future, is that from *solar energy*. Like the wind, solar energy is a non-polluting and plentiful source of energy but, being also an intermittent source of energy it suffers from the same drawbacks as wind energy. Unlike wind energy, however, it is still a very expensive way to generate electricity. With current technology, the production cost of hydrogen generated by photovoltaic systems is estimated at $28 kg^{-1} [91] – an order of magnitude higher than that based on fossil fuels, and also more expensive than that based on other renewable energy sources. Even with further development, including improved efficiency and the use of thin-film technology instead of crystalline silicon solar cells, the cost is estimated to remain above $5–6 kg^{-1} hydrogen [9]. Important technological breakthroughs would be needed to make any significant reductions in the costs of solar electricity and thus hydrogen produced by photovoltaic cells. Currently, there are new concepts at the research stage based on conductive organic polymers or nanostructured films, which could possibly be mass produced at lower costs than silicon-based photovoltaic cells. By avoiding the need to couple a photovoltaic device with an electrolyzer, the possibility of producing hydrogen directly from sunlight and water in a so-called photoelectrolysis (PE) device is also under development. Besides photovoltaics, electricity from thermal solar power plants could be used to produce hydrogen, but for the foreseeable future this also is too expensive. Experiments have also been conducted in France, Canada, Israel and other countries, to thermally split water into oxygen and hydrogen at high temperature (2000–2500 °C) using solar furnaces, although there has been limited success and no practical applications are in sight. Thermochemical water splitting using solar energy is another possibility that must be further explored. If solar energy is to become a low-cost option for the sustainable production of hydrogen in the future, then clearly further research is needed.

Electricity from *geothermal energy*, could possibly also be used in some geothermally active areas such as Iceland, The Philippines and Italy, to produce hydrogen. However, even if found economically viable, this process would only be of minor importance on the global scale, as the number of geothermal sources of the quality required for electricity generation is relatively limited.

Resource limitation, a lack of mature technologies, and difficulties of exploitation or environmental concerns also explain the limited development of energy extracted from oceans under various forms: tidal power, wave power or thermal energy, which are thus not expected to play any significant role in hydrogen production.

Hydrogen Production Using Nuclear Energy

As with renewable energy sources, nuclear power reactors do not emit any CO_2 or other pollutant gases into the atmosphere. Using off-peak periods, hydrogen could be produced by nuclear power through electrolysis, enabling a greater utilization of these plants. As mentioned earlier, higher temperatures improve both the thermodynamics and kinetics of the process: hydrogen can be generated more energy efficiently in less time. Most of the new generation IV reactors are planned to operate at much higher temperatures (700–1000 °C) than existing reactors, which operate between 300 and 400 °C [79]. These new reactors are thus well-suited for high-temperature electrolysis of steam for hydrogen production. The direct thermal decomposition of water is impractical as it requires temperatures in excess of 2000 °C, but thermochemical water splitting into hydrogen and oxygen can be achieved efficiently at temperatures of 800–1000 °C, by using chemical cycles. Among various processes, the currently most studied is the so-called iodine-sulfur cycle, in which SO_2 and iodine are added to water in an exothermic reaction to form sulfuric acid and hydrogen iodide [91] (Fig. 9.6). At temperatures above 350 °C, HI decomposes to hydrogen and iodine, the latter being recycled. Sulfuric

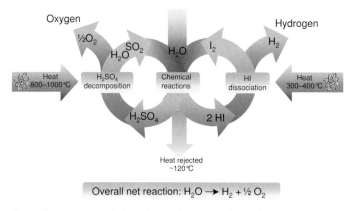

Figure 9.6 The sulfur-iodine thermochemical cycle for the production of hydrogen.

acid decomposes at temperatures in excess of 850 °C into SO_2 (which is also recycled), water, and oxygen. With SO_2 and iodine being continuously recycled, the only feeds to be used up in the process are water and high-temperature heat, giving the products hydrogen, oxygen, and low-grade heat. Considering its near-zero emission characteristics, nuclear power is particularly suited to the generation of hydrogen, and the development of such processes is being conducted at several locations, including the Japan Atomic Energy Research Institute (JAERI), the Oak Ridge National Laboratory, and the Commisariat à l'Energie Atomique (CEA). High temperatures generated by nuclear plants could also be used in other energy-intensive industrial applications. If used for methane steam reforming for example, the heat provided by a high-temperature nuclear reactor could significantly reduce the amount of methane required for hydrogen production and thus reduce CO_2 emissions.

Eventually, heat generated in fusion reactors could also be used in hydrogen production and in high-temperature applications proposed for advanced nuclear fission reactors.

The Challenge of Hydrogen Storage

Producing hydrogen is only the first step in the envisioned hydrogen economy (Fig. 9.7). The next is the storage of the generated hydrogen in a form that should be economical, practical, safe, and user-friendly. Because hydrogen is a very light gas, it contains much less energy per unit volume than conventional liquid fuels under the same pressure. Under normal conditions, hydrogen requires about 3000 times more space than gasoline for an equivalent amount of energy. Thus, hydrogen must be compressed, liquefied or absorbed on a solid material to be of any practical use for energy storage.

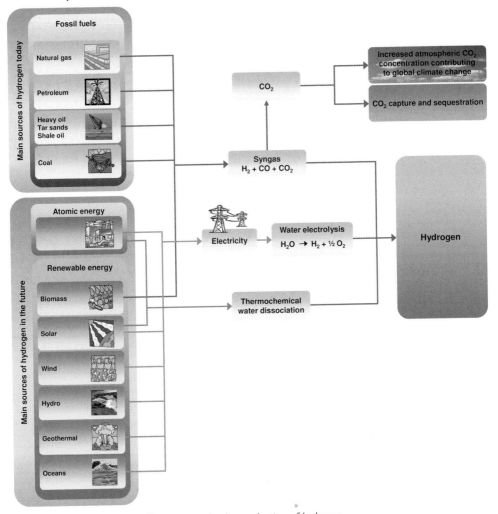

Figure 9.7 Different routes for the production of hydrogen.

Depending on the use for stationary or mobile applications, hydrogen will have very different storage characteristics. In stationary applications, including heating and air-conditioning of homes and buildings, electricity generation and varied industrial uses, hydrogen storage systems can occupy a relatively large space and their weight is not a major factor. In contrast, hydrogen storage in transportation such as in cars is limited by volume and weight which, to provide a driving range of some expected 500 km, must remain minimal. Hydrogen storage is therefore a key factor for the successful introduction of hydrogen as a transportation fuel, the presently considered major application for the hydrogen economy. The challenges faced by hydrogen storage in transportation are great and far from solved. With a driving range requirement of 500 km, a storage capacity of 5 to 10 kg of hydrogen will be needed even for a fuel cell-propelled vehicle. At the same time, refueling should take less than 5 minutes and should be as easy and safe as with hydrocarbon fuels today. The storage system should be also, of course, affordable.

Current technology involves the physical or chemical storage of hydrogen. Physical storage is effected in insulated or high-pressure containers in which hydrogen is stored as a liquefied or compressed gas. Chemical storage includes metals and other materials which absorb hydrogen. Each of these methods has its advantages, but also at the same time has serious drawbacks.

Liquid Hydrogen

On a weight basis, hydrogen has the highest energy content of any known fuel (almost three times that of gasoline). However, being the lightest gas, the density of liquid hydrogen is only 70.8 kg m^{-3}, corresponding on a volume basis to an energy content about a factor of three less than gasoline. Nevertheless, liquid hydrogen is a compact form of hydrogen, making it in principle an attractive candidate for hydrogen storage, especially in transportation. In fact, it is in this form that hydrogen is used as a propellant for space vehicles. However, hydrogen, which has a boiling point of −253 °C is, after helium, the most difficult gas to liquefy. Complex and expensive multi-stage cooling systems are necessary to obtain liquid hydrogen. Typically, in the first step, hydrogen is precooled with liquid ammonia to −40 °C and then to −196 °C using liquid nitrogen. In the following step, helium is used in a multi-stage compression-expansion system to obtain liquid hydrogen at −253 °C. The efficiency of this complex process increases with the plant's size and is thus more adapted for centralized production. The process is not only complex and expensive but also very energy intensive: about 30–40% of the energy content of the hydrogen is required for its liquefaction [96]. Moreover, liquid hydrogen storage systems inevitably lose hydrogen gas over time by evaporation or "boil off". The rate of loss is dependent on the amount stored and the tank's insulation, and is generally lower for larger quantities of liquid hydrogen. In the case of automobile tanks with small capacities, the result is that 1–5% of the hydrogen content would be released to the atmosphere each day in order to avoid pressure build-up [95]. Given the cost and energy invested into producing liquid hydrogen, this is unacceptable both from an economic and also from an environmental viewpoint. Guessing how much fuel is still in the storage tank after some days is certainly not something that most people would like to worry about! In addition to the cost of cryogenic storage being expensive liquid hydrogen, because of its extremely low temperature, must be handled with great care. But perhaps most important of all, hydrogen leaks can result in major safety hazards.

Compressed Hydrogen

In order to store sufficient amounts of energy in a given space, hydrogen compression is currently the preferred solution used in most of the hydrogen fuel cell-powered prototype cars. Because the same quantity of hydrogen can be stored in smaller tanks with increasing pressure, containers (tanks) were developed over the years able to withstand increasingly high pressures. Hydrogen can now be held under 350 or even 700 atm in tanks made from new lightweight materials, such as carbon-fiber-reinforced composites. However, even under these conditions, hydrogen has still a much lower energy content per volume than gasoline (4.6 times less than gasoline at 700 bar H_2; Fig. 9.8) and thus requires several-fold more voluminous tanks. In contrast to liquid fuel tanks which can adopt any shape and easily be adapted to any vehicle, compressed hydrogen tanks have a fixed cylindrical shape necessary to ensure their integrity under high pressure. Vehicle designers and engineers will need to pay great attention on how and where to integrate the pressure tanks. Although hydrogen compression is less energy-intensive than liquefaction, depending on the pressure, it still uses the equivalent of 10–15% of the energy contained in the hydrogen fuel [96]. Because of its small size, as mentioned, hydrogen is able to diffuse through many materials, including metals. During prolonged exposure to hydrogen, some metals can also become brittle. Consequently, because many parts of the fuel system in contact with hydrogen will be metallic, it is necessary to prevent material failure

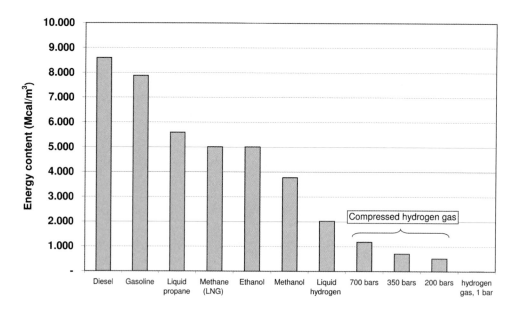

Figure 9.8 Volumetric energy content of hydrogen compared to other fuels.

which could have grave consequences, especially under high pressures. The risk of leaks is a major safety hazard as hydrogen is a highly flammable and explosive gas. This is of course an even greater concern in collision accidents. The high-pressure hydrogen storage systems made from high-tech materials are also complex and presently very expensive, and although technological improvements and larger-scale production will certainly reduce the costs, many of the concerns raised will remain.

Metal Hydrides and Solid Absorbents

An alternative to liquefaction and compression is to store hydrogen in solids, either physically absorbed or chemically bound. Most of the research in this field has concentrated on metals and metallic alloys which have the ability to absorb hydrogen, like a sponge, to form hydrides. In these materials, the metal matrix is expanded and filled with hydrogen. The process, depending on the nature of bonding, can be either reversible or irreversible. Hydrogen stored irreversibly in some materials (including chemical hydrides) can only be released by the chemical reaction of these compounds with another substance, such as water, producing byproducts which must be collected and reprocessed before they can be used again to store hydrogen. Thus, for practical purposes, using irreversibly formed hydrides is not an attractive means of hydrogen storage.

Reversible hydrides are generally solids, which can release the hydrogen contained in the metal in the form of molecular hydrogen, H_2. The uptake and release of hydrogen is typically controlled by temperature and pressure, and is different for every hydride. Some metals absorb hydrogen rapidly but release it slowly, while others require higher temperatures before release is possible. The release of the absorbed hydrogen may also be only partial. The storage and release of hydrogen in any case should take place at temperatures and on a time scale suitable for applications in the transportation field. Volume and temperature increases due to hydride formation should also to be taken into account for storage vessel design. Conventional metal hydrides, as well as their hydrogen-storage capabilities, have been well characterized, with most containing relatively heavy elements: TiFe, $ZrMn_2$, $LaNi_5$, etc. As only a few hydrogen atoms can be bound by each metal atom, this explains why typically only 1–3% by weight of these metal hydrides is actually usable hydrogen. In other words, in order to store 5 kg of hydrogen a tank will need to weigh 200 kg or more. So, unless compensated for by the use of lightweight materials in other parts of the vehicle, the additional weight of the tank would reduce fuel efficiency, which is one of the main goals of developing hydrogen-powered vehicles. Metal hydrides however, do have the advantage of being compact, requiring less space than compressed hydrogen for an equal amount of stored energy. Being only under moderate pressure, hydride tanks can also be shaped more freely, and this facilitates their integration into the vehicle body. Today, research into metal hydrides is focused mainly on lighter compounds such as $NaAlH_4$, Na_3AlH_6, $LiAlH_4$, $NaBH_4$, $LiBH_4$, and MgH_2, which offer higher hydrogen contents per unit of mass.

Besides metal hydrides, other solid absorbing materials for potential hydrogen storage are under investigation. Recently, fullerenes and, more significantly, carbon nanotubes have attracted much attention, but they are still a long way from finding their way into fuel tanks, partially because of their still exorbitant price and unproven potential.

Other Means of Hydrogen Storage

In seeking to overcome the problems associated with hydrogen storage and distribution, many investigators have set out to use liquids that are rich in hydrogen such as gasoline or methanol as a hydrogen source. In contrast to pure hydrogen, these are compact (they contain, on a volume basis, more hydrogen than even liquid hydrogen), and are easy to store and handle. The possibility of generating hydrogen with more than 80% efficiency by on-board reforming of gasoline has been demonstrated. As the reforming produces a mixture of H_2 and CO, the later must be separated and disposed of as CO_2, adding to atmospheric levels of greenhouse gases. Also, as CO will poison the fuel cells, every trace must be removed. The reforming process is expensive and also challenging, because it involves high temperatures and needs considerable time to reach steady-state operational conditions. The advantage is that the distribution network for hydrocarbon fuels already exists, though this would not solve the problems of diminishing oil and gas resources. Methanol reformers operating at much lower temperatures (250–350 °C) [95], though still expensive, are more adaptable for vehicle on-board applications. The absence of any C–C bonds in methanol also greatly facilitates its reforming to hydrogen, and it is the fuel used currently to generate hydrogen in DaimlerChrysler's Necar 5 fuel cell demonstration vehicle. As a liquid, methanol – much like gasoline – can be distributed through the existing infrastructure including filling stations, with only minor modifications. Furthermore, methanol can be made from different sources, including the recycling of CO_2 (see Chapter 11).

Hydrogen: Centralized or Decentralized Distribution?

If hydrogen is produced from gasoline or diesel fuels on board vehicles via reformers, then the existing distribution network could be used with essentially no modifications. If the reformer technology can be made economically viable, it would however, not solve our dependence on decreasing oil resources and would still produce considerably increased amounts of CO_2. In the long term, only reformers using methanol offer a perspective of sustainability.

In case hydrogen gas itself were to become the energy source of the future, it would need to be easily available anywhere at an affordable price, and its distribution should be safe and user-friendly – similar to today's hydrocarbon-based fuels. This implies the creation of a totally new infrastructure specifically designed for the transportation, storage and distribution of hydrogen. Hydrogen could also be produced directly in local fueling stations by natural gas reforming or via water electrolysis by electricity. A decentralized hydrogen production would not require a nationwide delivery system involving trucks or pipelines, but it would be very expensive and energy-consuming. Only some 60 hydrogen-producing and refueling stations are currently operating worldwide in Germany, the United States, Japan, and other countries, providing hydrogen for a small number of automobiles and buses. California, in its Hydrogen Highway program, is planning to construct 200 such fuel stations within a decade. Locally distributed hydrogen generation may be the preferred option as long as the number of hydrogen-powered vehicles and the demand for hydrogen remain low. However, even so there are drawbacks: hydrogen being generated in limited amounts has a high cost, and on-site production by reforming of natural gas emits CO_2 which, considering the scattered locations and small scale, cannot be economically captured or recycled with present-day technology. Electrolysis is only emission-free if the electricity used comes from renewable or atomic energy sources. Iceland for example, which produces electricity almost exclusively from rich geothermal and hydro resources, has the ambition to become the world's first hydrogen-based economy. In most countries, however, if electricity is taken from the existing grid, a significant part of it will have been generated by fossil fuel-burning power plants, thereby eliminating any possible benefits expected from the use of hydrogen. The local manufacture of leak-prone and explosive hydrogen in populated areas would also raise serious safety issues.

The other solution would be to produce hydrogen in large quantities in centralized plants and then to transport it to the local stations by road, rail, or pipelines. In such plants, the production costs would be lower, the efficiency higher, and CO_2 capture and sequestration (or recycling) easier to implement. Furthermore, hydrogen generation is not limited only to methane reforming or water electrolysis, but can also involve other technologies such as coal and biomass gasification or thermal splitting of water in high-temperature nuclear reactors. However, such means of hydrogen production require the establishment of an extensive storage and delivery system in order to service customers. Currently, the options taken in consideration for delivering hydrogen are by road transport using trailers containing hydrogen in its liquid or high-pressure form, and hydrogen gas pipelines.

The transportation by truck of hydrogen as a cryogenic liquid is today commonly used when delivering hydrogen to industries with limited needs, and where on-site generation would be uneconomical. It is by this means, for example, that the hydrogen required as propellant for Space Shuttle launches is transported from Louisiana to the NASA launch pads in Florida. Whereas the density of liquid hydrogen is about ten-fold lower than that of gasoline or diesel oil, it still has the advantage of being relatively compact compared to compressed hydrogen. Commercially available trailer-trucks can transport some 3500 kg hydrogen in liquid form, and this is energetically equivalent to about 13 000 L of gasoline [97]. Because the cryogenic trailers used must be double hulled and vacuum-insulated to avoid excessive blow-off, they are expensive. It has been suggested that transportation of hydrogen in its liquid form from centralized production centers to local distribution stations could play an important role in the initial transition phase to a hydrogen infrastructure. On the large scale necessary however, this solution is not suitable. As mentioned earlier, up to 40% of the energy content of the shipped hydrogen is consumed by its liquefaction, making the process far too expensive from both energetic and economic viewpoints. Hydrogen compression requires less energy than liquefaction. The steel cylinders and other on-board equipment needed for the safe handling of highly pressurized hydrogen is both heavy and expensive. Moreover, with road transport being weight-limited, and considering the extremely low density of hydrogen, this means that the current tube trailers used to transport hydrogen under high pressure (200 atm) can each deliver only about 300 kg of hydrogen. Even taking into account any expected future technical advances, a 40 000-kg truck would enable the delivery of only 400 kg of hydrogen, or about 1% of its dead weight [96]. In comparison, a similar truck could deliver some 26 tonnes of gasoline, containing more than twenty times more energy than the compressed hydrogen truck. So, instead of one driver and truck, more than twenty would be needed to deliver the equivalent amount of energy as gasoline; this in turn would generate higher costs and increase traffic congestion. The introduction of lightweight materials for high-pressure hydrogen storage, such as those currently under development for use in motor cars, could potentially be utilized, but the transportation capacity is expected to remain modest. Besides economic and energetic considerations, a substantial increase in the transportation of highly flammable and explosive hydrogen both in liquid and compressed form by road would imply considerable safety issues and risks.

The most commonly used system for the transmission of hydrogen in large quantities for the chemical and petrochemical industries is by pipelines. To date, worldwide, these have a combined length of only about 2500 km, of which 1500 km are located in Europe and 900 km in the United States [98]. Transport by pipeline allows a direct connection to be made between main hydrogen producers and users. Hydrogen pipelines require the use of special steels or metals, seals and pumps, and they are also expensive to build, maintain and operate. In contrast, there are many hundreds of thousand kilometers of existing pipelines for natural gas, oil and other hydrocarbon products around the world, though these are not suited for hydrogen transport. Most of the metal pipelines, when exposed to hydrogen, would allow it to diffuse through and themselves become brittle over time. Due to the small size of hydrogen, leakage – especially during transportation over long distances – is likely to occur and should therefore be carefully controlled in order to minimize significant losses and explosion hazards. The cost of hydrogen transport is also at least about 50% higher than that for natural gas and for the same volume [99], while hydrogen contains three times less energy than natural gas. Pipeline shipment and dispensing of hydrogen is estimated to cost some $1 kg^{-1} with current technology (ca. $0.7 kg^{-1} expected by future improvements). It is thus, much more expensive than the $0.19 per gallon currently paid to ship and distribute gasoline [91]. Nonetheless, the transport of hydrogen by pipelines may be the best solution to date, though the installation of a large hydrogen pipeline infrastructure would be highly capital-intensive. Indeed, it would only be an option for the long term, when the numbers of cars and other fuel cell-driven devices running on hydrogen would be sufficient to support the vast investments needed.

Safety of Hydrogen

As mentioned earlier, the chemical and petrochemical industries have been using hydrogen in their operations for many years, and the space industry for several decades. Within industrial settings, the production, storage and transportation infrastructures have been developed for the safe use of hydrogen. However, it must be borne in mind that hydrogen is a volatile, dangerous and explosive gas. Perhaps the most vivid image of this was the fire that destroyed the *Hindenburg* Zeppelin in 1937 while landing in Lakehurst, New Jersey. Initially, hydrogen used to keep the airship aloft was blamed for the disaster, but later investigations showed the real cause of the accident to be the extremely flammable lacquer (it had similar properties to rocket fuel) which was painted onto the outer hull of the airship and ignited due to electrostatic discharges. The ensuing fire helped by hydrogen burned the entire airship in about only 30 seconds.

In the past, hydrogen has also been used in many homes (the owners were most often unaware) in the form of town gas, which contained up to 60% hydrogen in a mixture with carbon monoxide. It was due in part to the high toxicity of CO, and not necessarily because of hydrogen, that town gas has been replaced by natural gas. Due to its unique physical properties compared to liquid and gaseous hydrocarbon fuels, the safety issues associated with the use of hydrogen are quite specific since, being small and light, hydrogen is a most leak-prone gas. Hydrogen itself is non-toxic, but it is explosive and flammable. Moreover, being colorless, odorless and tasteless, it is difficult to detect leaks. In the case of natural gas, which is also odorless, colorless and tasteless, sulfur compounds are added to make leaks readily detectable, but the addition of such odorants is impractical in the case of hydrogen. In addition, the odorants would leak at different rates compared to the extremely small hydrogen molecule. Consequently, it is necessary to use sensors for hydrogen detection, though even these have been found to be relatively ineffective. Additives could also contaminate and poison the hydrogen fuel cells. Hydrogen is flammable over a wide range of concentrations in air (from 4 to 75%), and the minimum energy necessary for its ignition (0.005 mcal) is about 20-fold lower than that for natural gas and gasoline. Common electronic devices such as a cell phone or even the friction of simply sliding over a motor car seat can cause ignition if the correct concentration of hydrogen in air is present [89,95]. Hydrogen burns with a scarcely, almost invisible, slightly bluish flame, which means that a person could actually step unknowingly into hydrogen flames. Hydrogen, as mentioned earlier, can also cause many metals (including steel) to become brittle over time, raising the risk of cracks and fractures that would result in failures with possible catastrophic consequences, especially in high-pressure systems. Hence, specialized materials or/and liners would be necessary for hydrogen storage.

Until now, the good safety record of hydrogen use in industry has been largely due to the numerous precautions, codes and standards required for hydrogen handling by trained professionals. It is also related to the fact that most hydrogen is produced on-site and so is not transported over long distances in large quantities. However, if hydrogen were to be handled by the wider public and by people with no formal training or awareness of its potential danger, then it would be vital that strict new safeguards be introduced. Such safety measures would most likely be very costly to introduce, and public compliance difficult to ensure.

Hydrogen in Transportation

The development of a hydrogen infrastructure, besides the difficulties discussed earlier, is facing the "chicken and egg dilemma". As long as there is no adequate hydrogen distribution infrastructure, there will be only a limited demand for hydrogen-powered cars and other applications, despite its obvious attractiveness as an energy storage material. On the other hand, there is no real incentive for investing hundreds of billions of dollars in a hydrogen infrastructure unless there is a solid and sustained demand for it. The question is: how can the demand for hydrogen be stimulated, and will the hydrogen economy become real? Today, the fate of the hydrogen economy seems to be tightly connected with the development of fuel cells which offer the promise of very efficient and zero-emission vehicles. Unfortunately, however, the questions raised about generation, handling and distribution of hydrogen are often neglected. Due to significant technical and economic challenges, the road to commercialization has still a long way to go. Currently, most prototype vehicles have very high price tags, and even using optimistic assumptions the U.S. Department of Energy has estimated that future fuel cell vehicles (FCV) would likely be 40–60% more expensive than conventional ones [95]. Thus, it may take decades before FCVs begin seriously to replace internal combustion engines (ICE) -powered cars and trucks. In aiming to accelerate the transition to a hydrogen economy, the use of hydrogen fuel in a conventional ICE has also been suggested. Except for replacing the fuel tank with a hydrogen tank, only minor and relatively inexpensive changes would be necessary to run an ICE vehicle on hydrogen. Safety precautions, however, must be seriously considered, and a number of automobile makers are heading down this pathway. BMW, in particular, has been conducting research on hydrogen-powered engines since 1978, testing the first prototype motor car a year later. The company's sixth-generation hydrogen-powered car, planned to be produced in limited series within a few years, has a bivalent engine which is able to run either on hydrogen or gasoline. This will allow the motor car to move on today's road network, where hydrogen filling stations are still few and far between. The liquid hydrogen tank, with a capacity of 170 L, will allow the car to cover a distance of about 300 km, with a secondary gasoline tank extending this range to 800 km (Fig. 9.9) [100]. In addition to BMW, Ford (hydrogen-powered model U concept car) and Mazda are also planning to introduce hydrogen ICE cars. These vehicles have the advantage of

Figure 9.9 The 745 h Hydrogen ICE car from BMW. (Courtesy of ©BMW AG.)

producing almost no pollutants (except for small amounts of NOx) and can be introduced relatively soon in the market compared to FCVs. In order to significantly increase its efficiency, the hydrogen ICE could also be coupled with an electric hybrid system, as found in Toyota's Prius (an ICE engine running on gasoline also charges batteries, which take over to provide an electric drive in slow city traffic). The efficiency, however, is expected to remain lower than for fuel cell vehicles. Despite the advantages of hydrogen ICE, the problem of on-board hydrogen storage, which presently limits the driving range, also remains. Besides fueling cars with hydrogen produced at central locations and in delocalized small units using electrolysis of water at filling stations, it should also be possible to fuel these cars with hydrogen produced by on-board reforming of a variety of hydrocarbon fuels. This would, however, bring no advantage over conventional hydrocarbon-burning ICEs. Carbon monoxide must be carefully separated from the generated syngas, so as not to poison the fuel cells. If the CO were to be oxidized to CO_2 and with no CO_2 sequestration, the CO_2 emission would only be relocated to the hydrogen-generating facilities or devices, but not eliminated.

Fuel Cells

History

Fuel cells are devices that convert the chemical energy of a fuel directly into electrical energy by electrochemical reactions. Fuel cells are considered to be one of the main solutions for the efficient utilization of fossil fuel-derived fuels. The concept of fuel cells was discovered by William R. Grove, a Scotsman, during the late 1830s. Grove discovered that by arranging two platinum electrodes with one end of each immersed in a container of sulfuric acid and the other ends separately sealed in containers of hydrogen and oxygen, a constant current would flow between the electrodes. The sealed containers held water as well as the gases, and Grove noted that the water level rose in both tubes as the current flowed. In 1800, William Nicholson and Anthony Carlisle in England had described the process of using electricity to decompose water into hydrogen and oxygen (the electrolysis of water). But combining hydrogen and oxygen to produce electricity and water was, according to Grove, "…a step further that any hitherto recorded." Grove realized that by combining several sets of these electrodes in a series circuit he might "…effect the decomposition of water by means of its composition." His device, which he named a "gas battery", was the first ever fuel cell.

However, the device remained a curiosity with no practical application in sight until, more than a century later, in 1953, Sir Francis T. Bacon constructed the first fuel cell prototype with a power output in the kW range. Bacon began experimenting with alkali electrolytes in the late 1930s, settling on potassium hydroxide (KOH) instead of using the acid electrolytes known since Grove's early discoveries. KOH performed as well as acid electrolytes and was not as corrosive to the electrodes. Bacon's cell also used porous "gas-diffusion nickel electrodes" rather than solid electrodes as Grove had used. Gas-diffusion electrodes increased the surface area in which the reaction between the electrode, the electrolyte, and the fuel occurred. Bacon also used pressurized gases to keep the electrolyte from "flooding" the tiny pores in the electrodes. Over the course of the following 20 years, Bacon made enough progress with the alkali cell to present large-scale fuel cell demonstration units. The U.S. space agency, NASA selected alkali fuel cells for the Space Shuttle fleet, as well as for the Apollo program, mainly because of power-generating efficiencies that approach 70%. Importantly, alkali cells also provided clean drinking water for the astronauts. The cells use platinum catalysts that are expensive (perhaps too expensive for commercial applications), but several companies are examining ways to reduce costs and improve the cells' versatility by using less-expensive cobalt catalysts. Most of these alkali fuel cells are currently being designed for transport applications.

Fuel Cell Efficiency

In contrast to heat engines (gasoline and diesel engines), the fuel cell does not involve conversion of heat to mechanical energy and the overall thermodynamic efficiencies can be very high.

The thermodynamic derivation of the Carnot cycle of a heat engine states that all the heat supplied to it cannot be converted to mechanical energy, and that some of the heat is rejected. The heat is accepted from a source at higher temperature (T_H in Kelvin), part of it is converted to mechanical energy, and remainder is rejected into a heat sink at lower temperature (T_S in Kelvin). The greater the temperature difference between the source and the sink, the greater the efficiency. The Carnot efficiency of a heat engine is given by Eq. (1). On the other hand, the fuel cell efficiency is related to the ratio of two thermodynamic properties, Gibbs free energy (ΔG^0) and the total heat energy or Enthalpy (ΔH^0) (Eq. (2)):

Maximum efficiency (Carnot), $\eta_{Carnot} = (T_H - T_S)/T_H$ (1)

Fuel cell efficiency, $\eta_{Fuel\ cell} = \Delta G^0/\Delta H^0$ (2)

The theoretical thermodynamic efficiency of a hydrogen-oxygen fuel cell is ~93% at ambient temperature. To achieve acceptable efficiencies, an internal combustion engine under ideal conditions must operate at a very high temperature. The variation of a hydrogen fuel cell theoretical efficiency versus the corresponding Carnot efficiency of a heat engine is shown in Figure 9.11.

The ambient temperature maximum thermodynamic intrinsic fuel cell efficiencies of different fuels can be very high. These data, along with reversible cell potentials of selected fuels, are listed in Table 9.2.

Fuel cells, therefore, are considered as very efficient electrical energy-producing devices with high power densities at relatively low temperatures. The possible applications of fuel cells are numerous, from micro fuel cells producing only a few

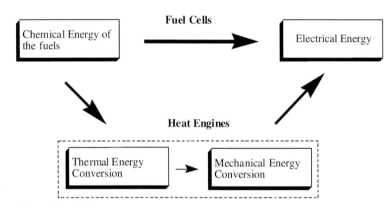

Figure 9.10 Fuel to energy conversions.

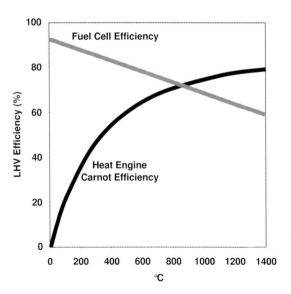

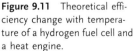

Figure 9.11 Theoretical efficiency change with temperature of a hydrogen fuel cell and a heat engine.

watts needed in cell phones, to on-board fuel cells for the automobile sector, and large units able to produce several MW to provide buildings with electricity. Major drawbacks to the widespread commercialization of fuel cells are mainly technological (reliability issues, material durability, catalyst utilization, mass transport,

Table 9.2 Theoretical reversible cell potentials (E^0_{rev}) and maximum intrinsic efficiencies for candidate fuel cell reactions under standard condition at 25 °C.

Fuel	Reaction	n	$-\Delta H^{o\ (a)}$	$-\Delta G^{o\ (a)}$	$E^{o\ (b)}_{rev}$	E [%]
Hydrogen	$H_2 + 0.5\ O_2 \rightarrow H_2O$ (l)	2	286.0	237.3	1.229	82.97
Methane	$CH_4 + 2\ O_2 \rightarrow CO_2 + H_2O$ (l)	8	890.8	818.4	1.060	91.87
Methanol	$CH_3OH + 1.5\ O_2 \rightarrow CO_2 + 2H_2O$ (l)	6	726.6	702.5	1.214	96.68
Formic acid	$HCOOH + 0.5\ O_2 \rightarrow CO_2 + H_2O$ (l)	2	270.3	285.5	1.480	105.62
Ammonia	$NH_3 + 0.75\ O_2 \rightarrow 0.5\ N_2 + 1.5\ H_2O$ (l)	6	382.8	338.2	1.170	88.36

^{a} kJ mol^{-1},
^{b} Volts

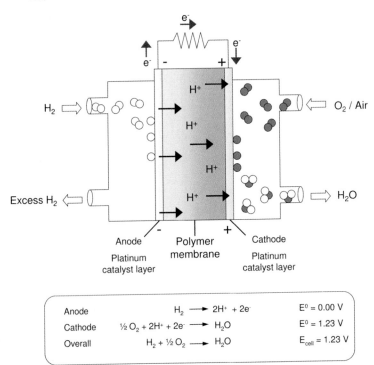

Anode	$H_2 \longrightarrow 2H^+ + 2e^-$	$E^0 = 0.00$ V
Cathode	$\frac{1}{2}O_2 + 2H^+ + 2e^- \longrightarrow H_2O$	$E^0 = 1.23$ V
Overall	$H_2 + \frac{1}{2}O_2 \longrightarrow H_2O$	$E_{cell} = 1.23$ V

Figure 9.12 The proton exchange membrane (PEM) hydrogen fuel cell.

etc.) and cost-related. Different types of fuel cells design exist, with some being more suited to certain applications than others. However, they all function on the same electrochemical principle. Fuel cells, in principle, can be built based on any exothermic chemical reaction.

Hydrogen-Based Fuel Cells

Hydrogen-based fuel cells produce electricity, heat and water by catalytically combining hydrogen with oxygen. They are composed of two electrodes, an anode (negatively charged) and a cathode (positively charged) separated by an electrolyte. This electrolyte can be made of a variety of materials from polymers to ceramics, which are in general ion (H^+, OH^-, CO_3^{2-}, O^{2-}, etc.) conductors. The nature of the electrolyte determines many of the fuel cell's properties, including the temperature of operation, and this is therefore used to categorize the different fuel cell types.

In a proton-exchange membrane (PEM) fuel cell (*vide infra*), hydrogen entering the fuel cell is split with help of a catalyst (generally platinum) on the anode side into electrons and protons (H^+). The electrons move along an external circuit to power an electric device, while the protons migrate through the electrolyte. At the cathode, by action of a catalyst, protons and electrons are recombined with oxygen of the air to produce water. The device is shown in Figure 9.12. Since every cell produces less than 1 V, many cells must be stacked together to produce higher voltages.

In an alkaline-based fuel cell, instead of protons, hydroxide ions (OH^-) move from cathode to anode (Fig. 9.13). Although, the alkaline fuel cells are utilized in space applications, their commercial use is hampered by their sensitivity to CO_2, which reacts with the alkali.

The most studied fuel cells designs are presently: phosphoric acid fuel cells (PAFC), molten carbonate fuel cells (MCFC), solid oxide fuel cells (SOFC), proton exchange membrane (PEM) fuel cells, and direct methanol fuel cells (DMFC). The latter will be discussed in more detail subsequently.

PAFC, MCFC and SOFC are generally designed to be used in stationary applications because they are heavy, bulky and require operating temperatures from around 200 °C for PAFC to about 650–1000 °C for MCFC and SOFC. PAFC, the most mature fuel cell technology, is commercially available from United Technologies Corp. Close to 300 units have been installed worldwide. As the name indicates, these fuel cells use liquid polyphosphoric acid as the electrolyte, the electrodes being made of carbon coated with finely dispersed platinum catalyst. The hydrogen required is obtained by methane (natural gas) reforming, and the overall efficiency from methane to electricity is 37–42%. With co-generation a heat efficiency approaching 80% can be achieved, which is comparable to conventional systems burning natural gas. At around $4500 kW^{-1} capacity [101], PAFCs also remain expensive compared to conventional fossil fuel-based technologies, with costs of less than $1000 kW^{-1} capacity. However, because fuel cells have no moving parts they are generally very reliable and require low operation and mainte-

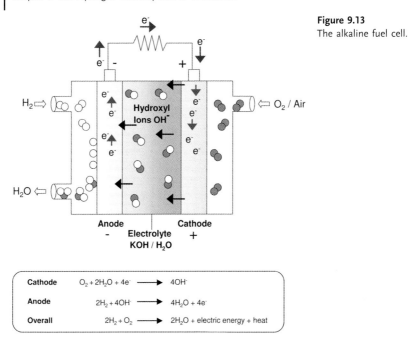

Figure 9.13
The alkaline fuel cell.

Cathode	$O_2 + 2H_2O + 4e^- \longrightarrow 4OH^-$
Anode	$2H_2 + 4OH^- \longrightarrow 4H_2O + 4e^-$
Overall	$2H_2 + O_2 \longrightarrow 2H_2O +$ electric energy + heat

nance costs. This explains why PAFCs have found their way only into "niche markets", mainly for consumers in need of a very stable, reliable and clean on-site electricity source such as banks , airports, hospitals, or military bases.

In MCFCs, the electrolyte is made of lithium-potassium carbonate salts heated to about 650 °C [101] (Fig. 9.14). At this high temperature the molten carbonate salts act as electrolytes and CO_2 formed by the reaction of carbonate with hydrogen is transported between the anode and cathode through carbonate ion. Because they operate at high temperature, natural gas (or other fuels including methanol and ethanol) can be converted into a hydrogen-rich gases directly inside the fuel cell in a process called internal reforming. Without the need for an external reformer, MCFCs can reach fuel to electricity efficiencies above 50% – much higher than the value of 37–42% obtained with PAFCs. The higher temperature also allows nickel to be used as a catalyst instead of expensive platinum at lower temperature because of its much higher reactivity than nickel. The first commercial unit was delivered by Fuel Cell Energy Inc. to a brewery in Japan in 2003, and today more than 50 units from that company and others are operating worldwide. Their price is in the same range as PAFCs. Molten carbonates are, however, highly corrosive, and this raises some concerns about the fuel cell's lifetime. Fuel-Cell Energy Inc. in collaboration with the U.S. Department of Energy, is also developing a hybrid system combining a MCFC with a gas turbine which could eventually lead to power plants with fuel to electricity efficiencies approaching 75%.

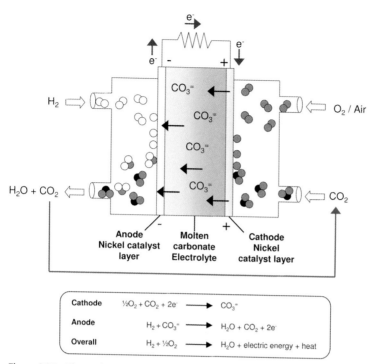

Cathode	$\frac{1}{2}O_2 + CO_2 + 2e^-$ ⟶ $CO_3^=$
Anode	$H_2 + CO_3^=$ ⟶ $H_2O + CO_2 + 2e^-$
Overall	$H_2 + \frac{1}{2}O_2$ ⟶ H_2O + electric energy + heat

Figure 9.14 The molten carbonate fuel cell (MCFC).

SOFC is the technology that currently attracts the most attention for stationary applications. These cells operate at high temperature (800–1000 °C) and thus, like MCFCs, do not require a reformer. However, in contrast to PAFCs and MCFCs, the SOFC uses a solid ceramic (usually Y_2O_3-stabilized ZrO_2) instead of a liquid as an electrolyte. O^{2-} ions are transported from cathode to anode, the latter being made of Co-ZrO_2 or Ni-ZrO_2 (Fig. 9.15). This feature allows the electrolyte to adopt different shapes, such as tubes or flat plates, giving a greater freedom in fuel cell design and also avoiding problems connected with the use of corrosive liquids. The efficiency of SOFCs is expected to be around 50–60%. The high temperature of the exhaust gases produced are ideal for co-generation and combined cycle electric power plants. In combination with gas turbines, efficiencies of 70% or more could thus be achieved. In co-generation units, the use of waste heat could bring overall fuel efficiencies to 80–85%. The U.S. Department of Energy has formed the Solid State Energy Conversion Alliance (SECA) involving companies, universities and national laboratories, with the goal of producing a highly efficient SOFC that would cost only about $400 kW^{-1}, allowing this technology to compete with diesel generators and natural gas turbines and rapidly to gain widespread market acceptance. The mass production of standardized basic ceramic modules, using manufacturing technologies similar to those developed for

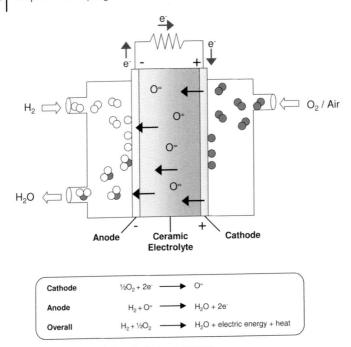

Cathode	$\frac{1}{2}O_2 + 2e^-$	$\longrightarrow$	$O^=$
Anode	$H_2 + O^=$	$\longrightarrow$	$H_2O + 2e^-$
Overall	$H_2 + \frac{1}{2}O_2$	$\longrightarrow$	$H_2O + $ electric energy $+$ heat

Figure 9.15 The solid oxide fuel cell (SOFC).

the production of electronic components, is believed to be the key to this ambitious cost reduction.

PEM Fuel Cells for Transportation

For transportation, PEM fuel cells are at present the most promising option to replace the current internal combustion engine. They operate like a refuelable battery, generating electricity for as long as they are supplied with hydrogen fuel and oxygen (from air). At the core of these fuel cells, a thin polymer film constitutes the electrolyte. Until now, this PEM has been based predominantly on a fluorocarbon polymer produced by DuPont and known under the commercial name of Nafion®; this polymer is permeable to protons when saturated with water, but does not conduct electrons. It is sandwiched between two platinum impregnated porous carbon (graphite) electrodes. PEM fuel cells can, in theory, convert more than 50% of the fuel into usable energy. They have a high power density, which translates into low weight and volume requirements. In addition, they operate at low temperature (generally 80 °C), which allows for fast start-ups, and they can very rapidly change their power output as a function of demand. Furthermore, they are safe, quiet, easy to operate, and of low maintenance. For all these reasons, PEM fuel cells are seen as the most suitable candidate for automotive applications, but they are also considered and being developed for small power applications. However, before PEM fuel cell-powered cars leave the assembly lines in the millions, numerous problems will have to be solved. The price of a prototype PEM fuel cell is today in the order of hundreds if not thousands of dollars per kW. A fuel cell with about 65–70 kW output (equivalent to ca. 90 horse power) is needed to power a small car. Thus, dramatic cost reductions are necessary to reach a price of less than $50 kW^{-1}, which would begin to make the PEM fuel cell competitive with ICEs.

The presently used perfluorinated membranes are way too expensive, representing about one-third of the cost of a fuel cell stack. New and inexpensive efficient materials, such as hydrocarbon membranes, will have to be developed to replace them. At the same time, they should be chemically and mechanically stable, have a great durability and a high tolerance to fuel impurities or reaction byproducts, such as carbon monoxide (CO).

Another key element for proper PEM fuel cell operation is the thin layer of platinum catalyst coated with electrically conductive graphite on both sides of the membrane. The use of platinum is necessary because of the low temperature of operation. Platinum however, is an expensive precious metal and extensive research is being conducted in order to reduce its content. These efforts include ways of increasing the catalyst's activity (e.g., by using nano-sized dispersed metals) so that less can be used for a same power output, or to seek alternative and cheaper catalysts. At a low operating temperature, platinum has also the disadvantage of binding strongly with CO, a common impurity in hydrogen obtained by reforming of fossil fuels. This reduces the platinum's availability for hydrogen chemisorption and electro-oxidation. At 80 °C, the platinum catalyst can tolerate only a few ppm of CO in hydrogen before its activity begins to subside. Thus, the hydrogen fuel used must be of very high purity, requiring additional purification steps such as oxidation of CO over gold or other catalysts – all of which will add to the cost of hydrogen obtained from hydrocarbons. In order to increase the catalyst's tolerance to CO, new bi-catalysts such as those based on platinum/ruthenium are currently being developed for use in fuel cells. PEM fuel cells that can potentially operate at temperatures 120 °C or higher will also alleviate the CO poisoning issue. However, high-temperature membranes with adequate water content and proton conductivity need to be developed to achieve this lofty goal.

Today, numerous PEM fuel cell-powered prototype vehicles are being tested worldwide by the major automobile companies, including Ford, General Motors, Honda, Toyota, Renault, and Volkswagen. DaimlerChrysler for example, has been working intensively on fuel cell technology since the early 1990s, and is now preparing a group of 60 of its latest hydrogen fuel cell car, the F-Cell, to be tested for their performance under everyday driving conditions. In addition, the company has also developed both a light duty vehicle and a bus which run on hydrogen. Small fleets of buses are presently driven on a daily basis for public transportation in Amsterdam, Luxembourg, and Reykjavik. Fleets of transit buses are ideal for the introduction of alternative fuels because they travel only short distances and generally refuel in a central depot. Furthermore, the installation of voluminous fuel tanks, which can be easily accommodated in these larger vehicles, is less of a problem than in a passenger car. In fact, transit buses is one area where alternative fuels have had the most success. Diesel buses in many cities have been progressively replaced by much less-polluting compressed natural gas (CNG) buses. To make fuel cell-powered vehicles affordable, however, major technological breakthroughs are clearly needed in order significantly lower production costs and also to mitigate safety issues. If hydrogen fuel has to be used to fuel these cars, the onboard hydrogen storage capacity must also to be substantially improved, and a massive and very expensive infrastructure for delivering needed hydrogen to the users must be built from scratch. Therefore, we need to look more carefully at other options. ICEs, for instance, have been continuously improved for more than a century, and are becoming increasingly efficient and ever less pollutant-emitting. Their combination with batteries and electric motors in hybrid vehicles increases their efficiencies even further. In order to compare various vehicles with different fuels, engines, electric motors, and drive trains, a well-to-wheel (WTW) analysis, representing the overall efficiency of an energy source from the time it is extracted from Earth or any other resource, to when it actually turns the wheels of the vehicle, is generally conducted. In the case of a gasoline vehicle operating with a regular ICE, the WTW gives an overall efficiency of only 14%. A gasoline ICE hybrid vehicle, however, can achieve a WTW efficiency of 28%, comparable to the present day 29% obtained with a fuel cell vehicle using compressed hydrogen generated from natural gas. The efficiency of a diesel hybrid ICE vehicle was found to be even higher than that for fuel cell motor car [93]. Fuel cells may be more efficient in converting hydrogen into electricity than through mechanical conversion, but the energy required to produce, handle and store hydrogen lowers the overall efficiency considerably. Considering greenhouse gas emissions, WTW analysis shows that hydrogen fuel cell vehicles offer only a slight advantage over hydrocarbon-based hybrid vehicles if hydrogen used in the fuel cell is derived from fossil sources.

The ICE is a proven and reliable technology, and its combination with a battery/electric motor has already been applied today in hybrid vehicles (Toyota Prius, Honda Insight, etc.). These are slightly more expensive than regular cars, but they allow a considerable reduction in fuel consumption and lower emissions. Fuel cell vehicles, on the other hand, are still in the developmental phase, with developers striving to bring down manufacturing costs and improve the lifetime and reliability of the fuel cell stacks. To reduce the consumption of petroleum and reduce CO_2 emissions, deploying hybrid cars on a large scale seems to be a better and more reasonable solution in the short term, than to rely on the still uncertain future advances in fuel cell technology, which however eventually might become economically viable for the transportation sector in the long term.

Regenerative Fuel Cells

If a hydrogen-oxygen fuel cell is designed to operate also in reverse as an electro-lyzer, then electricity can be used to convert the water back into hydrogen and oxy-gen. This dual-function system known as a regenerative fuel cell (also called uni-tized regenerative fuel cell, URFC), is lighter than a separate electrolyzer and gen-erator and is an excellent energy source in situations where weight is a concern.

Scientists at AeroVironment of Monrovia, California and NASA developed a propeller-driven aircraft called *Helios*, to be used for high-altitude surveillance, communications, and atmospheric testing. *Helios* was a $15 million dollar, solar-electric project. The unmanned aircraft had a wingspan of 75 m and was de-scribed by some as more like a flying wing than a conventional plane. In 2001, in a test flight, *Helios* reached an altitude of almost 29.4 km, an altitude considered by NASA to be a record for a propeller-powered, winged aircraft. *Helios* was de-signed for atmospheric science and imaging missions, as well as relaying tele-communications up to 30 km. The 5-kW prototype was powered by solar cells during the day and by fuel cells at night (a URFC device). Regretfully, in another test flight in 2003, *Helios* crashed.

For automotive applications, the Livermore National Laboratory and the Hamil-ton Standard Division of United Technologies have studied URFCs in great detail and found that, compared with battery-powered systems, the URFC is lighter and provides a driving range comparable to that of gasoline-powered vehicles. Over the life of a vehicle, the URFC was found to be more cost-effective because it does not require replacement. In the electrolysis (charging) mode, electrical power from a residential or commercial charging station supplies energy to pro-duce hydrogen by electrolyzing water. The URFC-powered motor car can also re-coup hydrogen and oxygen when the driver brakes or descends a hill. This regen-erative braking feature increases the vehicle's range by about 10%, and could re-plenish a low-pressure (about 14 atm) oxygen tank, the size of a football.

In the fuel-cell (discharge) mode, stored hydrogen is combined with air to gen-erate electrical power. The URFC can also be supercharged by operating from an oxygen tank instead of atmospheric oxygen to accommodate peak power demands such as entering a freeway. Supercharging allows the driver to accelerate the vehicle at a rate comparable to that of a vehicle powered by an ICE.

The URFC, in a motor car, must produce ten times the power of the *Helios* air-craft prototype, or about 50 kW. A car idling requires just a few kilowatts, highway cruising about 10 kW, and hill climbing about 40 kW, but acceleration onto a high-way or passing another vehicle demands short bursts of 60 to 100 kW. For this, the URFC's supercharging feature supplies the additional power. A URFC-pow-ered motor car must be able to store hydrogen fuel on board, but existing tank systems are relatively heavy, reducing the car's efficiency or range. Under the Part-nership for a New Generation of Vehicles – a government-industry consortium dedicated to developing vehicles with very low fuel consumption – the Ford Cor-poration provided funding to Lawrence Livermore National Laboratory, EDO Cor-poration, and Aero Tec Laboratories to develop a lightweight hydrogen storage tank (a pressure vessel). The team combined a carbon-fiber tank with a laminated, metalized, polymeric bladder (much like the ones that hold beverages sold in boxes) to produce a hydrogen pressure vessel that was lighter and less expensive than conventional hydrogen tanks. Equally important, its performance factor – a function of burst pressure, internal volume, and tank weight – was about 30% higher than that of comparable carbon-fiber hydrogen storage tanks. In tests where cars with pressurized carbon-fiber storage tanks were dropped from heights or crashed at high speeds, the cars generally were demolished while the tanks still held all of their pressure – an effective indicator of tank safety. Unlike other hydrogen-fueled vehicles in which refueling needs depend entirely on commercial suppliers, the URFC-powered vehicle carries most of its hydrogen infrastructure on board. Unfortunately, even a highly efficient URFC-powered vehicle needs periodic refueling, and until a network of commercial hydrogen suppliers is developed, an overnight recharge of a small motor car at home would generate enough energy for a driving range of about 240 km (150 miles), exceeding the range of present-day electrical vehicles. With the infrastructure in place, a 5-min fill up of a 350 atm (5000 psi) hydrogen tank would give a range of 580 km (360 miles). The commercial development of the URFC for use in automobiles is, however, at least five to ten years away.

Utilities are also looking at large-scale energy storage systems employing regenerative fuel cells. The proposed systems store or release electrical power through an electrochemical reaction between two liquid electrolytes such as sodium bromide and sodium polysulfide stored in tanks. Inside the cell, the two electrolytes are separated by a thin, ion-selective membrane. Inside this big rechargeable battery-like device, when subjected to current during charging, bromine is produced at the anode. The bromide ions in the electrolyte combine with bromine to give perbromide ions. In the discharge cycle, perbromide is converted to bromide ions, producing at the same time electric energy [102]. Systems based on vanadium salts or zinc/bromine are also being developed and are commercially available [103, 104]. Such regenerative fuel cell storage systems are expected to store up to a whopping 500 MW of energy for up to 12 h [105].

Outlook

Although hydrogen is widespread and abundant on Earth, its extraction from water or hydrocarbons requires much energy. Today, most efforts for efficient and economical hydrogen production are concentrated on natural gas and coal, because they are still the most inexpensive sources of hydrogen. One of the goals of the envisioned hydrogen economy is the mitigation of greenhouse gas emissions, but this would imply capture and sequestration of CO_2 on a very large scale. Alternatively, H_2 would be obtained by the electrolysis of water, but the required energy must be derived from non-fossil fuel sources (atomic and any form of alternative energy source). Even if this might become technologically and economically possible, the consequences of storing huge amounts of CO_2 underground or at the bottom of the seas are, at best, uncertain. In the long term, however, considering the finite amounts of fossil fuels, hydrogen will have to be produced increasingly using not only renewable energy sources such as wind and solar, but also nuclear energy. Once generated, the physical and chemical properties of hydrogen makes its storage, transportation and safe handling difficult and potentially hazardous. Whether the production is centralized or decentralized, due to its unique properties, a totally new and expensive infrastructure would have to be built to supply consumers with hydrogen. For vehicles, onboard hydrogen storage is likely to remain voluminous and costly, while the use of hydrogen as an automotive fuel from an energy efficiency and emissions viewpoint would be feasible only if used in combination with fuel cells. Efficient, reliable, and affordable fuel cells will almost certainly become a reality in the not too distant future, although unless an efficient and safe metal hydride or other storage system is developed it is also questionable to what degree people would feel safe, knowing that they are driving cars with a high-pressure tank filled with an explosive and highly flammable gas under their seats.

Other static applications of hydrogen fuel are feasible in suitable cases, and will be developed in time. The storage and transportation of energy in the form of hydrogen, as discussed here, has serious drawbacks and problems. Instead of the volatile and potentially explosive hydrogen gas, a new and feasible alternative in energy storage by its conversion with atmospheric CO_2 to liquid methanol is therefore proposed. In the near future, still-existing large natural gas reserves can be converted directly to methanol (without first conversion to syn-gas), thereby solving problems of transportation and shipping associated with LNG and allowing the gradual introduction of methanol-powered cars. What we now term the "Methanol Economy" (see Chapters 10 and 14) will eventually involve the recycling of CO_2 to methanol rather than its sequestration. This will provide an inexhaustible fuel source as well as a carbon source for synthetic hydrocarbons and their products while mitigating global warming caused by the generation of excess CO_2 from burning fossil fuels. In order to achieve its goals, the methanol economy will also require the production of hydrogen on a massive scale through water electrolysis, using electricity generated from any non-fossil sources (renewable energy, and also atomic energy). In this way, energy will be stored not as volatile hydrogen gas, but by its conversion with CO_2 into convenient and easy-to-handle liquid methanol.

Chapter 10
The "Methanol Economy": General Aspects

Oil and natural gas, the main fossil fuels besides coal, are not only still our major energy sources and fuels but also the raw materials for a great variety of man-made materials and products. These range from gasoline and diesel oil to varied petrochemical and chemical products including synthetic materials, plastics, and pharmaceuticals. However, what Nature provided as a gift, formed over the course of eons, is being used up rather rapidly. The expanding world population (now exceeding 6 billion and probably reaching 8–10 billion in the 21st century), and the increasing standards of living and demands for energy in developing countries such as China and India, is putting increasing pressure on our diminishing fossil fuel resources and making them even more costly. Whereas coal reserves may last for another two or three centuries, readily accessible oil and gas reserves – even considering new discoveries, improved technologies, savings and unconventional resources (such as heavy oil deposits, oil shale, tar sand, methane hydrates, coalbed methane, etc.) – may not last much beyond the 21st century.

In order to satisfy mankind's ever-increasing energy needs, all feasible alternative and renewable energy sources must be considered and used. These include biomass, hydro and geothermal energy as well as the energy of the Sun, wind, waves, and tides of the seas. In practical reality, nuclear energy will above all have to be further developed and utilized. Our discussion here, however, is not dealing with the question of energy generation, which we believe mankind must and will solve (as in the final analysis majority our energies are derived from the Sun, an enormous and permanent energy source), but rather with the challenges of how to store and to best use energy. Most of our energy sources are used primarily to provide heat and electricity. Electricity is generally a good way to transport energy over relatively short distances when a suitable grid exists, but it is very difficult to store on a large scale (batteries for example are still inefficient and bulky). Besides finding new energy sources, it is therefore necessary to identify and develop new and efficient ways to store and distribute energy from whichever source it is derived.

One approach that has been proposed and widely discussed recently is the use of hydrogen. This would be generated eventually by water electrolysis using any available energy source and subsequently used as a clean fuel (the so-called "hydrogen economy"; see Chapter 9). Hydrogen is clean in its combustion, produc-

Beyond Oil and Gas: The Methanol Economy. G. A. Olah, A. Goeppert, G. K. S. Prakash
Copyright © 2006 WILEY-VCH Verlag GmbH & Co. KGaA, Weinheim
ISBN 3-527-31275-7

ing only water, although its generation is less clean if the requisite energy is derived from fossil fuels with their attendant polluting effects. As we have seen in Chapter 9, hydrogen has certain desirable attributes for energy storage and as a fuel but, due to its extreme volatility and explosive nature, many difficult issues will need to be resolved if it is to be used on a massive scale as an everyday energy source and fuel. As the lightest element, hydrogen has serious limitations in terms of its storage, transportation and deliverance of energy. The handling of volatile and potentially explosive hydrogen gas necessitates special conditions (high-pressure technology, cryogenic tanks, special materials to minimize diffusion and leakage, etc.), as well as strict adherence to extensive safety precautions, all of which makes hydrogen use very costly. In addition to these difficulties, there is a need to develop a currently non-existent infrastructure for the "hydrogen economy", and although this may eventually be developed it seems at present economically prohibitive. In any case, hydrogen per se will be unable to solve our continuing need for hydrocarbons and their products; for this, new synthetic methods must be developed that use existing natural resources more efficiently, or they may be synthesized from a non-fossil hydrocarbon source.

Today, the field of transportation is the major user of oil, with liquid hydrocarbon fuel products (gasoline, diesel oil) being the fuels of choice. These are easy and relatively safe to handle, to transport and to distribute, mainly because a vast infrastructure already exists. Consequently, the preferred alternative fuels for transportation should be comparable liquids.

Some years ago, one of the authors suggested a new, viable alternative approach of how to use more efficiently available oil and natural gas resources and, eventually, to free humankind from its dependence on fossil fuels. This approach is based on methanol, which forms the basis of the "Methanol Economy". Methanol (CH_3OH), is the simplest, safest, and easiest to store and transport liquid oxygenated hydrocarbon. At present, it is prepared almost exclusively from synthesis gas (syn-gas, a mixture of CO and H_2) obtained from the incomplete combustion of fossil fuels (mainly natural gas or coal). Methanol can also be prepared from biomass (wood, agricultural byproducts, municipal waste, etc.), but these play only a minor role. As discussed in Chapter 12, the production of methanol (and/or dimethyl ether) is also possible by the oxidative conversion of methane, avoiding the initial preparation of syn-gas, or by reductive hydrogenative conversion of CO_2 (from industrial exhausts of fossil fuel burning power plants, cement plants, etc. and eventually the atmosphere itself). The hydrogen required (which is eventually generated from water using non-fossil fuel-based energy) is thus stored in the form of a safe and easily transportable liquid. The chemical recycling of excess CO_2 would, at the same time, also help to mitigate climate changes caused in a significant part by the excessive burning of fossil fuels.

Methanol is an excellent fuel on its own right, with an octane number of 100, and it can be blended with gasoline as an oxygenated additive. Alternatively, methanol can be used in today's ICEs with only minor modifications. Methanol can also be used to generate electricity in fuel cells. This is achieved by first catalytically reforming methanol to H_2 and CO; the H_2, after separation from CO, is then fed into the fuel cell. Methanol can also react directly with air in the Direct Methanol Fuel Cell (DMFC), without the need for reforming. The DMFC greatly simplifies fuel cell technology, making it readily available to a wide range of applications such as portable electronic devices (e.g., cell phones, laptops), and soon also for motor scooters and cars, or for electricity generators and emergency back-up systems in areas of the world where electricity is still not available from a grid.

Another potentially significant application of the direct conversion of natural gas (methane) to methanol is in its ready and safe transportation when pipelines are neither feasible nor available. Today, LNG is transported across oceans under cryogenic conditions by using very large tankers (>200 000 tonnes). The LNG is unloaded at terminals and fed into pipelines to satisfy increasing needs, or to serve as a substitute for diminishing local natural gas sources. LNG is potentially hazardous, however, due perhaps to accidents or to acts of terrorism, and a single supertanker exploding close to high-density population area might have the devastating effect of a hydrogen bomb. Whilst hoping that such a situation will never occur, realistically we must be prepared to find alternative, safe methods of transporting natural gas. In this respect, its conversion to methanol is a feasible alternative. The direct conversion of natural gas to liquid methanol, in contrast to its prior conversion to syn-gas, can be achieved relatively easily and without building very large plants (see Chapter 12). Methanol produced close to the natural gas sources, can be easily transported.

Besides its use as energy storage and fuel, methanol also serves as a starting material for chemicals such as formaldehyde, acetic acid, and a wide variety of other products including polymers, paints, adhesives, construction materials, synthetic chemicals, pharmaceuticals, and single-cell proteins. Methanol can also be conveniently converted via a simple catalytic step to ethylene and/or propylene (the methanol-to-olefin, MTO, process), which serve as the building blocks in the production of synthetic hydrocarbons and related compounds (see Chapter 13). Thus, hydrocarbon fuels and products currently obtained from fossil fuels can be obtained from methanol, which is in turn produced by the chemical recycling of atmospheric CO_2 (see Chapter 12). Methanol, therefore, has the ability to liberate mankind from its dependence on fossil fuels for transportation and hydrocarbon products, by allowing these to be produced via the hydrogenative recycling of CO_2.

The concept of the "Methanol Economy" has broad advantages and possibilities. It is suggested that methanol be used as: (i) a convenient energy storage medium; (ii) a readily transported and dispensed fuel, including uses in methanol fuel cells; and (iii) as a feedstock for synthetic hydrocarbons and their products, including polymers and single-cell proteins (for animal feed and/or human consumption). The carbon source will eventually be the air, which is available to all on Earth, while the required energy will be obtained from alternative energy sources, including atomic energy.

It should be emphasized that there is no preference for any particular energy source in the production of methanol. All sources, including alternative sources and atomic energy can be used in the most economical, safe and environmentally acceptable ways. Methanol is a most convenient way in which to store and distribute energy, a suitable fuel in its own right, and a raw material in the production of synthetic hydrocarbons and their related compounds. The "Methanol Economy" offers a new way in which convenient and safe reversible energy storage and transportation can be achieved in the form of a simple, easily to handle liquid chemical – methanol. The ready conversion of methanol to synthetic hydrocarbons and their products will ensure that future generations will have access to the essential products and materials that today form an integral part of our life. At the same time, the "Methanol Economy", by recycling excess atmospheric CO_2, will mitigate one of the major adverse effects on the Earth's climate caused by mankind, namely global warming.

The concept of the "Methanol Economy" has developed over a number of years, with the use of methanol as a fuel and gasoline additive only attracting interest during times of critical shortage. In fact, much more attention was (and is) paid to the use of ethanol obtained from agricultural sources, including fermenting corn (in the United States), sugar cane (in Brazil), or by bioconverting various other agricultural materials. The production and use of bioethanol as a transportation fuel was discussed in Chapter 8, and although this is feasible in some countries (e.g., Brazil, United States), it is able to satisfy only a small part of our overall transportation fuel requirements.

Although methanol and ethanol are chemically closely related (one and two carbon atom alcohols, respectively), when the public considers "alcohols" as transportation fuels they frequently fail to realize the significant differences between the two molecules. The fermentation of agricultural or natural products can be used to produce both alcohols, including "wood alcohol" (i.e., methanol) obtained from cellulose sources (primarily wood), though today methanol is produced mostly by synthetic processes. Industrial ethanol is produced by hydration of ethylene. Ethanol can also be prepared by fermentation, but whilst it represents a renewable, non-fossil fuel base, vast areas of suitable land are required to grow the sugar cane, corn, or wheat from which it is produced. The agricultural production of ethanol is also highly energy-demanding, and currently most of that energy comes from fossil fuels. Hence, the fundamental difference between the bioagricultural production of ethanol and of methanol is that the latter does not rely on agriculture or on diminishing fossil fuels.

In the interim, in the case of the methanol economy, still available natural gas (methane) can be efficiently converted to methanol by direct oxidative transformation, or from CO_2 in the exhausts of fossil fuel-burning power plants. Eventually, it can also rely on the chemical conversion of atmospheric CO_2, using hydrogen generated by the electrolysis of water (using any energy form, including alternative non-fossil fuel energies and atomic energy). In this way, mankind can produce methanol from chemical recycling of the CO_2 of the air, which is accessible to all and, together with water, are inexhaustible resources on Earth.

The Olah group has long been involved in the study of various new aspects of methanol chemistry, beginning in the 1970s with the superacidic selective oxidation of methane to methanol and the related condensation of methanol to higher hydrocarbons. During the 1980s, this was followed by the discovery of the bifunctional acid-base-catalyzed conversion of methanol or dimethyl ether to ethylene and/or propylene, and through them to both gasoline range aliphatic as well as aromatic hydrocarbons. These studies were conducted independently of the zeolite-catalyzed chemistry developed by Mobil (now ExxonMobil) and UOP for the conversion of syn-gas-based methanol to hydrocarbons (see Chapter 13).

As general realization has set in during recent years that our non-renewable fossil fuel resources are indeed diminishing, increasing efforts have been directed to finding solutions to counteract their depletion. A need for their more efficient and economic use is clear, and a variety of alternative energy sources and safer ways to use atomic energy, together with the proposed "hydrogen economy", have been pursued. It should be emphasized at this point, however, that in addition to finding better solutions to our overall energy needs in the post fossil fuel era, there will continue to be a lasting need for convenient and safe transportation fuels and for a multitude of hydrocarbon products that will necessitate the creation of vast quantities of synthetic hydrocarbons. Whilst the "hydrogen economy", as discussed, cannot itself fulfill these needs, it appears that the proposed broad concept of the "Methanol Economy" can indeed achieve this goal.

In summary, the "Methanol Economy" encompasses:

- New and more efficient ways of producing methanol (and/or derived dimethyl ether) from still-existing natural gas sources by their oxidative conversion, without prior production of syn-gas.
- Utilization of the hydrogenative recycling of CO_2 to methanol from industrial exhausts, but eventually from the air itself as the inexhaustible carbon source.
- The use of methanol and derived dimethyl ether as a convenient transportation fuel for both ICEs as well as in the new generation of fuel cells, including DMFC.
- The use of methanol as the raw material for producing ethylene and/or propylene to also provide the basis for synthetic hydrocarbons and their products.

The "Methanol Economy" offers a feasible means by which to liberate mankind from its dependence on diminishing oil and gas resources, while simultaneously utilizing and storing all sources of alternative energies (renewable and atomic). At the same time, by chemically recycling excess atmospheric CO_2, one of the major man-made causes of climate change – global warming – will also be mitigated. These points are discussed in more detail in Chapters 11 to 14.

Chapter 11
Methanol as a Fuel and Energy Carrier

Properties and Historical Background

Methanol, also called methyl alcohol or wood alcohol, is a colorless, water-soluble liquid with a mild alcoholic odor. It freezes at −97.6 °C, boils at 64.6 °C, and has a density of 0.791 at 20 °C. Methanol in its (relatively) pure form was first isolated in 1661 [106] by Robert Boyle, who called it "spirit of the box" because he produced it through the distillation of boxwood. Its chemical identity or elemental composition, CH_3OH, was described in 1834 by Jean-Baptiste Dumas and Eugene Peligot. They also introduced the word methylene to organic chemistry, from the Greek words *methu* and *hyle*, meaning respectively wine and wood. The term methyl, derived from this word was then applied to describe methyl alcohol, which was later given the systematic name methanol. Containing only one carbon atom, methanol is the simplest of all alcohols. Methanol is commonly referred to as wood alcohol because it was first produced as a minor byproduct of charcoal manufacturing, by destructive distillation of wood. In this process, one ton of wood generated, along other products, only about 10–20 L of methanol. Beginning in the 1830s, methanol produced in this way was used for lighting, cooking and heating purposes, but was later replaced in these applications by cheaper fuels, especially kerosene. Up until the 1920s, wood was the only source for methanol, which was needed in increasing quantities in the chemical industry. As hard as it may be to believe today, all the methanol required during World War I for example, was derived from charcoal furnaces along with acetone and other essential chemicals [107]. With the industrial revolution, wood was largely replaced by coal in many applications. Coal and coke gasification processes through the action of steam and heat were developed, by which gases containing carbon monoxide and hydrogen could be obtained to supply cities with town gas. Using hydrogen produced with this technology, Fritz Haber and Carl Bosch developed the technical hydrogenation of molecular nitrogen N_2 to ammonia at very high temperature and pressure. This breakthrough resulted in the development of a number of other chemical processes, necessitating similar severe conditions and feedstock, including methanol synthesis. In fact, from the earliest days, the synthesis of methanol and ammonia were so interrelated that they are often produced in the same plant. The synthetic route to methanol production, by reacting

Beyond Oil and Gas: The Methanol Economy. G. A. Olah, A. Goeppert, G. K. S. Prakash
Copyright © 2006 WILEY-VCH Verlag GmbH & Co. KGaA, Weinheim
ISBN 3-527-31275-7

carbon monoxide with hydrogen, was first suggested in 1905 by the French chemist Paul Sabatier [108]. In 1913, the Badische Anilin und Soda Fabrik (BASF), based on the investigations of A. Mittasch and C. Schneider, patented a process to synthesize methanol from syn-gas, produced from coal, over a zinc/chromium oxide catalyst at 300–400 °C and 250–350 atm [108, 109]. After World War I, BASF resumed its research into synthetic methanol and built, in 1923, the first commercial high-pressure synthetic methanol plant in Leuna, Germany. Between 1923 and 1926, F. Fischer and H. Tropsch reported from the Mühlheim Coal Research Laboratory their extensive studies of the production of hydrocarbons, including that of methanol, from syn-gas, a mixture of carbon monoxide and hydrogen, which was the basis of what is known as the Fischer–Tropsch synthesis [110, 111]. In 1927, in the United States, Commercial Solvents Corporation used the high-pressure technology to produce methanol from CO_2/H_2 mixtures obtained as fermentation byproduct gases [112]. At the same time, the DuPont company began the production of both methanol and ammonia in the same plant using syn-gas produced from coal. In the 1940s, steam reforming of natural gas began in the United States, based on developments of BASF in the 1930s. From then on, coal was slowly abandoned as a feedstock for syn-gas in favor of the cleaner, cheaper and plentiful natural gas. Over the years, other feedstock including heavy oil and naphtha have also been used, albeit at much lesser extent. The steam reforming process of methane, because of the very high purity of the syn-gas, opened the way to the technical realization of the low-pressure methanol process, introduced commercially in 1966 by Imperial Chemical Industries (ICI).

Table 11.1 Properties of methanol.

Synonyms	Methyl alcohol, wood alcohol
Chemical formula	CH_3OH
Molecular weight	32.04
Chemical composition (%)	
Carbon	37.5
Hydrogen	12.5
Oxygen	50
Melting point	−97.6 °C
Boiling point	64.6 °C
Density at 20 °C	791 kg m^{-3}
Energy content	5420 kcal kg^{-1}
	173.6 kcal mol^{-1}
Energy of vaporization	9.2 kcal mol^{-1}
Flash point	11 °C
Autoignition temperature	455 °C
Explosive limits in air	7–36%

This new process using a more active Cu/ZnO catalyst, and operating at 250 to 300 °C and 100 atm [113], put an end to the high-pressure methanol synthesis technology which operated under much more severe conditions. The use of these highly active catalysts was made possible by the lower content of catalyst poisons such as sulfur or metal carbonyls in the syn-gas feed. Not much later, Lurgi launched its own process with even lower operating temperature and pressure (230–250 °C, 40–50 atm). During the past 40 years, considerable further improvements have been made in methanol synthesis from carbon oxides (CO containing some CO_2) and hydrogen, making this technology a rather mature one. Using the low-pressure process, selectivity for methanol is now in excess of 99.8% with an energy efficiency of nearly 75%. Current research is aimed at developing new ways to synthesize methanol at even lower temperature and pressure from diverse origin carbon oxides/hydrogen feeds, as well as by direct oxidation of methane which is a superior method from an energetic viewpoint.

Today, almost all methanol worldwide is produced from syn-gas. However, as discussed in Chapter 12, new ways for its production directly from methane (natural gas) without going through syn-gas, as well as by hydrogenative chemical recycling of carbon dioxide are being developed.

Present Uses of Methanol

Today, methanol is mainly a primary feedstock for the chemical industry. It is manufactured in large quantities (over 32 million tons per year in 2004 [114]) as an intermediate for the production of a variety of chemicals (Fig. 11.1). Worldwide, almost 70% of the methanol production is used to produce formaldehyde (38%), methyl-*tert*-butyl ether (MTBE, 20%) and acetic acid (11%). Methanol is also a feedstock for chloromethanes, methylamines, methyl methacrylate, and dimethyl terephthalate, etc. [108]. These chemical intermediates are then processed to manufacture many products of our daily life, including paints, resins, silicones, adhesives, antifreeze, and plastics [115]. Formaldehyde, the largest consumer of methanol, is mainly used to prepare phenol-, urea- and melamine-formaldehyde and polyacetal resins as well as butanediol and methylenebis(4-phenyl isocyanate) (MDI). MDI foam is, for example, used as insulation in refrigerators, doors, and in motor car dashboards and fenders. The formaldehyde resins are then predominantly employed as adhesives in the wood industry in a wide variety of applications, including the manufacture of particle boards, plywood and other wood panels. The market for MTBE, an oxygenated gasoline additive and blending component, which became the second largest for methanol globally, grew strongly during the 1990s, especially in the United States where it accounted in 2001 for 37% of the methanol consumption. Because of its high octane rating it replaced the phased-out lead-based anti-knock compounds. At the same time this oxygenated compound, when added to gasoline, helped to reduce air pollution from motor cars. In recent years however, MTBE has come under serious environmental attacks, especially in California, due to MTBE contamination discovered

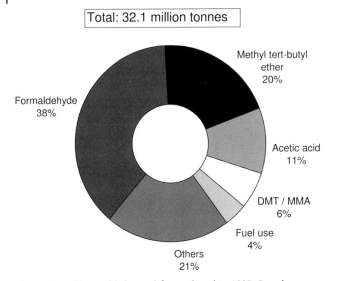

Total: 32.1 million tonnes

Methyl tert-butyl ether
20%

Formaldehyde
38%

Acetic acid
11%

DMT / MMA
6%

Fuel use
4%

Others
21%

Figure 11.1 The world demand for methanol in 2005. Based on data from *Chemical Week*.

in groundwater, primarily as the result of leaking underground storage tanks from local filling stations. Increased maintenance, stricter control or replacement of these tanks would certainly have gone a long way in solving this problem. Nevertheless, as a consequence of the contamination, the use of MTBE has been phased out in most of the United States and will probably also be phased out in other countries. Proposed substitutes include ethanol and ethyl-*tert*-butyl ether (ETBE). There is also the question of whether or not any oxygenated additives to mitigate air pollution are necessary in reformulated gasoline for today's internal combustion engines which, for the most part, use direct fuel injection systems and oxygen sensors allowing the effective control and substantial reduction of emissions, even without the addition of any oxygenated compounds. This of course, does not negate the fact that oxygenated additives provide superior performance and very clean-burning fuels. Methanol and derived dimethyl ether (DME) have excellent combustion characteristics which make them ideal fuels for today's ICE-driven vehicles [116] and diesel engines, respectively. Establishing an infrastructure for methanol fuels would also greatly ease the introduction of advanced fuel cell-based vehicles using methanol as a fuel either by onboard reforming to hydrogen or directly with Direct Methanol Fuel Cells. Thus, methanol as a transportation fuel will certainly play an increasing role in the future.

Use of Methanol and Dimethyl Ether as Transportation Fuels

Alcohol as a Transportation Fuel in the Past

The concept of using an alcohol (methanol or ethanol) as a fuel is as old as the ICE itself. Some of the early ICE models, developed at the end of the 19th century by Nicholas Otto and others, were actually designed to run on alcohol. By that time already, alcohol-powered engines had started to replace steam engines for farm machinery and train locomotives. Also used in automobiles, alcohol engines were advertised as less polluting than their gasoline counterparts. Most European countries with few or no oil resources were especially eager to develop ethanol as a fuel because it could be readily distilled from various domestic agricultural products. Germany for example, went from a production of almost 40 million liters of alcohol in 1887 to about 110 million liters in 1902 [117]. During the first decade of the 20th century, many races were held between alcohol- and gasoline-powered automobiles, and there were lively debates as to determine which fuel gave the best performances. On an economic basis, however, ethanol could hardly compete with gasoline, especially in the United States which had plentiful petroleum resources at the time and a very powerful opponent: the Standard Oil Trust, which was understandably reluctant to the introduction of any alternative fuel. World War I saw an important expansion of the ethanol industry not only for fuel uses but also for gunpowder and war gases manufacture. In the Soviet Union after the Bolshevik Revolution Lenin's first five-year plan had foreseen the wide use of agricultural grain-based alcohol for industrial uses (fuels and production of ethylene and its products). Due to opposition, however, to divert the Russian Vodka to these purposes, the plans were soon abandoned. In the United States, the prohibition – which became effective at the beginning of 1920 – outlawed the consumption of any alcoholic beverage and greatly complicated the production of alcohol, even for fuel purpose. The 1930s saw the introduction of Agrol, a blend of alcohol and gasoline which had some success in Midwestern States for a limited time but was vigorously opposed by oil companies. At the same time, European countries such as Germany, France and England – as well as Brazil and New Zealand, which were relying on imported oil – strongly encouraged the production of alcohol by offering subsidies, or even made the blending of alcohol with gasoline mandatory. It is by the way interesting to note that the very same Standard Oil of New Jersey (one of the companies resulting from the dismantling of the Standard Oil Trust) which had expressed itself against the use of alcohol fuels in the United States, was promoting its own alcohol blend in England under the name Discol. Nowhere however, was the effort of developing alternative fuels as ambitious as in Germany, where the Nazi government had the goal to achieve energy independence, mainly because of military considerations. Hitler was in fact convinced that Germany's failure in World War I was in a large part due to fuel shortages and that such a situation should never hinder the Nazi war machine. From the time Hitler came to power, the production of ethanol (mainly from potatoes) grew dramatically, reaching almost 1.5

million barrels in 1935 [117]. At the same time, methanol produced from coal via syn-gas using the process invented by BASF also expanded dramatically. Methyl and ethyl alcohol were blended with gasoline and sold under the name Kraftspirit. One of the difficulties with blending alcohols with gasoline, however, is the phase separation that can take place when moisture is present, leading to stalling of the engine. Besides methanol synthesis, the large coal reserves, especially in the Ruhr valley were also used to produce via syn-gas large amounts of synthetic gasoline following the Fischer–Tropsch process developed by I. G. Farben, the world's largest chemical company of the time. Shortly before World War II, all alternative fuels accounted for more than 50% of Germany's total light motor fuel consumption. During the war, a large part of the alcohol produced was diverted to other uses, including ammunition [118], medicines and synthetic rubber manufacture. Switzerland however, while struggling to remain neutral during the conflict, turned to methanol when its petroleum supplies were cut off. After the war the interest in alcohol-based fuels declined rapidly in United States and most other countries due to the ready availability of large quantities of cheap oil. Only with the oil crises of the 1970s and concerns about pollution, would the interest in alcohol fuels grew again. The large-scale development of alcohol fuel was most successful in Brazil, which launched in 1975 its National Alcohol Program (PNA). More than 20 years later, the production of ethanol, mainly from sugar cane and its residues amounted to about 15 million m^3 per year in 1997/1998 (equivalent to some 220 000 barrels of oil per day [60]), allowing millions of vehicles to run on alcohol. With production costs averaging about US$35–45 per barrel of oil equivalent, the discovery of large oil reserves off the Brazilian shore and the low oil prices of the 1990s however, the share of alcohol-fueled motor cars in total sale of news cars fell from 96% in 1985 to 0.07% in 1997! The PNA project was kept afloat by blending up to 24% of ethanol into gasoline [60]. With increasing oil prices and the introduction of flexible fuel vehicles (FFV), able to run on any mixture of gasoline and ethanol, the interest in ethanol has been recently revived. The utilization of ethanol fuel in developed countries, where relatively cheap and efficient sources such as sugar cane are generally not available, is limited. The production of ethanol from corn is practiced on a relatively large scale in the United States but, compared to the size of the automobile fuel market, it covers only a very small fraction of the demand. Ethanol is, however, increasingly in demand as an oxygenated additive replacing MTBE. Although Brazil's government and agricultural producers in United States fought to establish ethanol as an alternative fuel, attention on alcohol fuels in the automotive sector also focused on methanol. This is due to the fact that methanol is a very flexible fuel which can be obtained from a wide variety of both renewable and fossil fuel resources: natural gas, coal, wood, agricultural and municipal waste, etc., at a cost generally lower than that for ethanol. The idea of using methanol as an automotive fuel was revived in the 1970s. Thomas Reed, a researcher at the Massachusetts Institute of Technology (MIT) was one of the first to advocate methanol as a fuel in the United States, publishing in 1973 a paper in *Science* magazine explaining some of its advantages. In passing, Reed also mentioned its use in a methanol economy, without further elaborating on it or even using further the name [118]. He stated that adding 10% methanol to gasoline improved performance, gave better mileage, and reduced pollution. Similar results were obtained in Germany by Volkswagen (VW), with the support of the West German government. In 1975, VW began an extensive test, involving a fleet of 45 vehicles using a 15% blend of methanol in gasoline. After minimal modifications to existing engines, VW was able to operate these vehicles efficiently on methanol blends, with only minor problems [117]. At the time, the methanol blends were described as already competitive with gasoline. Furthermore, methanol (like ethanol) acted as an octane booster, the methanol/gasoline blend delivering more power than the pure gasoline. Five vehicles running on pure methanol were also tested by VW. Cold start problems due to the lower volatility of methanol were successfully solved by using small amounts of additives such as butane or pentane. The use of methanol significantly improved the cars' performance. Methanol, being considered also safer than gasoline, has been the fuel of choice at the Indianapolis 500 races since the mid- 1960s [119]. In the United States, most oil companies were first at best apathetic if not clearly opposed to the introduction of methanol as an alternative automotive fuel. The interest of American motor car manufacturers for methanol-fueled cars was also very limited. In California, however, which was trying to reduce its significant air pollution problems and reliance on imported fuel, a research program on methanol was started in 1978 at the University of Santa Clara, where a Ford vehicle running on pure methanol was extensively tested. This test was followed by several fleets of Ford and VW vehicles operated in various state and local agencies. The 84 vehicles [120], which accumulated a total mileage of over 2 million km, showed good fuel economy and engine durability which was comparable to that of gasoline vehicles. At about the same time in 1980, the Bank of America, based in San Francisco, decided to convert most of its vehicle fleet to methanol fuel in response to high oil prices. During their lifetime, the more than 200 methanol-fueled vehicles accumulated over 30 million km on the roads. The Bank of America concluded that, compared to gasoline engines, the use of neat methanol was found to be cheaper, increased the engine's lifespan, and greatly decreased exhaust pollutants. Nevertheless, the bank's interest in methanol declined rapidly with the sharp drop in oil prices of the mid-1980s. The state of California however continued its efforts on promoting methanol, mostly in the form of a blend composed of 85% methanol and 15% gasoline and called M85. A small fleet of cars using this methanol blend driven on a daily basis at the Argonne National Laboratory near Chicago also showed that, even in the frigid climate of northern Illinois, the cars had no problems with cold starting [121]. At the end of the 1980s, automobile companies began to develop cars powered by alternative fuels because they were concerned about meeting the new air pollution standards. These vehicles were first introduced in California, which has one of the most stringent emission control standards of the nation. General Motors, Ford, Chrysler, Volvo, Mercedes and others transformed exiting models to run on methanol with an additional cost of, at most, a few hundreds of dollars. Because of methanol's higher octane rating, Ford found that acceleration from 0 to 100 km h^{-1} was up to one second faster with the methanol-fuelled models compared to the gasoline versions [121]. Due to the limited number of filling stations dispensing the methanol blended fuel, most of the models were designed as Flexible Fuel Vehicles (FFVs), being able to run on gasoline alone in case the M85 methanol blend was not available. This concept enabled bypassing of any lack of availability standing in the way of most alternative fuels in the early stages of their introduction. The number of methanol-fueled vehicles in use in the United States (mostly in California) reached a maximum in 1997 of still minuscule 20000 units [122]. During the 1990s, different technological advances were achieving wide acceptance in the automobile industry: direct fuel injection, three-way catalytic converters, reformulated gasoline, etc., reducing dramatically emissions problems associated with gasoline-powered vehicles and decreasing at the same time the interest in methanol-based fuels. However, the recent dramatic increase in oil prices, combined with growing concerns about human-caused climate changes is reviving the interest for alternative fuels, among which methanol plays an important role.

Although the flexibility of FFVs represent a powerful means to circumvent the fuel supply conundrum, and also a way to build up the demand for methanol, it must be borne in mind that this is only a compromise, and it does not offer the best performance achievable in either emissions or fuel economy. In the long term, the use of cars optimized to run only on methanol (M100) would be preferable, and would also greatly facilitate the transition to methanol-powered fuel cell vehicles.

Looking back at the history of methanol in the automotive sector, it must be observed that the fate of methanol fuel is extremely dependent on economic aspects, and especially oil prices. Resistance to the widespread introduction of methanol by special interest groups (some of which favor agricultural ethanol), energy security issues, governmental energy and emission policies and other political considerations also play an important role. With diminishing oil and gas reserves, a new realization for the need of finding alternative solutions is finally achieving a foothold, and the future of methanol as a transportation fuel is entering a new period. Methanol is also easily dehydrated to dimethyl ether, which is an effective fuel particularly in diesel engines due to its high cetane number and favorable properties. Haldor Topsoe first promoted its use as an efficient diesel fuel in the 1990s. Interest in DME is rapidly growing

Methanol as Fuel in Internal Combustion Engines (ICE)

In contrast to gasoline, which is a complex mixture containing many different hydrocarbons and some additives, methanol is a simple chemical. It contains about half the energy density of gasoline, which means that 2 L of methanol contains the same energy as 1 L of gasoline. Even though methanol's energy content is lower, it has a higher octane rating of 100 (average of the research octane number (RON) of 107 and motor octane number (MON) of 92) which means that the fuel/air mixture can be compressed to a smaller volume before it is ignited by the sparkplug. This allows the engine to run at a higher compression ratio (10–11 to 1 against 8–9 to 1 for gasoline engines) and thus also more efficiently than a gasoline-powered engine. Efficiency is also increased by methanol's higher "flame speed" which enables a faster and more complete fuel combustion in the cylinders. These factors explain why, despite having half the energy density of gasoline, less than double the amount of methanol is necessary to achieve the same power output. This is true even in engines that are only modified gasoline engines and not specifically designed for methanol's properties. Methanol-specific engines however provide even better fuel economy [116]. Methanol also has a latent heat of vaporization which is about 3.7 times higher than gasoline, so that methanol can absorb a much larger amount of heat when passing from the liquid to gaseous state. This helps to remove heat away from the engine so that it may be possible to use air-cooled radiators instead of heavier, water-cooled systems. For similar performance to a gasoline-powered car, a smaller, lighter engine block, reduced cooling requirements, better acceleration and mileage are to be expected from methanol-optimized engines in the future [116]. In addition, methanol vehicles have low overall emissions of air pollutants such as hydrocarbons, NO_x, SO_2, and particulates.

Some problems remain, however, which arise mainly from the chemical and physical properties of methanol, and these need to be addressed. Methanol, not unlike ethanol, is miscible with water in all proportions. It has a high dipole moment as well as a high dielectric constant, making it a good solvent for ionizable substances such as acids, bases, salts (contributing to corrosion problems) and some plastic materials. Gasoline on the other hand, as already mentioned, is a complex mixture of hydrocarbons, the majority of which have low dipole moment, low dielectric constant and are non-miscible in water. Gasoline is therefore a good solvent for non-polar, covalent materials.

Due to the different chemical characteristics of gasoline and methanol, some of the materials used in gasoline distribution, storage, devices and connectors are predictably often incompatible with methanol. Methanol, consequently, can corrode some metals, including aluminum, zinc and magnesium, though it does not attack steel or cast iron [123]. Methanol can also react with some plastics, rubbers and gaskets, causing them to soften, swell or become brittle and fail, resulting in eventual leaks or system malfunction. Therefore, systems built specifically for methanol use must be different from those used for gasoline, but are expected to be only marginally, if at all, more expensive. Specific lubricating engine oils and greases that are compatible with methanol but already exist must be further developed.

With pure methanol, cold start problems can occur because it lacks the highly volatile compounds (butane, isobutane, propane) generally found in gasoline, which provide ignitable vapors to the engine even under the most frigid conditions [108]. The addition of more-volatile components to methanol is usually the preferred solution. In FFVs using M85 for example, the 15% gasoline provides enough vapors to allow the motor to start even in the coldest climates. Another possibility is to add a device to vaporize or atomize methanol into very small droplets which are easier to ignite.

Technical problems have to be expected during the development of any new technology. The technological difficulties facing methanol as a blending component or substitute for gasoline in ICE vehicles however, are relatively easy to solve and, indeed, the majority of them have already found solutions.

Methanol and Dimethyl Ether as Diesel Fuels Substitute in Compression Ignition Engines

Methanol, when combusted, does not produce smoke, soot or particulates. Particulate matter – whether carcinogenic compounds are absorbed onto them, or not – have been identified as a significant health hazard, especially in large cities. Diesel fuel generally produces particles during combustion. This, and the fact that methanol produces very low emissions of NOx because it burns at lower temperatures, makes methanol attractive as a substitute for diesel fuel [123].

Diesel engines are quite different from gasoline engines. Instead of using sparkplugs to ignite the fuel/air mixture in the engine's cylinders, diesel motors rely on the self-ignition properties of the fuel to ignite under specific high-temperature and high-pressure conditions. While a typical gasoline engine has a compression ratio of about 8–9 to 1, a diesel engine has generally a compression ratio in excess of 17 to 1. In the past, such engines were used mainly in heavy-duty vehicles such as buses, tractors, trucks, locomotives and ships. However, in the early 1970s – due to their better fuel economy compared to gasoline engines – diesel engines were increasingly used to power personal automobiles. Today, in western Europe for example, diesel-fueled cars represent about 50% of all cars in operation.

Like gasoline, diesel fuel is composed of many hydrocarbons that have a wide boiling range. The physical and chemical properties of diesel fuel are, however, quite different. Whereas gasoline contains predominantly branched alkanes with three to 10 carbon atoms as well as aromatic compounds to ensure a high octane rating, diesel fuel is mainly composed of straight-chain alkanes with 10 to 20 carbon atoms. A fuel's propensity to self-ignite under high heat and pressure is measured by the "cetane" rating. While diesel fuel has cetane ratings in the range from 40 to 55, methanol's cetane rating is only about 3. With methanol and diesel fuels being practically non-miscible, the possibility of using any blends of methanol and diesel fuel in diesel vehicles must be excluded. In order to overcome the low cetane rating of methanol, diesel motors must be changed and adapted. Additives can be added to increase the cetane rating of methanol to levels close to diesel fuels. These ignition improvers, added in the order of a few percent to methanol, are typically composed of nitrogen-containing compounds such as octyl nitrate and tetrahydrofurfuryl nitrate [123], although many of them are toxic and/or carcinogenic. Non-toxic cetane enhancers based on peroxides and higher alkyl ethers have also been developed. If neat methanol is used, ignition through spark plugs or glow plugs is necessary.

The energy content of methanol on a volume basis is about 2.2 times less than that of diesel fuel [123], which means that the fuel tank must be about twice the size of a conventional diesel fuel tank to provide the same amount of energy. Methanol has a significant vapor pressure compared to diesel fuel. The higher volatility allows heavy-duty engines to start easily in the coldest weathers, thereby avoiding the white smoke which is typical of cold-starts with conventional diesel engines.

As for many other alternative fuels, transit buses have been the main test-field for methanol-powered diesel engines. The Detroit Diesel Corporation (DDC) in particular developed a methanol version of its 6V-92TA diesel engine which in the early 1990s was the lowest emission heavy-duty diesel engine certified by the Environmental Protection Agency (EPA) and the California Air Resource Board (CARB) [123]. Using methanol instead of diesel fuel also allowed a dramatic reduction in particulate as well as NOx emissions. Methanol contains no sulfur; hence, SO_x emissions that lead to acid rain are also nearly eliminated. Fleets of buses equipped with methanol-fueled diesel engines were tested in different places around the United States including Los Angeles, Miami and New York [124]. The Metropolitan Transit Authority (MTA) of Los Angeles in particular operated a large fleet of some 330 methanol-powered transit buses for some years in the 1990s (Fig. 11.2). Compared to conventional diesel buses, higher maintenance costs due to some technical problems in the fuel system and engine were experienced using methanol. Similar operating problems were also encountered when these buses were converted to run on ethanol. Most of the difficulties experienced with alcohol use in diesel fuels are believed to be connected with the fuel delivery system and the use of non-compatible materials. Nevertheless, these technical problems are not insurmountable and can be solved with additional studies. Methanol, however, must also compete with other alternative fuels. When considering transit buses, for example, where the bulkiness of fuel tanks is not a major issue, compressed natural gas (CNG) has now become the preferred fuel because of its low emissions, price and widespread availability. Improved diesel engines enabling cleaner operation, particle filters as well as advanced diesel fuels containing less-polluting impurities are also being developed.

Another possibility is to use dimethyl ether, a superior and more calorific fuel than methanol for diesel engines, which can be easily obtained by dehydration of methanol.

Figure 11.2 Methanol-powered regional transit bus in Denver, Colorado. (Source: Gretz, Warren DOE/NREL.)

Dimethyl ether, DME, the simplest of all ethers, is a colorless, non-toxic, non-corrosive, non-carcinogenic and environmentally friendly chemical that is mainly used today as an aerosol propellant in various spray cans, replacing banned CFC gases. DME has a boiling point of $-25\,°C$, being a gas under ambient conditions. DME however is generally handled as a liquid and stored in pressurized tanks, much like LPG (liquefied petroleum gas), which contains primarily propane and butane, commonly used for cooking and heating purposes. The interest in DME as an alternative transportation fuel lies in its high cetane rating of 55–60, compared with 40–55 for conventional diesel fuel and much higher than that of methanol. Therefore, DME can be effectively used in diesel engines, as has been introduced by Haldor Topsoe. Like methanol, DME is clean-burning, produces no soot, black smoke or SO_2, and only very low amounts of NO_x and other emissions even without exhaust gas after-treatment (Tables 11.2 and 11.3).

Today, DME is produced exclusively by the dehydration of methanol. Being directly derived from methanol, DME can be produced from a large variety of feedstocks: coal, natural gas, biomass, etc. or reductive CO_2 recycling (see Chapter 12). A method to synthesize DME directly from syn-gas by combining the methanol synthesis and dehydration steps in a single process has also been developed [125]. The direct synthesis of DME from CO_2 and H_2 has also been studied [126]. The global demand for DME is currently only about 150 000 tons per year, but this could be considerably increased if large quantities of DME were to be needed as fuels.

Table 11.2 Properties of dimethyl ether (DME).

Chemical formula	CH_3OCH_3
Molecular weight	46.07
Appearance	Colorless gas
Odor	Slightly sweet smell
Chemical composition (%)	
Carbon	52
Hydrogen	13
Oxygen	35
Melting point	$-138.5\,°C$
Boiling point	$-24.9\,°C$
Density of liquid at 20 °C	$668\ kg\ m^{-3}$
Energy content	$6880\ kcal\ kg^{-1}$
	$317\ kcal\ mol^{-1}$
Cetane number	55–60
Flash point	$-41\,°C$
Autoignition temperature	$350\,°C$
Flammability limits in air	3.4–17%

Table 11.3 Comparison of the physical properties of DME and diesel fuel.

	DME	Diesel fuel
Boiling point (°C)	−24.9	180–360
Vapor pressure at 20 °C (bar)	5.1	–
Liquid density at 20 °C (kg m^{-3})	668	840–890
Heating value (kcal kg^{-1})	6880	10 150
Cetane number	55–60	40–55
Autoignition temperature (°C)	235	200–300
Flammability limits in air (vol. %)	3.4–17	0.6–6.5

DME vehicles are being developed in many areas of the world. In Europe, according to Volvo, which is running tests on a DME-powered bus and truck, DME is one of the most promising fuels for substituting conventional diesel oil (Fig. 11.3) [127, 128]. It can be used in an ordinary diesel engine equipped with a new fuel injection system producing the same performance, whilst dramatically reducing emissions. Japan, which widely uses imported LPG for domestic applications, has an extensive LPG infrastructure that can be easily adapted to DME. Realizing that future LPG supplies may not meet demand, Japan is studying the DME option not only for the transportation sector but also for electric power generation as well as household and industrial uses [129]. General Electric showed DME to be an excellent fuel for gas turbines, with emissions and performances comparable to those of natural gas [130]. Due to similar combustion characteristics, cooking stoves designed for natural gas can use DME without any modification [131]. Road tests with several DME-powered Japanese trucks and buses led to conclusions comparable to those obtained by Volvo. Isuzu, a well-known diesel engine manufacturer, is taking part in these tests and is confident that diesel oil can be supplanted in the future by improved technologies, including the use of DME. Two major Japanese consortia composed of leading compa-

Figure 11.3 DME-fuelled Volvo bus developed in Denmark. (Courtesy: Danish Road Safety and Transport Agency.)

nies are currently assessing the economical viability of DME. One consortium, led by the giant NKK Corp, created DME International Corp and DME Development Company to investigate the economics and facilitate the introduction of DME as a fuel with commercial production in the 850 000 to 1 650 000 tonnes per year range, envisaged to start in late 2006. The other consortium, Japan DME, led by Mitsubishi Gas Chemical, is planning to build a DME plant in Papua New Guinea which could produce up to 3 000 000 tonnes of DME per year. The Japanese government, through substantial financial support from the Ministry of Economy, Trade and Industry (METI) is also committed to the development of mass produced, low-cost DME.

Developing countries in Asia, such as China and India, are also very interested in DME, given their rapidly increasing needs and growing demands for diesel and LPG, as well as their deteriorating air quality. China in particular, because of its enormous reserves of coal, is interested in coal to DME liquefaction technology. In Shandong Province, the construction of a plant to produce 1 million tons of DME per year, and which will also produce 1.5 million tonnes methanol per year, based on coal is under construction. At the Shanghai Jiao Tong University, researchers are developing a transit bus powered by DME, whilst the Chinese Ministry of Science has offered subsidies to produce 30 such buses.

Countries which have low-cost natural gas reserves but are distant from important consuming centers, in the Middle East, Australia, Trinidad and Tobago and others, are also interested in DME as a convenient way to transport energy to markets in highly populated areas. A plant with a production capacity of 800 000 tonnes per year of DME for fuel uses is currently under construction in Iran. In the Middle East [3], BP is also seeking for partners for a planned DME plant producing 1.8 million tons per year [132].

Besides DME, dimethyl carbonate (DMC), which has also a high cetane rating, can be blended into diesel fuel at a concentration of up to 10%, thereby reducing the fuel viscosity and improving emissions. In China, plans to produce DMC (mainly from coal or some natural gas resources) on a commercial scale as a diesel additive are under way [132]. Due to a melting point of 3 °C, however, neat DMC is not an ideal fuel because of its expected freezing problems at lower operating temperatures. The commercial route to DMC has used the reaction of methanol with phosgene. However, phosgene – being highly toxic – was replaced by the oxidative carbonylation of methanol, developed by EniChem and other companies [133].

Biodiesel Fuel

Another way to use methanol in diesel engines and generators is through biodiesel fuels. These can be made from a large variety of vegetable oils and animal fats which are reacted with methanol in a transesterification process to produce compounds known as fatty acid methyl esters, which compose biodiesel. Biodiesel can be blended without major problems with regular diesel oil in any proportion. It is a renewable, domestically produced fuel which also reduces emissions of unburned hydrocarbons, carbon monoxide, particulate matter, sulfur compounds as well as CO_2 (one of the main greenhouse gases). The use of biodiesel has grown substantially during the past few years, mainly in Europe and the United States. However, as pointed out earlier in Chapter 6, the feedstocks for biodiesel are limited, and consequently biodiesel can cover only a relatively small portion of our energy needs. Alone, biodiesel will be unable to replace diesel fuel obtained from hydrocarbons in the quantities required for our transportation systems.

Advanced Methanol-Powered Vehicles

Methanol or its derivatives (DME, DMC, biodiesel) can already be used as substitutes for gasoline and diesel fuel in today's ICE-powered cars, with only minor modifications to existing engines and fuel systems. ICE is a much-proven and reliable technology, which has been continuously improved and perfected since its invention over a hundred years ago. Fuel economy compared to generated power is now better, and emissions lower than ever before. Hybrid cars, combining an ICE with an electric motor are commercialized by a growing number of companies (Toyota, Honda, Ford, etc.) and can reduce even further fuel consumption and emissions. In these vehicles too, gasoline and diesel fuel can be easily substituted by methanol or its derivatives. Their use on a large scale is realizable in the relative short term. In the foreseeable future, however, in order to further increase efficiency and lower emissions, fuel cell technology will be the best alternative to ICEs in the transportation field. Much effort and financial resources are currently being invested by major motor car manufacturers and governments to make fuel cell vehicles (FCV) an affordable and viable option for consumers in the foreseeable future. FCVs promise to be much quieter, cleaner and to require less maintenance than ICE because of fewer moving parts. Proton exchange membrane fuel cells (PEMFC) are currently the favored type of fuel cell to power cars because of their relatively light weight, low operating temperature, and high power output. As described in Chapter 9, these vehicles operate on hydrogen which can be stored in liquid, gaseous or solid metal hydride forms, or even reformed on-board from different liquid fuels, including gasoline and methanol.

Hydrogen for Fuel Cells from Methanol Reforming

In seeking to overcome the problems associated with hydrogen storage and distribution, numerous approaches have set out to use liquids rich in hydrogen such as gasoline or methanol as a source of hydrogen via on-board reformers. In contrast to pure hydrogen-based systems, they are compact (containing on a volume basis more hydrogen than even liquid hydrogen) and easy to store and handle without pressurization. The possibility of generating hydrogen with more than 80% efficiency by the on-board reforming of gasoline has been demonstrated. However, the process is expensive and challenging, because it involves high temperature and needs considerable time to reach a steady operational state. The advantage is that the distribution network for gasoline already exists, though this would not solve the problems of diminishing oil resources and dependence from oil-producing countries. On the other hand, methanol steam reformers operating at much lower temperature (250–350 °C) [95], albeit still expensive, are more adapted for on-board applications. The absence of C–C bonds in methanol, which are difficult to break, greatly facilitates its transformation to high-purity hydrogen with 80–90% efficiency [134]. Methanol, furthermore contains no sulfur, a contaminant for fuel cells. With the reformer operating at low temperature, no nitrogen oxides are formed. The use of an on-board reformer enables the rapid and efficient delivery of hydrogen from a liquid fuel that can be easily distributed and stored on the vehicle. To date, methanol is the only liquid fuel that has been processed and demonstrated on a practical scale in fuel cells for transportation applications. The disadvantages of this system are, however, the added weight, complexity and cost of the overall system, as well as trace emissions that may be produced from the reformer when it burns some of the methanol to provide the necessary heat for hydrogen production [119].

The potential for on-board methanol reformers to power FCVs has been demonstrated by several prototypes constructed and tested by various automobile companies. In 1997, DaimlerChrysler presented the first methanol-fueled FCV, the Necar 3, a modified A-Class Mercedes-Benz compact vehicle equipped with a 50-kW PEM fuel cell and a driving range of 400 km. In 2000, an improved version with a 85-kW fuel cell, the NECAR 5 was introduced (Fig. 11.4) [135]. In this vehicle, which was described by the company as being fit for practical use [119], the entire fuel cell and reformer system has been accommodated in the underbody of the car, enabling five passengers and their luggage to be transported at a maximum speed of 150 km h^{-1} and a driving range approaching 500 km [135]. In 2002, this FCV was the first to complete a coast-to-coast trip across America, from San Francisco to Washington D.C., or a distance of more than 5000 km, with methanol being refueled every 500 km [136]. DaimlerChrysler also presented in 2000, a fuel cell/battery hybrid Jeep Commander SUV powered by methanol. Based on a Ford Focus, Ford constructed TH!NK FC5 [137], a methanol-fueled FCV with a fuel cell/reformer system beneath the vehicle's floor and characteris-

Figure 11.4 DaimlerChrysler's methanol-fueled NECAR 5 fuel cell vehicle (introduced in 2000, Courtesy of DaimlerChrysler).

tics similar to NECAR 5. Other companies which have developed methanol-powered FCVs include General Motors, Honda, Mazda, Mitsubishi, Nissan, and Toyota. However, most of these companies, including Daimler-Chrysler, have recently concentrated their efforts on vehicles with on-board storage of pure hydrogen.

Georgetown University of Washington DC has been in the forefront in the development of transportation fuel cells for some 20 years. Supported by the U.S. Federal Transit Administration, the University has developed several fuel cell transit buses running on methanol [138]. In 1994 and 1995, Georgetown produced three buses that were the world's first FCVs able to operate on a liquid methanol fuel (Fig. 11.5). These methanol buses, each powered by a 50-kW Phosphoric Acid Fuel Cell (PAFC) combined with a methanol steam reformer, are still operating today. In 1998, an improved second-generation bus using a more powerful 100 kW PAFC provided by UTC Fuel Cells was introduced, followed in 2001 by the first urban transit bus powered by a liquid-fueled 100 kW PEMFC system manufactured by Ballard Power System, a major fuel cell developer. Batteries provide surge power and a means to recover braking energy by regeneration. These two buses, each able to seat 40 passengers, meet all the requirements of the transit industry, are much more quiet than their ICE-powered counterparts, and have a driving range of some 560 km between refueling. The use of methanol allows the refueling to be as quick and easy as with diesel buses. PM and NO_x emissions are virtually eliminated, and other emissions are well below even the cleanest CNG buses on the road with the most stringent clean air standards [138]. Capitalizing on previous experience, a third generation fuel cell hybrid bus fueled by methanol is presently under development.

Besides on-board methanol reforming, methanol is also seen as a convenient way to produce hydrogen in fueling stations to refuel hydrogen FCVs. Mitsubishi Gas Chemical has developed a process to produce high-purity hydrogen by steam reforming of methanol using a highly active catalyst which allows operation at relatively low temperature (240–290 °C) and enables rapid start-up and stop, as well as flexible operation. These methanol-to-hydrogen (MTH) units, which range in production capacity from 50 to 4000 m^3 H_2 per hour, are already used by a variety of customers in the electronic, glass, ceramics, and food processing industries [139]. They provide an excellent reliability, prolonged life service, and minimal

Figure 11.5 Methanol fuel cell buses developed at Georgetown University in front of the U.S. Capitol (2002) (Source: Georgetown University).

maintenance [134]. Based on the technology developed by Mitsubishi Gas Chemical, the first fueling station to supply hydrogen by methanol reforming was constructed in Kawasaki, Japan as part of the Japan Hydrogen & Fuel Cell Demonstration Project (JHFC), which is studying a large array of possible feedstocks for hydrogen generation. According to JHFC, methanol is the safest of all materials available for hydrogen production [140]. Operating at relatively low temperature, the MTH process has a clear advantage over the reforming of natural gas and other hydrocarbons, which needs to be carried out above 600 °C. A smaller amount of energy is necessary to heat methanol to the appropriate reaction temperature. At the Kawasaki station, methanol and water are evaporated and reacted over a catalyst. After purification and separation, the hydrogen produced is compressed and stored to provide FCVs with high-pressure hydrogen [140]. Although methanol reforming is a convenient and attractive way of producing hydrogen, it does not solve the problems associated with the costly and difficult on-board storage of hydrogen.

Significant investigations aimed at further improving methanol reforming to hydrogen, whether focused on on-board or stationary applications, are under way. Hydrogen obtained by methanol reforming in current processes always contains more than 100 ppm CO, a poison for PEM fuel cell catalysts operating below 100 °C. At present, reformed gas has thus to be cleaned to remove CO, lowering the total efficiency of the process. At the Brookhaven National Laboratory, new catalysts have been designed which produce hydrogen with high yield, generating by the same time only marginal amounts of CO. Using a process known as oxidative steam reforming, which combines steam reforming and the partial oxidation of methanol, and different novel catalyst systems, the National Industrial Research Laboratory of Nagoya has also achieved the production of high-purity hydrogen with either zero or only trace amounts of CO, at high methanol conversion and temperatures as low as 230 °C. Oxidative steam reforming of methanol also has the advantage of being – contrary to steam reforming – an exothermic reaction, minimizing energy consumption. The exothermicity of the reaction however, can also be a drawback since the generated heat and consequently the reactor's temperature may be difficult to control. In an ideal case, the reaction should therefore only produce enough energy to sustain itself. This is the principle of autothermal reforming. The autothermal reforming of methanol, which combines steam reforming and partial oxidation of methanol in a specific ratio, is an idea first developed in the 1980s by Johnson-Matthey. It is neither exothermic nor endothermic, and thus does not require any external heating once the reaction temperature has been reached. For a fast start-up, the methanol/oxygen feed ratio can be varied, as has been shown for example in Johnson-Matthey's "Hot-Spot" methanol reformer.

Direct Methanol Fuel Cell (DMFC)

In contrast to hydrogen fuel cells, direct methanol fuel cells are not dependent upon hydrogen generation by processes such as electrolysis of water, or natural gas or hydrocarbon reforming. As mentioned earlier, the storage and distribution of hydrogen fuel will require an entirely new infrastructure or a complete overhaul of existing systems, all of which constitutes a major barrier for entry into the commercial market. Methanol, on the other hand, is a clear liquid fuel (b.p. = 64.7 °C, d = 0.791 g mL^{-1}) that does not require special cooling at ambient temperature, and will fit into existing storage and dispensing units with only small modification.

Methanol has a relatively high volumetric theoretical energy density compared to other systems such as conventional batteries and the H_2-PEM fuel cell (Fig. 11.6). This is of basic importance for small portable applications, as battery technology may not be able to keep up with the demand for laptops and mobile phones that are lightweight and have extended operating time [141, 142].

There is more hydrogen in 1 L of liquid methanol than in 1 L of pure cryogenic hydrogen (98.8 g of hydrogen in 1 L of methanol at room temperature compared to 70.8 g in liquid hydrogen at –253 °C). Therefore, it transpires that methanol is a safe carrier fuel for hydrogen.

In the past, methanol-based PEM cells have used a separate reformer to release the hydrogen from liquid methanol, after which the pure hydrogen is fed into the fuel cell stack. However, since 1990 researchers at the Jet Propulsion Laboratory, and the authors' group at the University of Southern California [143, 144], have developed a simple DMFC that consists of two electrodes separated by a proton exchange membrane (PEM) and connected via an external circuit that allows the conversion of free energy from the chemical reaction of methanol with air to be directly converted into electrical energy (Fig. 11.7).

The anode is exposed to methanol water mixture fed by flow from an external container, where it is oxidized to produce protons that travel through the PEM by ionic conduction, and electrons that travel through the external circuit by electronic conduction. The cathode containing platinum as catalyst is exposed to oxygen or air, which may be either ambient or pressurized. The PEM is coated on both sides with layers of catalyst (1:1 Pt–Ru catalyst at the anode and Pt catalyst at the cathode), usually supported by a gas diffusion electrically conductive carbon (graphite) electrode at the cathode and a liquid feed-type carbon electrode struc-

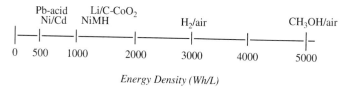

Energy Density (Wh/L)

Figure 11.6 Theoretical energy density of batteries, H_2-PEM fuel cells, and DMFCs.

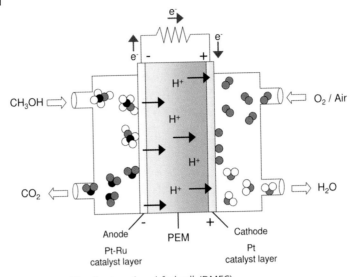

Figure 11.7 The direct methanol fuel cell (DMFC).

ture at the anode that facilitates reduction of oxygen and oxidation of methanol, respectively Eqs. (1) and (2).

Anode reaction

$$CH_3OH + H_2O \longrightarrow CO_2 + 6H^+ + 6e^-$$

Eq. (1)

Cathode reaction

$$1.5\,O_2 + 6H^+ + 6e^- \longrightarrow 3\,H_2O$$

Eq. (2)

Overall reaction

$$CH_3OH + 1.5\,O_2 \longrightarrow CO_2 + 2\,H_2O$$

Eq. (3)

At room temperature, the overall reaction (Eq. (3)) gives a theoretical open circuit voltage of 1.21 V with a theoretical efficiency close to 97%. Although fundamentally simple, DMFCs presently still perform well below their theoretical Nernstian potential, even under open-circuit conditions because of sluggish redox kinetics and fuel crossover. Advances in both catalysts and membranes have – and are being – made and promise to conquer these issues, in terms of both performance and cost.

PEMs intended for H_2-PEM fuel cells are not good candidates for DMFCs due to the issue of methanol crossover from anode to cathode. High crossover rates have a deleterious effect on DMFC performance. As oxygen is reduced to produce a cathodic current, oxidation of crossover methanol simultaneously produces an anodic current that results in a mixed potential and overall reduced cathode potential. Methanol may poison the cathode catalyst (Pt) and block cathode catalyst sites, further reducing its ability to efficiently reduce oxygen, and necessitating an increase in oxygen flow above and beyond stoichiometric requirements. Furthermore, the chemical oxidation of methanol also produces excessive water that hampers cathode performance due to flooding. Instead of meaningful electrical energy, waste heat is generated and fuel utilization efficiency is lowered. Methanol permeates through the PEM by two methods: (i) by simple diffusion due to a concentration gradient; and (ii) by electro-osmotic drag from proton migration when the cell is under an applied current. New membranes based on hydrocarbon/hydrofluorocarbon materials with reduced cost and cross-over characteristics have been developed that allow room temperature efficiency of 34% [145, 146].

With the advances made in all aspects of DMFC research, many companies are now actively developing low-power DMFCs for portable devices such as cellular phones and laptop computers [147]. The success of these initial devices is pivotal to transition away from rechargeable batteries, which can in theory deliver only 600 W·h kg^{-1} at best. Currently, rechargeable commercial lithium ion batteries deliver a power density anywhere between 120 and 150 W·h kg^{-1}. Consumers will soon enjoy the benefits of DMFC-powered devices, including longer cellphone talk time, extended usage time on laptop computers, rapid rechargeability, and lighter weight contribution of the power source.

Many issues arise when single fuel cells are assembled into practical stack assemblies such as temperature and pressure control, resistance, and water management. Various materials and designs have been employed to deal with these issues. The Jet Propulsion Laboratory [148] has developed a small, six-cell DMFC stack with total active area of ~48 cm^2 (anode and cathode catalyst loading of 4–6 mg cm^{-2}) using ambient air with 1 M methanol at room temperature, having a power density from 6 to 10 mW cm^{-2}. The cell is a flat array, where the cells are externally connected in series sharing a single membrane. The cathode catalyst, when applied to teflonized carbon supports, imparted excellent water-removal properties to the system, although the design also provides increased ohmic resistance. Three two-flat pack arrays are necessary to power a mobile phone, and a 10-h operating time is estimated before methanol replenishment.

Los Alamos National Labs, in conjunction with Motorola [149], have also developed a stack for mobile phones utilizing ceramic fuel plates with microfluidic channels for efficient delivery of methanol and water and removal of CO$_2$. Their four-cell stack with total active area ~60 cm^2 gives a power density between 12 and 27 mW cm^{-2} with catalyst loading of 6–10 mg cm^{-2} at room temperature.

The Korea Institute of Energy Research (KIER) has developed a 10-W DMFC stack [150]. The cell has a bipolar plate design with six cells each of 52 cm^2 active area. Most notably, the stack was operated with 2.5 M methanol, and achieved 6.3 W at room temperature using ambient oxygen flow. In addition, the Korea Institute of Science and Technology (KIST) has developed and assembled a six-cell monopolar stack [151] with a total active area of 27 cm^2, and which produced a power density of 37 mW cm^{-2} using 4 M methanol and ambient air.

Toshiba has developed a promising DMFC prototype stack for laptop computers. The stack has an average output of 12 W and may be continuously used for 5 h with a 50-mL methanol storage cartridge. To minimize the size of the cartridge, the cell collects output water for recombination with methanol. Sensors are hooked up directly to the PC to tell users when the cartridge needs replacing. NEC has developed similar stacks and, within the next few years, anticipates the stack to have a 40-h operation time.

Both stationary and portable DMFC stacks are now available for consumer purchase. For instance, The Fuel Cell Store offers its SFC A25 Smart Fuel Cell [152], capable of 25 W continuous output, and can cover four days worth of energy demand using only 2 kg of fuel. A larger 50-W model is also available.

In the transportation area, Daimler-Chrysler, working on DMFC for automotive purposes, constructed a one-person go-cart prototype vehicle powered by a 3-kW DMFC (Fig. 11.8) [153]. Recently, a similar vehicle, equipped with a 1.3-kW DMFC known as "JuMOVe", has been developed in Germany by the Julich Research Center [154]. In Japan, Yamaha presented in 2003 its FC06 prototype, the first two-wheeler motor cycle powered by a DMFC with an output of 500 W. Equipped with a 300 W AC outlet, this bike can also serve as an electric power source for outdoor activities or during emergencies [155]. An advanced version of this motor bike, the FC-me, is presently in practical use on a lease basis in Japan (Fig. 11.9). In the United States, Vectrix plans to commercialize in 2006

Figure 11.8 DaimlerChrysler DMFC go-cart (Courtesy of DaimlerChrysler).

Figure 11.9 Yamaha FC-me two-wheeler powered by direct methanol fuel cell (DMFC) (Courtesy © Yamaha Motor Co.).

a hybrid fuel cell scooter powered by a 800-W DMFC attached to a rechargeable battery [156, 157]. The fuel cell continuously recharges the battery, which powers the electric motor. Regenerative braking technology also captures the energy usually dissipated during braking to provide additional battery charging. With a maximum speed in excess of 100 km h^{-1} and a range of about 250 km at cruising speed, it has characteristics comparable to conventional scooters.

Considerable development efforts are still needed to make larger DMFCs practical, for example to be able to power motor cars, but ongoing progress is impressive. DMFCs offer numerous benefits over other proposed technologies in the transportation sector. By eliminating the need for a methanol steam reformer, the vehicle's weight, cost and the system's complexity can be significantly reduced, thereby improving fuel economy. DMFC systems also come much closer to the simplicity of direct hydrogen-fueled fuel cell, without the cumbersome problem of either on-board hydrogen storage or hydrogen-producing reformers. By emitting only water and CO_2, other pollutant emissions (NO_x, PM, SO_2, etc.) are eliminated. As methanol will eventually be made by recycling atmospheric carbon dioxide, CO_2 emissions will not be of any concern and there will be no dependence on fossil fuels.

Methanol as a transportation fuel has a variety of important advantages. In contrast to hydrogen, methanol does not need any energy-intensive procedures for pressurization or liquefaction. Because it is a liquid, it can be easily handled, stored, distributed and carried on board vehicles. Methanol is already used today in ICE vehicles. Through on-board methanol reformers it can act as an ideal hydrogen carrier for FCVs, and be used in the future directly in DMFC vehicles. Employing the same fuel from present ICE to advanced vehicles equipped with DMFCs will allow a smooth transition between existing and new technologies. With the progressive phase-out of MTBE as a gasoline additive, substantial methanol production over-capacity is immediately available to be used as a transportation fuel.

Over the next decade new innovations such as novel proton-conducting materials, membrane-less fuel cells, and cheaper and more efficient catalysts may lead DMFC technology away from the traditional cell structure and design. DMFC is poised to play a critical role in electricity production from methanol in the overall methanol economy structure.

Fuel Cells Based on Other Fuels and Biofuel Cells

Direct oxidation fuel cells based on other fuels such as ethanol, formaldehyde, formic acid, DME, dimethoxymethane, and trimethoxymethane, have been studied in laboratories worldwide. However, none of these has shown so far the promise of either the H_2-PEM fuel cell or DMFC, although application of fuel mixes is feasible.

Biofuel cells use biocatalysts for the conversion of chemical energy into electrical energy. As most organic materials undergo combustion with the evolution of energy, the biocatalyzed oxidation of organic substances by oxygen or other oxidizers at two-electrode interfaces provides a means for the conversion of chemical to electrical energy. Abundant organic raw materials such as ethanol, hydrogen sulfide, organic acids or glucose can be used as substrates for these oxidation process, while molecular oxygen or H_2O_2 can be reduced. Intermediate formation of hydrogen as a potential fuel is also possible. Biofuel cells can use biocatalysts, enzymes or even whole-cell organisms. The power produced in such devices are miniscule (microwatt to nanowatt range), although such devices have potential uses as chemical and biological sensors.

Regenerative Fuel Cell

A regenerative fuel cell concept based on methanol/formic acid fuel cells has also been proposed (Fig. 11.10) [158]. The key to the success of such an approach is efficient capture of CO_2 and its electrochemical reduction to either HCOOH or CH_3OH in high current efficiencies. Intense research to achieve efficient electrochemical reduction of CO_2 is currently under way in many laboratories.

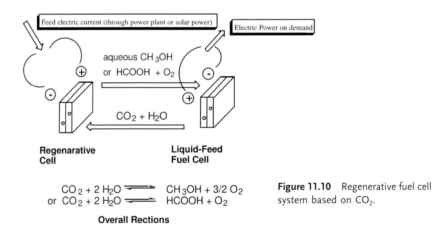

Overall Rections

$$CO_2 + 2\,H_2O \rightleftharpoons CH_3OH + 3/2\,O_2$$
$$\text{or } CO_2 + 2\,H_2O \rightleftharpoons HCOOH + O_2$$

Figure 11.10   Regenerative fuel cell system based on CO_2.

Methanol for Static Power and Heat Generation

Transportation and other mobile applications are not the only areas where methanol can be used as a fuel; indeed, it is also an attractive fuel for static applications. It can be used directly as a fuel in gas turbines to generate electric power. Gas turbines use typically either natural gas or light petroleum distillate fractions as fuels. Compared to these fuels, tests conducted by many institutions beginning in the 1970s, have shown that methanol can achieve higher power output and lower NO_x emissions due to lower flame temperatures. Since methanol does not contain sulfur, SO_2 emissions are also eliminated [159, 160]. Operation on methanol offers the same flexibility as on natural gas and distillate fuels, including the ability to start, stop, accelerate and decelerate rapidly, following the electric power needs. Existing turbines, designed originally for natural gas and other fossil fuels, can be relatively easily and inexpensively modified to run on methanol. For this application, fuel-grade methanol with lower production costs than higher purity chemical-grade methanol can be used. Considering increasing natural gas prices, methanol produced at low-cost from remote natural gas resources in the Middle-East or other regions and shipped much more easily and less expensively than LNG, offers also an alternative for power generation in large consuming centers such as North America, Europe, or Japan.

For static uses, the size and weight of fuel cells are of lesser importance compared to mobile applications. Besides PEM fuel cells and DMFC, phosphoric acid, molten carbonate and solid oxide fuel cells (PAFC, MCFC and SOFC), all of which are ill-suited for automobiles, can also be used for static power and heat generation. As described in Chapter 9, these fuel cells are already being used in the production of electricity in facilities sensitive to power outages such as airports, hospitals, military complexes, and banks. Whereas the present cost of these installations is still high, their price is expected to decrease with further development and the number of units produced. General Electric, for example, is developing a fuel cell unit called the HomeGen 7000 with a 7-kW capacity, and which is intended for the production of electricity for the residential home market [161]. For such static applications, liquid methanol, which is easy to handle, deliver and store, would be the fuel of choice.

In developing countries, methanol has been proposed as a substitute cooking fuel in place of wood and expensive and inconvenient kerosene. The consumption of large quantities of wood for cooking purposes by more than 2.5 billion people is in fact one of the major causes of deforestation and all the ecological (desertification, excessive erosion, land-slides, etc.) and socio-economical problems associated with it in the developing areas of the world. Wood-burning stoves in use in these countries are generally also very inefficient, producing also much smoke, fumes and soot, all of which are serious health hazards. In eliminating these drawbacks, stoves designed specifically for methanol have been developed [162, 163].

Methanol Storage and Distribution

In parallel to the development of methanol-fueled vehicles, a widespread distribution network for methanol will have to be established to make it as easily available for the consumer as are petroleum-based fuels today. While the passage from ICE to fuel cell-powered methanol vehicles represents a radical technological change, the development of a fueling infrastructure to fuel them is not. Refueling stations dispensing methanol will be almost identical to today's fueling stations, reflecting very little change to consumers' habits. Rather than gasoline or diesel fuel, they will simply fill their tanks at the local service station with a different liquid fuel.

The installation of methanol storage tanks and distributing pumps in existing facilities or specifically designed stations is quite straightforward, and is in any case no more difficult than the installation of their gasoline counterparts. Starting in the late 1980s, a network of almost a hundred methanol refueling stations was built in California to fuel the private and state-owned 15 000 or so methanol-powered vehicles. Most of these were Flexible Fuel Vehicles (FFV), able to run on any mixture of methanol and gasoline, and usually fueled with M85 (85% methanol and 15% gasoline), though others were designed specifically to run on pure methanol (M100). Other methanol pumps were also installed across the United States and Canada [164].

For retail stations, the conversion costs are minimal. Converting existing doubled-walled underground gasoline or diesel fuel storage tanks and installing new piping and dispenser pumps compatible with methanol is quite trivial. For some $20 000, an existing 40 000 L tank can be cleaned and the remainder of the system equipped with methanol-compatible elements. The complete operation takes only about one week. The cost of adding a new double-walled underground methanol storage with a 40 000 L capacity and methanol-compatible piping, dispensers, valves, etc. to an existing service station is around $60 000–65 000. In rural areas, or where space is available and local codes allow, an above-ground storage tank can be installed and the overall cost reduced to about $55 000 [164]. This means that in the United States, an investment of about $1 billion would enable 10% of the 180 000 service stations to distribute methanol, and for less than $3 billion, methanol pumps could be added to one-fourth of the service stations [119, 153]. This amounts only to a fraction of the more than $12 billion that have been spent by the oil industry to introduce reformulated gasoline to United States service stations [153].

Methanol fueling stations are also much less capital intensive than an infrastructure based on hydrogen, which would need special equipment and materials to handle high pressures or very low temperatures. General Motors has estimated that in order to build 11 700 new hydrogen fueling stations, $10–15 billion would have to be invested [165], which represents about $1 million per station. Besides the high cost, the technology to dispense hydrogen is presently still immature and has not yet reached the degree of convenience and safety that consumers have come to expect with conventional liquid fuels. Numbers of regulations also stand in the way of hydrogen. In the United States, the National Fire Protection Association (NFPA) currently prohibits the placing of hydrogen fueling equipment within 25 m of gasoline pumps [153]. This makes hydrogen pumps difficult – or even impossible – to install in most existing fueling stations, especially in cities. Further higher costs are thus expected if hydrogen has to be dispensed in hydrogen-only stations. Ease of delivering liquid methanol from production centers to local stations avoids all the difficulties encountered for hydrogen transportation whether under high pressure or in cryogenic form.

Today already, methanol is a widely available commodity with extensive distribution and storage capacity in place. More than 500 000 tons of methanol are presently transported each month to diverse and scattered users in the United States alone [134], by rail, boat, and trucks. Overland, transport by railway – where methanol is moved in rail cars each holding about 100 tons – is the preferred option for the long-distance transportation of bulk quantities. The railroad system in the United States, Europe, Japan and other major consuming countries is generally very comprehensive, enabling methanol shipments to be made to all major markets. For smaller volumes and distribution to local markets, tanker trucks with capacities of up to 30 tons are generally used. Where inland waterborne shipment through rivers and canals is possible, methanol can be transported by barges which typically contain some 1250 tons (10 000 barrels) of methanol. This is the largest inland transportation method, and is especially adapted to deliver methanol to large consumers and inland methanol hubs for regional redistribution [166]. Another means of transporting large quantities of liquids, and one which is also used extensively for oil, natural gas and their products, is via pipelines. At present, methanol pipelines are only viable in regions where major methanol producers and users are concentrated in close proximity, such as on the Texas Gulf Coast between Houston and Beaumont. For long-distance transportation, the volumes of methanol to be shipped are generally insufficient to justify the high investments needed to build a pipeline. In the future, however, if methanol has to be increasingly used as a fuel, the much larger amounts of methanol to be moved overland will improve the economics and make transportation through pipelines not only viable but also indispensable. Technically, transporting methanol through pipelines does not pose any problems, as has been demonstrated successfully in two test runs conducted in Canada. One demonstration used the Trans Mountain crude oil pipeline running from Edmonton, Alberta to Barnaby, British Columbia over a distance of 1146 km; the other involved the Cochin pipeline, primarily used for LPG, over a distance of nearly 3000 km [167]. In both cases the quantity of methanol shipped was the same (4000 tons), and the quality of the delivered product was well suited for fuel applications. When methanol is produced in remote locations where cheap natural gas is available, it is shipped throughout the world by dedicated methanol ocean tankers which range in size from 15 000 to almost 100 000 dead weight tons (DWT) in the case of the latest super tanker used by Methanex (Fig. 11.11) [134, 168], one of the world leaders in methanol production. When transported in such large vessels, the costs of shipping methanol will become similar and ultimately equal

Figure 11.11 Millennium explorer methanol tanker (Courtesy Mitsui O. S. K. Lines).

to that of crude oil. Once delivered, methanol can be easily stored in large quantities, much like petroleum and its products, in tanks with capacities exceeding 12 000 tons. Such tanks can be constructed from a variety of materials, including carbon steel and stainless steel, which are compatible with methanol.

Methanol Price

Since 1975, the average wholesale price for methanol has been around $175 t^{-1}, but has fluctuated roughly between $100 and $275 t^{-1} (Fig. 11.12). As with any other commodity, methanol is subject to fluctuations depending on offer and demand. The high prices experienced in 1994–1995, which reached more than $350 t^{-1} for example, were due to an increase in demand for major methanol derivatives such as MTBE, formaldehyde and acetic acid, coupled with production problems at methanol plants. However, with increased capacity and competition, as well as concerns about the use of MTBE as a gasoline additive, prices decreased rapidly. In recent years, high natural gas prices – especially in North America – have driven the price of methanol (which is mainly produced from natural gas) to higher levels. As methanol production in North America is gradually phased out, new methanol production facilities are being constructed in regions rich in natural gas but far from main consuming centers, such as the Middle-East. The construction of extremely efficient mega-methanol plants with very low production costs in these parts of the world will allow the price of methanol to remain at a relatively low level as long as sufficient natural gas reserves are available. The production cost for methanol in mega-methanol plants has been estimated to be well below $100 t^{-1} (equal to less than ¢8.5 per liter, or ¢30 per gallon)

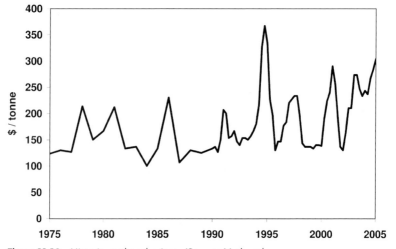

Figure 11.12 Historic methanol prices. (Source: Methanol Institute, Methanex and *Chemical Week*.)

[132]. Even considering its relatively lower energy content (half that of gasoline), methanol will then be quite competitive with gasoline and diesel fuels. At present oil prices of $50 to $75, a liter of crude oil costs already between ¢31 and ¢47 (from $1.2 to $1.8 a gallon), without including costs of further processing in refineries to create a suitable fuel from the raw material. Methanol production from feedstocks other than natural gas (in particular coal) are generally higher because of the added cost of generating and purifying the syn-gas necessary for the methanol synthesis. In regions rich in coal, such as United States or China, the production of methanol from coal on a large scale would however bring the costs down, providing an alternative domestic route to methanol.

The cost of producing methanol from sources other than fossil fuels and, most importantly, from the reaction of CO_2 with hydrogen, is still difficult to evaluate with precision at this stage. As with any other synthetic material or fuel, the price of methanol from alternative routes is, however, expected to be more costly than existing pathway via fossil fuels. It is much more difficult and energy-intensive to manufacture a convenient fuel that is not directly derived from natural sources. In many ways, fossil fuels should be considered as a gift from nature, which have allowed mankind to reach unprecedented levels of development. They served us well, but now – due to their finite nature – must be replaced by more sustainable sources of energy.

Methanol Safety

Methanol, as mentioned earlier, is a colorless liquid with a mild alcoholic odor. It is widely used as a chemical intermediate and solvent by industries and is present in a variety of consumer products. This includes, for example, the blue windshield washer fluid that most motor car owners are familiar with, which is in fact composed in large part of methanol. The use of methanol not only as a windshield washer fluid but also as a deicing fluid, antifreeze or even fuel for camping cooking vessels, implies that albeit almost every house contains methanol. Even if vigilance is always required, no significant problems have been associated with its use by the general public. With its widespread usage as an automotive fuel, exposure to methanol is likely to increase. As shown by a number of studies, the risks encountered by the consumer will however remain minimal and, in any case, not greater than those associated with the use of gasoline.

Methanol, like all other motor fuels, is toxic to the human body and should be handled with the same care as gasoline or diesel fuel with regard to its adverse effects on human health. Methanol is readily absorbed by ingestion, inhalation, and more slowly by skin exposure. The ingestion of 25 to 90 mL of methanol [169] may be fatal if not treated in time (compared to 120–300 mL of gasoline). Shortly after exposure, methanol causes a temporary effect on the brain, of similar nature but of lesser strength, than that of ethanol. Methanol is eliminated from the body by metabolism (i.e., enzymatic conversion) in the liver to formaldehyde, and then to formic acid which can be excreted in the urine or further me-

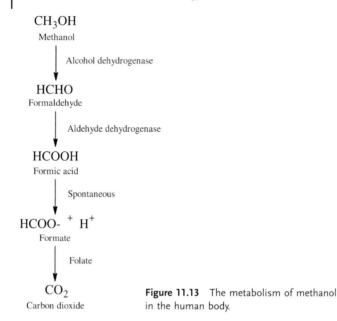

CH_3OH
Methanol

Alcohol dehydrogenase

$HCHO$
Formaldehyde

Aldehyde dehydrogenase

$HCOOH$
Formic acid

Spontaneous

$HCOO^-$ + H^+
Formate

Folate

CO_2
Carbon dioxide

Figure 11.13 The metabolism of methanol in the human body.

tabolized to CO_2 (Fig. 11.13). As methanol is metabolized, the most severe effects are delayed for up to 30 h, these being caused mainly by the formic acid produced, which humans metabolize very slowly. Higher concentrations of formic acid, when dissociated into formate and hydrogen ions, leads to increased acidity in the blood. Symptoms may include weakness, dizziness, headache, nausea, and vomiting, followed by abdominal pain and difficulties in breathing. In severe cases methanol poisoning may progress to coma and death. Another well-known symptom associated with methanol poisoning is that of visual impairment, which ranges from blurring to total loss of vision, and is caused by formic acid affecting the optic nerve.

Several treatments can be applied to combat methanol poisoning, and these generally lead to complete recovery if administered in timely manner. Early treatment with sodium carbonate counters the higher blood acidity and prevents or reverses vision impairment. Dialysis is effective in removing both methanol and formate from the bloodstream. In addition, 4-methylpyrazole (Antizol®, fomepizol) [169, 170], an antidote approved by the US Food and Drug Administration (FDA), and operating in the same manner for ethanol ingestion, though without its side effects (it is also effective against ethylene glycol poisoning), can be administered either intravenously or orally.

Although overexposure to methanol can be dangerous to human health, it is also important to realize that both methanol and formate are naturally present in our bodies from the diet, and also as a result of metabolic processes. Methanol is ingested when eating fresh fruits, vegetables and fermented foods and beverages. Aspartame, a widely used artificial sweetener included in many diet foods and soft drinks is also partially converted to methanol during the digestion process. According to the FDA, a daily intake of up to 500 mg methanol is safe for an adult's diet [171]. Methanol and formate are naturally present in the blood in concentrations of approximately 1–3 and 10 mg L^{-1}, respectively; moreover, formate is an essential building block for many biological molecules, including components of DNA [169]. Methanol is not considered to be either a carcinogenic or a mutagenic hazard; this is in contrast to gasoline, which contains a number of chemical compounds that are considered to be hazardous, including (amongst others) benzene, toluene, xylene, ethylbenzene and *n*-hexane, some of which are known to be carcinogenic.

Refueling a motor car with methanol at a service station equipped with an usual refueling system is only expected to result in low-dose exposures (23–38 ppm during the refueling process [169]) to the general public. By inhalation, a small oral intake of 2–3 mg of methanol is thus expected during a typical refueling. For comparison, this is much less than drinking a single 0.35-L can of diet soda containing 200 mg of aspartame, which will produce some 20 mg of methanol via the body's digestive system. Using vapor recovery systems, exposure to methanol during refueling can be further reduced to the 3 to 4 ppm level , adding only insignificantly to the methanol balance of the body. Even considering a worst-case scenario involving a malfunctioning vehicle in an enclosed garage where the methanol concentration is estimated to reach 150 ppm, a 15-min exposure would only add some 40 mg to the body's intake, or the equivalent of drinking 0.7 L of diet soda. Exposure to methanol can be readily minimized through the correct design of fueling systems and fuel containers.

In order to avoid spills, spill-free nozzles have been developed which makes it virtually impossible for the consumer to come into contact with methanol during refueling. Nevertheless, in case of contact with the skin, the affected area should be washed thoroughly with water and soap. To avoid accidental ingestion of methanol, the addition of distinct taste and odor agents should be considered. A dye may also be added to give methanol fuel a distinctive color. As indicated, however, by the 35 000 annual cases of gasoline ingestion in the United States alone (mostly through mouth siphoning when fuel is transferred from one tank to another), the unpleasant taste and smell may not be enough to prevent accidents. Thus, refueling systems should also be designed to make siphoning impossible and to allow only the vehicles themselves to be refueled with methanol. Other containers not meant specifically to contain methanol, the improper labeling of which could lead to mistakes or misuses, should be prohibited. In order to prevent the accidental ingestion of methanol by some heavy drinkers or the ill-informed public, the name "methyl alcohol" should also be avoided for methanol in order to minimize possible confusions with ethyl alcohol (i.e., ethanol). Above all, commonsense precaution in handling methanol should make its use safe.

Fire and explosion are major hazards associated with the use of transportation fuels, and these are also of concern for methanol safety. Compared to gasoline, methanol's physical and chemical properties significantly reduce the risk of fire. Combined with its lower volatility, methanol vapor in air must be four times more concentrated than gasoline for ignition to occur. If it does ignite, methanol burns about four times slower than gasoline and releases heat at only one-eighth the rate of gasoline fires. Because of the low radiant heat output, methanol fires are less likely to spread to surrounding ignitable materials. In tests conducted by the EPA and the Southwest Research Institute [108], two cars – one fueled by methanol and the other by gasoline – were allowed to leak fuel on the ground adjacent to an open flame. Whilst the gasoline ignited rapidly, resulting in a fire that consumed the entire vehicle within minutes, methanol took three times longer to ignite and the resulting fire damage affected only the rear part of the car. The EPA has estimated that switching fuels from gasoline to methanol would reduce the incidence of fuel-related fires by 90%, saving annually in the United States more than 700 lives, preventing some 4000 serious injuries, and eliminating property losses extending to many millions of dollars [172]. Methanol has been the fuel of choice for Indianapolis-type race cars since the mid-1960s because, in addition to achieving superior performances, it is one of the safest fuels available. Unlike gasoline fires, methanol fires can be quickly and easily extinguished even by simply pouring water on them. Methanol burns with little or no smoke, reducing the risks of injuries associated with smoke inhalation and allowing a better visibility around the fire, enabling easier fire fighting. Methanol's combustion generates a light blue flame that is visible in most situations, but may not be easily seen in bright sunlight. In a majority of fires, however, the burning of materials other than fuel, such as upholstery, engine oil and paint, would impart the color of the flames, making them visible in any situation. In confined areas such

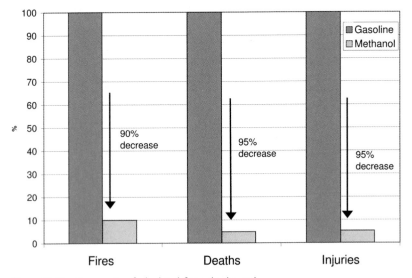

Figure 11.14 Comparative fuel-related fires, deaths and injuries. (Source: U.S. Environmental Protection Agency, EPA 400-F-92-010.)

as fuel tanks and reservoirs, an ignitable methanol/air mixture can form at ambient temperature. This property of methanol is however unlikely to lead to fires or explosions even in the event of a collision, and has been addressed by simple fuel tank modifications or addition of a volatile compound which makes the vapor space in the tank too rich to ignite [120].

In summary, compared to gasoline, methanol fires are far less likely to occur, and are much less damaging when they do (Fig. 11.14).

Emissions from Methanol-Powered Vehicles

Transportation-associated air pollution is a major problem in large metropolitan areas. CO, NO_x, volatile organic compounds (VOCs), SO_2 and particulate matter (PM) emitted by automobiles, trucks and buses can have serious effects on the population's health, especially in children, elderly and other sensitive persons. As described earlier, the use of clean-burning methanol in ICEs could immediately help to reduce these emissions [173]. Included in the VOCs is formaldehyde, an air-toxic and ozone precursor (it is present naturally in low concentrations in the atmosphere) that is produced in small quantities by the incomplete combustion of not only gasoline and diesel but also methanol, and is classified as a possible carcinogen. The issue of formaldehyde formation in methanol-powered ICEs has been successfully addressed by the development and use of a highly effective catalytic muffler to remove the relatively reactive formaldehyde by catalytic oxidation. One should bear in mind that, although methanol is an inherently cleaner fuel, gasoline- and diesel-fueled ICE vehicles have made – and will continue to make through improved technology – considerable progress in emission control, and thus effectively compete with alternative fuels. In ICE cars however, due to the sophistication of the systems needed to keep emissions low, a lack of proper maintenance and regular inspection can easily lead to dramatically higher emissions as the vehicle ages.

In the long term the use of methanol-powered FCVs offers the promise almost to eliminate all current air pollutants from vehicles. Furthermore, unlike ICE vehicles the emission profile of methanol-powered vehicles will remain almost unchanged as they age. Emissions of methanol FCVs equipped with an onboard methanol reformer are expected to be even much lower than the already stringent limits set by the State of California for Super Ultra Low Emission Vehicles (SULEV) (Fig. 11.15). Direct methanol FCVs are expected to be virtually zero emission vehicles (ZEV) [119]. Tests conducted with Georgetown University's reformed methanol fuel cell bus have shown that it is almost a ZEV, releasing only negligible amounts of carbon monoxide and hydrocarbons, and no NO_x or PM [174].

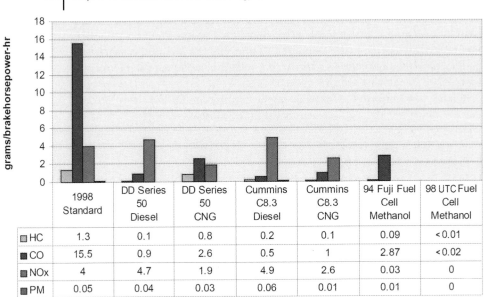

	1998 Standard	DD Series 50 Diesel	DD Series 50 CNG	Cummins C8.3 Diesel	Cummins C8.3 CNG	94 Fuji Fuel Cell Methanol	98 UTC Fuel Cell Methanol
▫ HC	1.3	0.1	0.8	0.2	0.1	0.09	< 0.01
▪ CO	15.5	0.9	2.6	0.5	1	2.87	< 0.02
▪ NOx	4	4.7	1.9	4.9	2.6	0.03	0
▪ PM	0.05	0.04	0.03	0.06	0.01	0.01	0

Figure 11.15 Pollutant emission from methanol-powered fuel cell buses. (Source: Georgetown University.)

Methanol and the Environment

Methanol is emitted into the atmosphere from a number of natural sources, including volcanoes, vegetation, microbes, insects, animals and decomposing organic matter [169]. The anthropogenic (human-caused) release of methanol into the environment is presently mainly due to its use as a solvent through evaporation. Releases to the water and ground are less significant. In the environment, methanol is readily degraded by photooxidation and biodegradation processes. This is, beside other reasons, why methanol is widely used for example in windshield washer fluids. Methanol is rapidly degraded under both aerobic (in the presence of air) and anaerobic conditions (in the absence of air) in fresh and salt water, groundwater, sediments and soils with no evidence of bioaccumulation. It is a regular growth substrate (source of carbon and energy) for many microorganisms which are found in the ground and are able to degrade methanol to CO_2 and water. Methanol is of low toxicity to aquatic and terrestrial organisms, and effects due to environmental exposure to methanol are unlikely to be of consequence under normal conditions [175]. In fact, methanol is used for the denitrification of wastewater in sewage treatment plants. The addition of methanol in this process accelerates the conversion of nitrates to harmless nitrogen gas by anaerobic bacteria. If discharged in large quantities, nitrates can accumulate in rivers, lakes, oceans and have devastating effects on the water ecosystem. Excess nitrogen from nitrates causes an overgrowth of algae, and this prevents oxygen and sunlight from penetrating deeply into the water; the result is often suffocation of the fish and all aquatic life below. Currently, more than 100 wastewater treatment plants across the United States use methanol in their denitrification process. The Blue Plains Wastewater Treatment Facility, which serves the Washington D.C. area, is one of the largest such treatment plants in the country, and by using methanol denitrification avoids the release each day of some 10 tonnes of nitrogen into the Potomac River [176].

The accidental release of methanol into the environment during its production, transportation and storage – although possible – would cause much less damage than a corresponding crude oil or gasoline spill, due to the above-mentioned favorable chemical and physical properties of methanol. A large accidental methanol spill into surface water would have some immediate impact on the ecosystem in close proximity to the spill. However, since the methanol is totally miscible with water it would rapidly be diluted and dissipated into the environment by wave action, winds or tides to non-toxic levels. It has been calculated that the release of 10 000 tonnes of methanol into the open sea would result in a concentration of only 0.36% within the first hour of the spill, and this would be much less during the following hours [177]. At this point, biodegradation through microorganisms would take care of the dilute methanol in a matter of days. A similar accident involving petroleum oil, which is not miscible with water, covers large surfaces and does not easily dissipate into the environment, would most likely lead to an ecological disaster. Methanol, in leaving no residues, also avoids the long and fastidious cleaning of beaches, shorelines, birds and wild life needed after a crude oil spill. On land, the accidental release from a tank truck or rail car transporting methanol, or from an underground storage tank, are possible scenarios. Depending on the size and location of the spill, different situations might be encountered, but in most cases the miscibility of methanol with water should allow it to be diluted and the effects to be dispersed rapidly, allowing also rapid biodegradation by microorganisms present in the ground. It should also be pointed out that the behavior of methanol in the environment is very different from that of MTBE, which is not easily degraded, and explains the reason for its ban as an oxygenated additive for transportation fuels [119]. Methanol leaks are also less hazardous than gasoline leaks because the latter contains many toxic and carcinogenic compounds (e.g., benzene) which biodegrade slowly and persist for a longer time in the environment. To further minimize the risk of leaks in underground storage, the use of double-walled storage tanks and leak detectors are preferable for all fuels.

Compared to gasoline or diesel fuel, methanol is clearly environmentally much safer and less toxic.

Methanol and Issues of Climate Change

Today, methanol is manufactured almost exclusively from syn-gas produced by catalytic reforming of natural gas or coal (i.e., from fossil fuel sources). In contrast to natural gas, our coal reserves are extensive. However, because of coal's deficiency in hydrogen, its conversion to methanol produces the most CO_2 compared to all other fossil fuels, and especially natural gas. Currently, on a large scale the cleanest, most efficient and most economical way to produce methanol is from natural gas-generated syn-gas. In the area of ongoing research and development, however, much progress has been made in the direct oxidative conversion of methane to methanol (see Chapter 12). As long as natural gas is still quite abundant, it would seem inevitable that it will be used to produce methanol. It is, however, feasible to develop technologies to convert it directly to methanol without passing first through the generation of syn-gas. It should also be recognized that (as discussed in Chapter 4) there are large resources of methane hydrates tied up in vast areas of the subarctic tundras and under the seas in the areas of the continental shelves. All of these resources will eventually be utilized, but they will only extend the inevitable exhaustion of our natural hydrocarbon resources.

Using methanol produced from natural gas in traditional ICE vehicles will only moderately – if at all – reduce the CO_2 emissions compared to their gasoline and diesel equivalents. Real benefits in terms of greenhouse gas emissions will be obtained only with more efficient FCVs powered by hydrogen produced from methanol reforming or the direct use of methanol in DMFC.

In order to further reduce CO_2 emissions and to provide a sustainable alternative source for the long term, methanol will have to be produced from feedstocks other than fossil fuels. Production from biomass is one possibility, but this could provide only a minor portion of our growing energy needs. Methanol, however can also be obtained from CO_2 by catalytic reduction with hydrogen or by electrochemical reduction in water (see Chapter 12). The flue gases of coal- and other fossil fuel-burning power plants, as well as from facilities such as cement and steel factories, contain high concentrations of CO_2, and these will increasingly need to be captured and disposed of. Instead of sequestration, the chemical recycling of CO_2 into methanol, besides providing useful fuels and a source for synthetic hydrocarbons will also mitigate human-caused climate changes. Eventually, the CO_2 content of the atmosphere itself will be similarly recycled, freeing humankind from its dependence on fossil fuels, assuming that we will be able to produce the required energy from non-fossil fuel sources (renewable energy sources and safe atomic energy) for the production of hydrogen and conversion to methanol.

Chapter 12
Production of Methanol from Syn-Gas to Carbon Dioxide

In 2004, the worldwide demand for methanol stood at 32 million tonnes. Although virtually any hydrocarbon source (coal, petroleum, naphtha, coke, etc.) can be converted to methanol via derived syn-gas, methane from natural gas accounted for some 90% of the feedstock used for methanol manufacture. Most existing plants have production capacities ranging from 100 000 to 800 000 t per year. In the past, plants were generally constructed close to large methanol-consuming centers in the United States or Europe. However, with decreasing local reserves and increasing prices of natural gas in these areas, production has shifted to countries which still have large natural gas resources but lie far from centers of major consumption, such as Chile, Trinidad and Tobago, Qatar, and Saudi Arabia (Fig. 12.1). In the United States, for example, the annual methanol production capacity peaked at some 7.4 million tonnes in 1998, supplying 70% of the domestic requirements. In 2005, after the progressive phase-out of methyl-*tert*-butyl ether (MTBE, a reformulated gasoline additive), together with the closure of aging plants and high natural gas prices, the annual production capacity fell to only about 2 million tonnes, covering merely 20–25% of the United States domestic demand [178]. Plans were also announced to close down altogether methanol production in North America. Not only the geographical location, but also the size of new methanol plants has changed. Facilities with capacities of 1 million tonnes per year or more – termed "mega methanol plants" – are becoming the norm. Two such mega methanol plants, each able to produce almost 1 million tonnes of methanol per year, are currently operating in Qatar and Saudi-Arabia [132]. Methanex recently started up its 1.7 million tonnes per year Atlas plant in Trinidad and Tobago [179], and further plants are either planned or are already under construction in Qatar, Saudi-Arabia, Iran, Chile, Malaysia, and other countries. Designs for plants with capacities up to 3.5 million tonnes per year are now offered by major methanol technology developers (Synetix, Lurgi and Haldor-Topsoe). The economy of scale brought by the construction of ever larger and more efficient syn-gas-based plants is expected to lower considerably the cost of methanol production, and is starting to open up the market for methanol as an alternative fuel and petrochemical feedstock (Fig. 12.2). New methods for the direct oxidative conversion of existing natural gas (methane) sources to methanol without going through syn-gas, as well as the hydrogenative

Beyond Oil and Gas: The Methanol Economy. G. A. Olah, A. Goeppert, G. K. S. Prakash
Copyright © 2006 WILEY-VCH Verlag GmbH & Co. KGaA, Weinheim
ISBN 3-527-31275-7

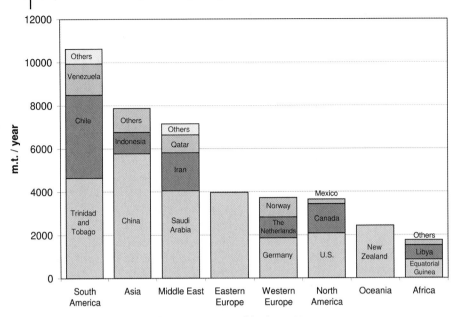

Figure 12.1 Methanol production capacity worldwide in 2004.
Source: Adapted from data in *Chemical Week*, June 22, 2005.

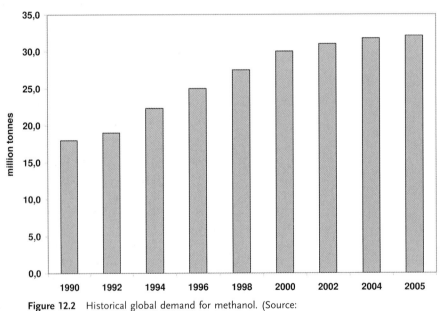

Figure 12.2 Historical global demand for methanol. (Source:
Methanex and *Chemical Week*.)

conversion of CO_2 (to be discussed subsequently) are essential new technologies for the proposed "Methanol Economy".

There are today more than 600 million private motor cars and some 200 million light and heavy trucks registered worldwide. Together, they represent 80% of the energy consumed in the transportation field. Air, maritime, rail and pipeline transport account for the remaining 20% or so. In 2002, the transportation sector consumed some 1.75 billion tonnes of oil, yet with increasing demand – especially from developing countries – this figure is expected to grow substantially to reach more than 3 billion tonnes in 2030, with 1.3 billion vehicles on the road [180]. Given these numbers, it is evident that any alternatives to oil-based fuels will have to be produced on a very large scale. Methanol, as discussed earlier, has for the same volume only about half the energy content of gasoline or diesel fuel. Replacing only 10% of our current energy needs for transportation would require the production of some 350 million tonnes of methanol per year (about 11 times today's world production). However, as methanol FCVs are expected to be twice as energy-efficient as present ICE vehicles, this would decrease the amount of fuel necessary by about half. With the widespread use of fuel cell technology, the amount of methanol required for the transportation sector would thus become comparable to that of oil-based fuels. For methanol to become a major fuel of the future, a significant scale-up in production capacity and new technologies will be needed.

Although conventional natural gas reserves are currently the preferred feedstock for the production of syn-gas-based methanol, unconventional gas resources such as coalbed-methane, tight sand gas and eventually vast methane hydrate resources will also be used. So will, besides methane, any other available fossil fuel source – and particularly coal, with its vast reserves, widespread availability and low cost. Coal, at present, is thus the best alternative to natural gas (methane) -based syn-gas. Large domestic coal supplies in major energy-consuming countries such as the United States or China offer price stability and enhanced independence from geopolitically unstable regions. Coal however, must be mined and has the drawback of releasing large amounts of pollutants into the atmosphere. Considering climate change, it also produces more CO_2 compared to methane and petroleum. In order to minimize CO_2 releases, biomass represents another possible source for methanol but, due to the enormous energy needs of our modern society, biomass will be able to cover only a fraction of the demand. To lessen our dependence on fossil fuels and also to address the CO_2 emission problem, new and improved methods of methanol production are needed. Besides directly converting methane (natural gas) to methanol without going through syn-gas, the combination of hydrogen, made with use of renewable or nuclear energy sources, with CO_2 from fossil fuel-powered electric plants, other industrial sources and eventually the atmosphere itself allow independence from diminishing fossil fuel resources.

Methanol from Fossil Fuels

Production via Syn-Gas

Today, methanol is almost exclusively produced from syn-gas, which is a mixture of hydrogen, carbon monoxide and CO_2 over a heterogeneous catalyst according to the following equations:

$$CO + 2H_2 \rightleftharpoons CH_3OH \qquad \Delta H_{298K} = -21.7 \text{ kcal} \cdot \text{mol}^{-1} \qquad \text{Eq. (1)}$$

$$CO_2 + 3H_2 \rightleftharpoons CH_3OH + H_2O \qquad \Delta H_{298K} = -11.9 \text{ kcal} \cdot \text{mol}^{-1} \qquad \text{Eq. (2)}$$

$$CO_2 + H_2 \rightleftharpoons CO + H_2O \qquad \Delta H_{298K} = 9.8 \text{ kcal} \cdot \text{mol}^{-1} \qquad \text{Eq. (3)}$$

The two reactions in Eqs. (1) and (2) are exothermic, with heats of reaction equal to -21.7 kcal mol^{-1} and -11.9 kcal mol^{-1}, respectively. They both result in a decrease of volume as the reaction proceeds. According to Le Chatelier's principle, the conversion to methanol is therefore favored by increasing pressure and decreasing temperature. Equation (3) describes the endothermic reverse water gas shift reaction (RWGSR) which also occurs during methanol synthesis, producing carbon monoxide which can then further react with hydrogen to produce methanol. In fact, Eq. (2) is simply the sum of the reactions in Eqs. (1) and (3). Each of these reactions is reversible, and thus limited by thermodynamic equilibria depending on the reaction conditions, mostly temperature, pressure and composition of the syn-gas.

Synthesis gas for methanol production can be obtained by reforming or partial oxidation of any carbonaceous material such as coal, coke, natural gas, petroleum, heavy oils and asphalt. Economic considerations dictate the choice of raw materials. The long-term availability of raw materials, energy consumption and environmental aspects will however also play increasingly important roles.

The composition of syn-gas is generally characterized by the stoichiometric number S. Ideally, S should be equal to or slightly above 2. Values above 2 indicate an excess of hydrogen, whereas values below 2 mean a hydrogen deficiency relative to the ideal stoichiometry for methanol formation.

$$S = \frac{(\text{moles } H_2 - \text{moles } CO_2)}{(\text{moles } CO + \text{moles } CO_2)}$$

Synthesis gas from coal has less than the optimum hydrogen to carbon content. Treatment of the gas before methanol synthesis or addition of hydrogen is therefore needed to avoid the formation of undesired byproducts. Reforming of feeds with a higher H/C ratio such as propane, butane, or naphthas leads to S values in the vicinity of 2, which is ideal for conversion to methanol. Steam reforming of methane on the other hand, yields syn-gas with a stoichiometric number of 2.8 to 3.0. In this case, the addition of CO_2 can lower S close to 2. Excess hydrogen can also be used in an adjacent plant to produce ammonia.

The production of methanol from syn-gas on an industrial scale using high pressures (250–350 atm) and temperatures (300–400 °C) was first introduced by BASF during the 1920s. From then until the end of World War II, most methanol was produced from coal-derived syn-gas and off-gases from industrial facilities such as coke ovens and steel factories. The use of such feedstocks containing high levels of impurities, was made possible by the design of a catalyst system consisting of zinc oxide and chromium oxide, which was highly stable to sulfur and chlorine compounds. After World War II, the feedstock for methanol synthesis shifted rapidly to natural gas, which became widely available at very low costs, particularly in the United States. Whereas 71% of the methanol in the United States was still derived from coal in 1946, by 1948 almost 77% was obtained from natural gas. Natural gas is still the preferred feedstock for methanol production because it offers, besides a high hydrogen content, the lowest energy consumption, capital investment and operating costs. Furthermore, natural gas contains fewer impurities such as sulfur and halogenated compounds or metals, which can poison the needed catalysts. When present, these impurities (mostly sulfur in the form of H_2S, COS or mercaptans) can, however, also be removed relatively easily. Lower levels of impurities in the syn-gas allowed the use of more active catalysts, operating under milder conditions. This led, during the 1960s, to the development by ICI (now Synetix) of a process using a copper-zinc-based catalyst allowing the conversion of syn-gas to methanol at pressures of 50–100 atm and temperatures of 200–300 °C. This low-pressure route is the basis for most current processes for methanol production. The formation of byproducts (dimethyl ether, higher alcohols, methane, etc.) associated with the old high-pressure technology was also drastically reduced or even eliminated [109]. Production using the high-pressure process is no longer economical, and the last plant based on this technology closed in the 1980s. Worldwide, the production capacity for methanol is dominated by processes from only a few companies, with Synetix accounting for some 60% of the installed capacity and Lurgi for 27% (Fig. 12.3).

All present processes use copper-based catalysts which are extremely active and selective, and are almost exclusively used in gas-phase processes. They mostly dif-

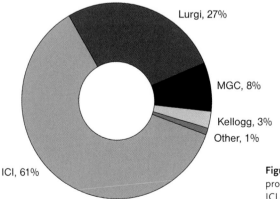

Lurgi, 27%

MGC, 8%

Kellogg, 3%
Other, 1%

ICI, 61%

Figure 12.3 Worldwide methanol production by process. (Source: ICI, now Synetix.)

fer only in the type of reactor design and catalyst arrangement (in fixed beds, tubes, suspension, etc.). Control of the temperature is very important, and overheating of the catalyst should be avoided as it will rapidly decrease its activity and shorten its lifetime [108]. In one pass over the catalyst only a part of the syn-gas is converted to methanol. After separation of methanol and water by condensation at the reactor's outlet, the remaining syn-gas is recycled to the reactor. Recently, liquid-phase processes for methanol production have also been introduced. Air Products in particular, developed the Liquid Phase Methanol Process (LPMEOH) in which a powdered catalyst is suspended in an inert oil, providing an efficient means to remove the heat of reaction from the catalyst and control the temperature (Fig. 12.4). The syn-gas is simply bubbled into the liquid. This process promotes also a higher syn-gas to methanol conversion, so that a single pass through the reactor is generally sufficient [181].

Generally, modern methanol plants have selectivities to methanol higher than 99%, with energy efficiencies above 70% (Fig. 12.5). Crude methanol leaving the reactor contains, however, water and small amounts of other impurities, the distribution and composition of which will depend on the gas feed, reaction conditions, and type and lifetime of the catalyst. Impurities that may be present in methanol include dissolved gases (methane, CO, CO_2), dimethyl ether, methyl formate, acetone, higher alcohols (ethanol, propanol, butanol) and long-chain hydrocarbons. Commercially, methanol is available in three grades of purity: fuel grade, "A" grade, generally used as a solvent, and "AA" or chemical grade. Chemical grade has the highest purity with a methanol content exceeding 99.85% and is

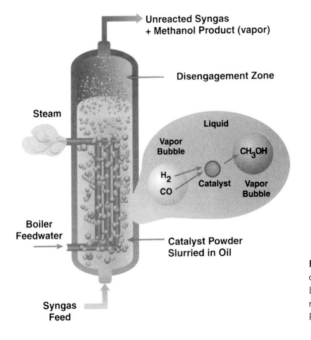

Figure 12.4 Synthesis of methanol in the LPMEOH™ slurry bubble reactor developed by Air Products (Source: DOE).

Figure 12.5 Atlas mega-methanol plant in Trinidad. (Courtesy of Methanex).

the standard generally observed by the industry for methanol production. Depending on the amount of impurities and purity desired, methanol will have therefore to be purified by distillation systems using one or more distillation columns.

Asides from the methanol synthesis step, the most crucial part of present methanol plants is the syn-gas generation and purification system, which will depend on the nature and purity of the feedstock used. Although natural gas is generally the preferred feedstock due to the simplicity of obtaining an adequate syn-gas with low levels of impurities, other routes are also used under given circumstances. Regions rich in coal or heavy oils, but having limited natural gas sources, in view of the high natural gas prices, can turn to these resources to produce methanol despite higher cost for the needed syn-gas purification system.

Syn-Gas from Natural Gas

Methane Steam Reforming

In methane steam reforming, methane is reacted in a highly endothermic reaction with steam over a catalyst, typically based on nickel, at high temperatures (800–1000 °C, 20–30 atm) [182] to form CO and H_2. A part of the CO formed reacts consequently with steam in the water gas shift (WGS) reaction to yield more H_2 and also CO_2. The gas obtained is thus a mixture of H_2, CO and CO_2.

$$CH_4 + H_2O \rightleftharpoons CO + 3H_2 \qquad \Delta H_{298K} = 49.1 \text{ kcal / mol}$$

$$CO + H_2O \rightleftharpoons CO_2 + H_2 \qquad \Delta H_{298K} = -9.8 \text{ kcal / mol}$$

The concentration of the various components depends on the reaction conditions: temperature, pressure and H_2O/CH_4 ratio. Syn-gas generation from methane however increases with increasing temperature and decreasing pressure. With increasing temperatures, the WGS reaction becomes also less dominant and the main products are CO and H_2 [182]. Since the overall methane steam reforming process is highly endothermic, heat must be supplied to the system by burning a part of the natural gas used as feedstock. Steam reforming is the most widely used technology to produce not only syn-gas for methanol synthesis but also ammonia and other bulk chemicals. The stoichiometric number S obtained by steam reforming of methane is close to 3, far from the desired value of 2. This can generally be corrected by addition of CO_2 to the steam reformer's exit gas or use of excess hydrogen in some other process such as ammonia synthesis.

Methane reforming can also be affected by a process called "coking"; this involves the formation of carbon, which may deposit in the form of soot or coke on the catalyst (greatly reducing its activity) as well as all the internal parts of the reformer and downstream equipment, resulting in possible clogging. Carbon may be formed by CH_4 decomposition or CO disproportionation (Boudouard reaction).

$$CH_4 \rightleftharpoons C + 2H_2 \qquad \Delta H_{298K} = 18.1 \text{ kcal} \cdot \text{mol}^{-1}$$

$$2CO \rightleftharpoons C + CO_2 \qquad \Delta H_{298K} = -41.2 \text{ kcal} \cdot \text{mol}^{-1}$$

In practice, the undesired formation of carbon is largely prevented by the use of excess steam and short residence times in the reactor. It can, however, be more problematic for the partial oxidation process operating at higher temperature than methane reforming.

Partial Oxidation of Methane

Partial oxidation is the reaction of methane with insufficient oxygen, which can be performed with or without a catalyst. This reaction, which is exothermic and operated at high temperature (1200–1500 °C) [132], generally yields syn-gas with an H_2/CO ratio of 2, which is ideal for methanol synthesis. The problem with partial oxidation is that the products, CO and H_2, can be further oxidized to form undesired CO_2 and water in highly exothermic reactions, leading to S values typically below 2. The production of large amounts of energy in the form of heat is also not desirable and wasteful if no immediate use for it can be found in an other process.

$$CH_4 + 1/2\,O_2 \rightleftharpoons CO + 2H_2 \qquad \Delta H_{298K} = -8.6 \text{ kcal} \cdot \text{mol}^{-1}$$

$$CO + 1/2\,O_2 \rightleftharpoons CO_2 \qquad \Delta H_{298K} = -67.6 \text{ kcal} \cdot \text{mol}^{-1}$$

$$H_2 + 1/2\,O_2 \rightleftharpoons H_2O \qquad \Delta H_{298K} = -57.7 \text{ kcal} \cdot \text{mol}^{-1}$$

Autothermal Reforming and Combination of Steam Reforming and Partial Oxidation

In order to produce syn-gas without either consuming or producing much heat, modern plants are usually combining exothermic partial oxidation with endothermic steam reforming in order to have an overall thermodynamically neutral reaction while obtaining a syn-gas with a composition suited for methanol synthesis (S close to 2). In this process, termed "autothermal reforming", the heat produced by the exothermic partial oxidation is consumed by the endothermic steam reforming reaction. Partial oxidation and steam reforming can be conducted simultaneously in the same reactor by reacting methane with a mixture of steam and oxygen. Having only one reactor lowers the costs and complexity of the system. However, because the two reactions are optimized for different temperature and pressure conditions, they are generally conducted in two separate steps. After the steam reforming step, the effluent from the reformer's outlet is fed to the partial oxidation reactor where all the residual methane is consumed [183]. The oxygen required for the oxidation step means that an air separation plant is needed, but to avoid the construction of such a unit the use of air rather than oxygen is also possible. The produced syn-gas will however contain a large amount of nitrogen and require special processing before conversion to methanol. Thus, most modern methanol plants use pure oxygen.

Syn-Gas from CO$_2$ Reforming

Syn-gas can also be produced by the reaction of CO$_2$ with natural gas or other aliphatic hydrocarbons, generally termed CO$_2$ or "dry" reforming, because it does not involve any steam. With a reaction enthalpy of $\Delta H = 70.7$ kcal mol^{-1} [182], this reaction is more endothermic than steam reforming (49.1 kcal mol^{-1}).

$$CO_2 + CH_4 \rightleftharpoons 2CO + 2H_2$$

The syn-gas produced has also an H$_2$/CO ratio of 1; much lower than the values of around 3 obtained with steam reforming. Whilst this is a disadvantage for methanol synthesis, it makes a suitable feed gas for other processes, especially iron ore reduction and Fischer–Tropsch synthesis. Several industrial processes, taking advantage of high CO and low water contents of synthesis gas produced by CO$_2$ reforming, have thus been applied for the production of iron. For methanol production, however, hydrogen generated from other sources would have to be added to the obtained syn-gas. Combination of steam and CO$_2$ reforming, which can be performed using the same catalysts, to reach adequate syn-gas composition is also possible.

Syn-Gas from Petroleum and Higher Hydrocarbons

Natural gas is not the only hydrocarbon source used for syn-gas generation for methanol production. Although on a smaller scale, liquefied petroleum gas and the different fractions obtained during oil refining (and especially naphtha) are also employed to produce syn-gas for the manufacture of ammonia, methanol and higher alcohols (propanol, butanol, etc.). Crude oil, heavy oil, tar and asphalt can all be transformed to syn-gas. The methods used are similar to those for natural gas: steam reforming and partial oxidation or a combination of both.

$$C_nH_m + nH_2O \rightleftharpoons nCO + (n + m/2) H_2$$

$$C_nH_m + n/2 O_2 \rightleftharpoons nCO + (m/2) H_2$$

The problem with higher carbon-rich feedstocks, is generally their higher content in impurities, and especially sulfur compounds or chlorine which can very rapidly poison the catalysts used for steam reforming and subsequent methanol synthesis. Therefore, considerably more capital must be invested in the purification steps, and other catalysts which are more resistant to poisoning may be employed. Heavy oils, tar sands and other hydrocarbon sources containing large and complex aromatic structures are also poor in hydrogen, resulting in a syn-gas that will be rich in CO and CO_2 but deficient in hydrogen.

Syn-Gas from Coal

Coal has been the feedstock used originally in the industrial production of syn-gas. It is still widely used in South Africa and China for the production of ammonia and methanol, and was expected for years to become the preferred route for large scale syn-gas generation in the United States, because of existing huge domestic coal reserves. Syn-gas is produced from coal by gasification, a process combining partial oxidation and steam treatment, according to the following reactions:

$$C + 1/2 O_2 \rightleftharpoons CO$$

$$C + H_2O \rightleftharpoons CO + H_2$$

$$CO + H_2O \rightleftharpoons CO_2 + H_2$$

$$CO_2 + C \rightleftharpoons 2 CO$$

Different coal gasification processes have been developed and commercialized over the years. The selection of a particular design depends greatly on the characteristics of the coal used: lignite, sub-bituminous, hard coal, graphite, water content, ash content, levels of impurities, etc. Due to the low H/C ratio of coal, the obtained syn-gas is rich in carbon oxides (CO and CO_2) and deficient in hydrogen.

Before being sent to the methanol unit, the syn-gas must thus be subject to the WGS reaction to enhance the amount of hydrogen formed. Some of the CO_2 produced must also be separated and any H_2S removed to avoid poisoning of the very sensitive methanol synthesis catalyst. Whilst sulfur must be removed to protect the catalyst, some of the cost of this operation may be offset by its commercialization for the preparation of chemicals such as sulfuric acid.

Economics of Syn-Gas Generation

The investment in the syn-gas generation unit accounts generally for more than half of the total investment for natural gas-based plants producing methanol. For plants using coal as a feedstock, this represents even more, generally 70–80% [183]. The remaining 20–50% accounts for the capital costs involved in the actual production of methanol. The price of methanol from coal is thus more dependent on the technology than the cost of the feedstock, which remains much cheaper and less fluctuating than natural gas, due to still large and easily accessible coal deposits to be found all around the globe. Despite the higher initial investment for the construction of a methanol plant, increasing natural gas prices created a new interest in the production of methanol from coal. Many plants in China are already using the still abundant and inexpensive coal sources to produce methanol. In the United States, only one plant – operated by the Eastman Corporation in Kingsport, Tennessee – presently produces methanol by coal gasification. The co-production of methanol and electricity in the same plant is also an attractive option.

Methanol through Methyl Formate

In order to reduce the pressure and temperature needed for the current methanol production process, and also to improve its thermodynamic efficiency, alternative routes to convert CO/H_2 mixtures to methanol under milder conditions have been developed. Among these, the most notable is the synthesis of methanol via methyl formate, first proposed in 1919 by Christiansen [184–186].

$$CH_3OH + CO \longrightarrow HCOOCH_3$$

$$HCOOCH_3 + 2\,H_2 \longrightarrow 2\,CH_3OH$$
$$\overline{}$$
$$CO + 2\,H_2 \longrightarrow CH_3OH$$

This methanol synthesis route consists of two steps. Methanol is first carbonylated to methyl formate, which is subsequently reacted with hydrogen to produce twice the amount of methanol. The carbonylation reaction is carried out in the liquid phase using sodium or potassium methoxide ($NaOCH_3$ or $KOCH_3$) as a homogeneous catalyst. It is a proven and commercially available technology used in the production of formic acid. High activities for methanol carbonylation have also been shown recently with Amberlyst and Amberlite resins used as heterogeneous catalysts. The subsequent reaction of methyl formate with hydrogen (hydrogenolysis) to synthesize methanol can be conducted either in the liquid or gas phase using generally a copper-based catalyst (copper chromite, copper supported on silica, alumina, magnesium oxide, etc.). Carbonylation and hydrogenolysis can be carried out in two separate reactors, but are preferably combined in a single reactor. In order to run the carbonylation and hydrogenation simultaneously in a single reactor, different combinations of catalysts have been investigated; in particular CH_3ONa/Cu and CH_3ONa/Ni. Nickel-based systems are very active and selective, but due to the volatility and high toxicity of the $Ni(CO)_4$ that may be formed during the reaction this process is considered difficult and hazardous to use in industrial settings. Copper systems are thus preferred because they offer similar activities and selectivities, without the toxicity problems associated with the nickel systems.

Overall, by the methyl formate route, methanol is produced from syn-gas, but at lower temperature and pressure than that employed in conventional methanol production processes. Patents from Mitsui Petrochemicals, Brookhaven National Laboratories and Shell using the methyl formate route claim to produce methanol at 80–120 °C and pressures of 10 to 50 atm. Another report states that continuous operation in a bubble reactor at 110 °C and a pressure of only 5 atm is also possible.

The problem with this process is the presence of CO_2 and water in the syn-gas, which will react with sodium methoxide, deactivating the catalyst and forming undesirable byproducts. In order to minimize this deactivation, CO_2 and water therefore need to be removed from the gas feed. Catalysts more tolerant to these poisons (such as the recently reported $KOCH_3$/copper chromite systems) are also being developed [186]. Although further improvements are still needed, the synthesis of methanol via methyl formate at low temperature and pressure, could lead to an attractive alternative to current processes operating at 200–300 °C and 50–100 atm. At the same time as methyl formate is also made by the dimerization of formaldehyde, the methyl formate to methanol route could also play a role in the secondary treatment of the oxidative conversion of methane to methanol, without going through syn-gas.

Methanol from Methane Without Syn-Gas

As long as methane, the main component of natural gas is still quite abundant, it will inevitably be used to produce methanol, and through it synthetic hydrocarbons and their products. It should also be recognized that (as discussed in Chapter 4) there are also other large methane resources such as unconventional natural gas sources as well as methane hydrates tied up in vast areas of subarctic tundras and under the seas in the areas of the continental shelves. All of these resources will eventually be utilized to produce methanol as our traditional natural gas resources are increasingly depleted. New ways to convert methane to methanol more efficiently, and without going through the currently used syn-gas-based processes, are much needed [187–190]. Extensive research has been conducted in this direction, and much progress has been made in recent years, especially in the oxidative direct conversion of methane to methanol.

Selective Oxidation of Methane to Methanol

A major disadvantage of the present process of producing methanol through syn-gas is the large energy requirement of the first highly endothermic steam reforming step. The process is also inefficient in the sense that it first transforms methane in an oxidative reaction to carbon monoxide (and some CO_2) which, in turn, must be reduced again to methanol. The direct selective transformation of methane to methanol is therefore a highly desirable goal, but this is difficult to accomplish in a practical way (with high conversion and selectivity). It would, however, eliminate the processing step for producing syn-gas, increase the amount of methanol to be obtained, and save on capital costs in commercial plants. The main problem associated with the direct oxidation of methane to methanol is the higher reactivity of the oxidation products themselves (methanol, formaldehyde and formic acid) compared to methane, giving eventually CO_2 and water – that is, the thermodynamically favored complete combustion of methane.

$$
CH_4
\begin{cases}
\xrightarrow{0.5\,O_2} & CH_3OH & \Delta H = -30.4\ \text{kcal}\cdot\text{mol}^{-1} \\
\xrightarrow{O_2} & CH_2{=}O\ +\ H_2O & \Delta H = -66.0\ \text{kcal}\cdot\text{mol}^{-1} \\
\xrightarrow{1.5\,O_2} & CO\ +\ 2H_2O & \Delta H = -124.1\ \text{kcal}\cdot\text{mol}^{-1} \\
\xrightarrow{2\,O_2} & CO_2\ +\ 2H_2O & \Delta H = -191.9\ \text{kcal}\cdot\text{mol}^{-1}
\end{cases}
$$

Finely tuned reaction conditions to achieve high conversion without complete oxidation producing CO_2 are thus explored. At present, however, no process has yet succeeded in achieving the combination of high yield, selectivity and catalyst stability that would make direct oxidative conversion competitive with conventional syn-gas-based methanol production methods.

A number of ways exist for the oxidation of methane. These include homogeneous gas phase oxidation, heterogeneous catalytic oxidation, and photochemical and electrophilic oxidations [191].

Catalytic Gas-Phase Oxidation of Methane

In homogeneous gas-phase oxidation, methane is generally reacted with oxygen at high pressures (30 to 200 atm) and high temperatures (200 to 500 °C). Optimum conditions for the selective oxidation to methanol have been extensively investigated. It was observed that selectivity to methanol increases with decreasing oxygen concentration in the system. The best result (75–80% selectivity in methanol formation at 8–10% conversion) was achieved under cold flame conditions (450 °C, 65 atm, less than 5% O_2 content), using a glass-lined reactor, which seemed to suppress secondary reactions. Most other studies obtained under their best reaction conditions (temperatures of 450–500 °C and pressures of 30–60 atm), selectivities of 30–40% methanol with 5–10% methane conversion [189]. At high pressure, the gas-phase radical reactions are predominant, limiting the expected favorable effects of solid catalysts. Selectivity to methanol can only be moderately influenced by controlling factors affecting the radical reactions such as reactor design and shape and residence time of the reactants.

At about ambient pressure (1 atm), the catalysts can play a crucial role in the partial oxidation of methane with O_2. A large number of them, primarily metal oxides and mixed oxides, have been studied. Metals were also tested but they tend to favor complete oxidation. In most cases the reactions were performed at temperatures from 600 to 800 °C, and formaldehyde (HCHO) was obtained as the predominant (and often only) partial oxidation product. Silica itself exhibits a unique activity in the oxidation of methane to formaldehyde. Higher methane conversions were however obtained with silica-supported molybdenum (MoO_3) and vanadium (V_2O_5) oxide catalysts. The yield of formaldehyde produced remained nevertheless in the range of 1 to 5%. An iron molybdate catalyst, Fe_2O_3 $(MoO_3)_{2.25}$, was found to be most active catalyst for partial oxidation of methane, with a reported formaldehyde yield of 23%. Reaction over a silica-supported PCl_3-$MoCl_5$-R_4Sn catalyst gave 16% yield. More recently, a high selectivity of 90% to oxygenated compounds (CH_3OH and HCHO) was obtained at methane conversion of 20–25% in an excess of steam, using a silica-supported MoO_3 catalyst. In most cases, however, such high yields were difficult to reproduce by other research groups, and yields of oxygenates generally do not exceed 2–5%. At the high temperatures used for partial oxidation of methane, methanol formed on the catalyst's surface is rapidly decomposed or oxidized to formaldehyde and/or carbon oxides, which explains its absence from the obtained product mixture.

As methanol is formed together with formaldehyde and formic acid in the oxidation of methane, it has recently been found by Olah and Prakash that this mixture can be further processed in a second treatment step without the need of prior separation, resulting in a substantial increase in methanol content and making the overall process more selective and practical for the production of methanol [192].

The initial oxidation of methane to a mixture of methanol, formaldehyde and formic acid can utilize any known catalytic oxidation procedure. The formation of CO_2, however, is to be kept to a minimum. In order to minimize overoxidation, the overall amounts of oxygenated products are kept at relatively low levels (20–30%), with unreacted methane being recycled.

The subsequent treatment of the formaldehyde and formic acid formed in the oxidation step, without any separation can be carried out in different ways. One is by dimerizing the formed formaldehyde over TiO_2 or ZrO_2 [189] to give methyl formate. Formaldehyde can also undergo conversion over solid-base catalysts such as CaO and MgO, a variation of the so-called Cannizaro reaction, giving methanol and formic acid. These readily react with each other to form methyl formate.

$$2 \text{ HCHO} \xrightarrow{\text{TiO}_2 \text{ or ZrO}_2 \text{ catalyst}} \text{HCOOCH}_3$$

$$2 \text{ HCHO} \xrightarrow{\text{H}_2\text{O}} \text{CH}_3\text{OH} + \text{HCO}_2\text{H}$$

$$\text{CH}_3\text{OH} + \text{HCO}_2\text{H} \xrightarrow{-\text{H}_2\text{O}} \text{HCOOCH}_3$$

The methyl formate obtained can then be catalytically hydrogenated or electrochemically reduced using suitable electrodes (made from copper, tin, lead, etc.), giving two molecules of methanol with no other byproduct.

$$\text{HCOOCH}_3 + \text{H}_2 \longrightarrow 2 \text{ CH}_3\text{OH}$$

Formic acid formed during the oxidation can itself serve as a hydrogen source to be reacted with formaldehyde in aqueous solutions at 250 °C or over suitable catalysts to produce methanol and CO_2.

$$\text{HCHO} + \text{HCO}_2\text{H} \longrightarrow \text{CH}_3\text{OH} + \text{CO}_2$$

Suitable combinations of these reactions allow the oxidative conversion of methane to methanol with overall high selectivity and yield, considering that CO_2 formed as byproduct is to be recycled to methanol (*vide infra*). As methanol is readily dehydrated to dimethyl ether, the oxidative conversion of methane to methanol is also well suited to produce dimethyl ether for fuel or chemical applications.

Avoiding any additional steps, and producing directly methanol from methane remains a most desirable goal. Catalysts able to activate methane at lower temperatures are much sought after for the direct and selective synthesis of methanol from methane and air (oxygen).

A new development for the oxidation of methane is the use of O_2-H_2 gas mixtures. With $FePO_4$ as a catalyst, methanol is formed as the main product in the presence of O_2-H_2 at temperatures below 400 °C [189]. Nitrous oxide (N_2O) was also reported to be effective in producing both methanol and formaldehyde in the presence of silica-supported MoO_3 and V_2O_5 catalysts. Iron-containing catalysts also exhibited very unique properties in the partial oxidation of methane. Using the above-mentioned $FePO_4$ catalyst, partial oxidation of methane with N_2O, gave methanol with a remarkable selectivity close to 100% in the presence of H_2 at 300 °C. The yield is however low, with only 3% methanol obtained at 450 °C. The O^- ion formed by the decomposition of N_2O at the catalyst's surface was postulated to be the active species initiating the activation process of methane. Even if N_2O turns out to be a suitable oxidizing agent, its application on an industrial scale for methanol production is unlikely due to the costs associated with its production and its high activity as a greenhouse gas. Oxygen of the air is, and will remain, the most affordable, ubiquitous – and therefore preferred – oxidizing agent.

Liquid-Phase Oxidation of Methane to Methanol

In order to minimize the formation of side products and increase the selectivity to methanol, the use of lower reaction temperatures ($<250\,°C$) is preferable. Currently used catalysts for methane oxidation are, however, not sufficiently active at lower temperatures. The development of a new generation of catalysts which could produce selectively methanol directly from methane in high yields at lower temperatures is therefore highly desirable. As discussed, gas-phase reactions have met with limited success and require generally higher temperatures ($>400\,°C$). This has prompted much investigation of liquid-phase reactions operating at more moderate temperatures.

During the 1970s it was first observed by Olah and co-workers that methanol could be obtained through electrophilic oxygenation (i.e., oxygen functionalization) of methane. When reacted in superacids (acids many million times stronger than concentrated sulfuric acid) at room temperature with hydrogen peroxide (H_2O_2), methane produces methanol in high selectivity. From the results obtained, it was concluded that the reaction proceeds in strong acids through the insertion of protonated hydrogen peroxide, $H_3O_2^+$ into the methane C–H bond. The formed methanol in the strong acid system is in its protonated form, $CH_3OH_2^+$ and is thus protected from further oxidation; this accounts for the high selectivity observed. The use of H_2O_2 is, however, not suited for the large-scale production of methanol, as is generally the high cost of liquid superacids [191]. Studies are continuing with cheaper peroxides and strong acid systems.

The concept of chemically protecting the methanol formed in the oxidation of methane was successfully further developed by Periana and co-workers, using predominantly metals and metal complex catalysts dissolved in sulfuric acid or oleums [193]. Several of these homogeneous catalytic systems, able to activate the otherwise very unreactive C–H bonds of methane at lower temperatures and with surprisingly high selectivity, have been developed. At temperatures around $200\,°C$, the conversion of methane to methanol by concentrated sulfuric acid using a $HgSO_4$ catalyst was found to be an efficient reaction. The oxidation of methane produces methyl hydrogen sulfate (CH_3OSO_3H), which can be hydrolyzed to methanol in a separate step. At a conversion of 50%, an 85% selectivity to methyl hydrogen sulfate was achieved [194]. To complete the catalytic cycle, Hg^+ is reoxidized to Hg^{2+} by H_2SO_4. Overall, the process uses one molecule of sulfuric acid for each molecule of methanol produced. SO_2 generated during the process can be easily oxidized to SO_3 which, upon reaction with H_2O, will give H_2SO_4 that can be recycled.

$$CH_4 + 2H_2SO_4 \xrightarrow{Hg^{2+}} CH_3OSO_3H + 2H_2O + SO_2$$

$$CH_3OSO_3H + H_2O \longrightarrow CH_3OH + H_2SO_4$$

$$SO_2 + 1/2\,O_2 + H_2O \longrightarrow H_2SO_4$$

$$CH_4 + 1/2\,O_2 \longrightarrow CH_3OH \qquad \text{(Overall reaction)}$$

The cleavage of methyl hydrogen sulfate to methanol and its separation from sulfuric acid media is, however, an energy-consuming process. Furthermore, the use of poisonous mercury makes the process also somewhat unattractive. Interest has therefore shifted towards other less toxic metals. Systems incorporating platinum, iridium, rhodium, palladium, ruthenium and others have been tested for homogeneous methane oxidation. Recently, even gold was found to be capable of catalyzing the oxidation of methane to methanol [195, 196]. However, the best results were obtained with a platinum complex in H_2SO_4.

With this system, methane was converted into methyl hydrogen sulfate, and subsequently, methanol with yields in excess of 70% and selectivities to methanol over 90% [193].

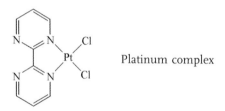 Platinum complex

Water formed during the reaction accumulates progressively in sulfuric acid, decreasing its acidity. This makes the process problematic, as the activity not only of the platinum but also of mercury and most other systems tested, is rapidly decreasing at lower acidities and can be inhibited by high water content. For this reason and others, the conversion is advantageously conducted in oleum (a mixture of H_2SO_4 and SO_3). SO_3 reacts with the water formed to give H_2SO_4, avoiding a decrease in acidity. The use of a catalyst system exhibiting high activity, even at lower acidities, would be preferable however. A gold-based system using selenic acid as the oxidant has also been developed and shows promise [193].

In order to replace the high-cost platinum and other noble metal-based catalysts, the development of a less expensive, but still effective and selective catalyst would also be desirable.

Although our understanding of the reaction mechanism for the direct oxidation of methane in homogeneous systems has greatly improved over the years, many key questions have still to be solved. There is also need to develop more active, selective and stable catalysts before a commercial process based on this technology becomes a reality. Today, however, this goal appears increasingly achievable.

Methanol Production through Mono-Halogenated Methanes

Another possible alternative for the selective conversion of methane to methanol proceeds through the intermediate catalytic formation of methyl chloride or bromide (CH_3Cl, CH_3Br), which are then hydrolyzed to methanol (or dimethyl ether) with the byproduct HCl or HBr being reoxidized [197, 198].

$$CH_4 \xrightarrow{X_2} HX + CH_3X \xrightarrow{H_2O} CH_3OH \ (or \ CH_3OCH_3) \ + \ HX$$

$$2HX \xrightarrow{1/2 \ O_2} X_2 + H_2O$$

$$X = Cl, Br$$

The chlorination of paraffins, discovered by Dumas in 1840, is the oldest known substitution reaction and is practiced on a large scale in industry. It is usually a free radical process initiated either thermally or photochemically (by heat or light). The major problem with free radical reactions, also experienced in oxidative methane conversion, is the lack of selectivity. In the case of methane chlorination, all four chloromethanes (CH_3Cl, CH_2Cl_2, $CHCl_3$ and CCl_4) are generally obtained.

$$CH_4 \xrightarrow{Cl_2} CH_3Cl \xrightarrow{Cl_2} CH_2Cl_2 \xrightarrow{Cl_2} CHCl_3 \xrightarrow{Cl_2} CCl_4$$

As methyl chloride is chlorinated more rapidly than methane itself under free radical conditions, a high ratio of methane to chlorine of at least 10:1 is required to obtain methyl chloride as the main desired product. Chlorination of methane has also been reported over catalysts such as active carbon, Kieselguhr, pumice, alumina, kaoline, silica gel and bauxite, but showed limited selectivity due to the free radical nature of the reaction. In industry, the strongly exothermic chlorination reaction is generally conducted in the absence of a catalyst at 400–450 °C without external heating under slightly elevated pressures.

In the 1970s, it was observed by Olah et al. that chlorination of methane with superacidic SbF_5/Cl_2 at low temperature gives methyl chloride in high selectivity with only small amounts of methylene chloride and no chloroform or carbon tetrachloride [199,200]. The high selectivity obtained in the electrophilic reaction contrasted strongly with the low selectivity for radical chlorinations. As, however, the conversions were low, no adaptation of the reaction for practical use was made.

In extending the electrophilic halogenation of methane to catalytic heterogeneous gas-phase reactions during the 1980s, it was shown that methyl chloride could be formed with high selectivity and acceptable yields [197]. The catalytic monohalogenation (chlorination or bromination) of methane was achieved over either solid superacids such as $TaF_5/Nafion-H$, $SbF_5/graphite$, supported metals such as Pt/Al_2O_3 and $Pd/BaSO_4$, or supported oxychloride and oxyfluoride such as $ZrOF_2/Al_2O_3$ and GaO_xCl_y/Al_2O_3. The reactions were carried out at temperatures between 180 and 250 °C, giving 10 to 60% conversion with selectivities to methyl chloride (bromide) generally exceeding 90%. Methylene halides (CH_2Cl_2, CH_2Br_2) were the only higher halogenated methanes formed and no haloforms ($CHCl_3$, $CHBr_3$) or carbon tetrahalides (CCl_4, CBr_4) were observed.

In the halogenation of methane, hydrogen halides (HCl or HBr) are formed as equimolar byproducts. Similarly hydrolysis of methyl halides also gives hydrogen halides. Their recycling is essential to utilize halogens only as a catalytic agent in the overall conversion of methane into methanol. The oxidation of hydrogen chloride to chlorine is industrially possible (Deacon process, Kellogg's improved Kel-Chlor process, Mitsui Toatsu Chemicals MT Chlor process), but it remains technologically difficult. An improved way to effect the reaction was also found by Benson and co-workers [201, 202]. In contrast, HBr is readily oxidized to bromine by air. Its reoxidation and continuous recycling thus is more easily achieved.

Combination of the selective monohalogenation (preferentially bromination) of methane with subsequent catalytic hydrolysis to methanol/dimethyl ether was shown by Olah and co-workers to be an attractive alternative route to the preparation of methyl alcohol without syn-gas. Methyl halides and/or methyl alcohol/dimethyl ether obtained directly from methane also offer a way to convert methane via zeolite or bifunctional acid-base-catalyzed condensation into ethylene and propylene, starting materials for synthetic hydrocarbons and their products (see Chapter 13). Related processes were developed by a research group at U.C. Santa Barbara [203] and Dow Chemicals [204].

In order to produce methanol from methane in a single step, combining halogenation, hydrolysis and reoxidation of hydrogen halide byproduct using a $Br_2/H_2O/O_2$ or $Cl_2/H_2O/O_2$ mixture can also be achieved, although to date the conversions remain modest.

Iodine was recently found by Periana to be a suitable catalyst for the selective conversion of methane in oleum [205]. At about 200 °C, methyl hydrogen sulfate was formed with up to 45% yield and 90% selectivity, and subsequently hydrolyzed to methanol. The intermediate formation of methyl iodide, followed by conversion to methyl hydrogen sulfate and reoxidation of HI to I_2 can be assumed, not unlike the chlorine- and bromine-catalyzed conversions described above.

$$CH_4 + 2 SO_3 \xrightarrow{I_2} CH_3OSO_3H + SO_2$$

The use of iodine for the conversion of methane to methanol through methyl iodide is consequently also of interest.

Methanol production from fossil fuels, as long as they are readily available, will remain a major source of methanol (Fig. 12.6). Improved methods of methane conversion without going through syn-gas are developed and will be used. Besides natural gas, methane hydrates and other sources will also be utilized. Biological production of methanol and chemical reductive recycling of CO_2 into methanol (vide infra) will, however, increasingly become important.

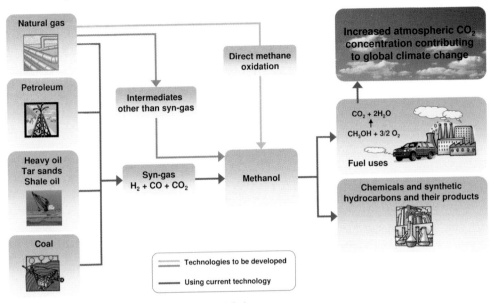

Figure 12.6 Methanol from fossil fuel resources.

Microbial or Photochemical Conversion of Methane to Methanol

Certain bacteria, known as methanotrophs, can obtain all the energy and carbon they need for life from methane [187]. The key step for their utilization of methane is its selective conversion to methanol using oxygen. In subsequent biological processes, methanol is further oxidized to formaldehyde, which in turn can be either incorporated into biomass or oxidized to CO_2, thereby providing the energy needed by the bacteria. Nature's catalyst for the conversion of methane to methanol is an enzyme called methane monooxygenase (MMO), which can operate in aqueous solution at ambient temperature and pressure [206]. In such a system composed of a number of imbricated proteins, the reactivity is controlled by molecular recognition and the regulation characteristics of the enzyme, allowing a virtually complete selectivity to methanol. The conversion is achieved by reductive activation of O_2 with a reductant known as NADH.

$$CH_4 + NADH + H^+ + O_2 \longrightarrow CH_3OH + NAD^+ + H_2O$$

Two varieties of MMO systems have been found in methanotrophic organisms. The first, which contains copper, is a membrane-bound enzyme which has proven difficult to isolate and eluded full characterization. Most attention has thus been focused on the soluble, iron-containing MMO, especially from the species *Methylosinus trichosporium* and *Methylococcus capsulatus*. The active site in this enzyme which is directly responsible for methane oxidation contains a pair of iron atoms. Given the ability of MMO to activate methane at room temperature and ambient pressure, much research as been devoted in trying to reproduce such activity for the large-scale production of methanol. Our understanding of the reaction mechanisms taking place in MMO has greatly improved over the years, and simpler catalyst systems which try to model the behavior of MMO have been developed and tested. However, because of their complexity, the direct use of MMO enzymes has proven difficult to apply for the practical production of methanol. Besides MMO systems and their ability to oxidize specifically the simplest alkane, methane, enzymes of the cytochrome P-450 family, with a less complicated structure than MMO, have been found to catalyze the oxidation of many types of hydrocarbons to alcohols [207]. At the heart of the cytochrome P-450 is a porphyrin system containing at its center a metal atom, most commonly iron, catalyzing the oxidation reaction. Although to date no P-450 system has been shown capable of oxidizing methane, significant advances have been made in that direction through the genetic engineering of enzymes based on P-450, specifically designed for high activity towards small alkanes. In attempting to approach Nature's ability to activate methane under mild conditions, methanol has recently also been obtained by reacting methane with oxygen and H_2O_2 in water at temperatures below 75 °C, using a vanadium-based catalyst [208]. The finding of a practical, highly efficient and selective catalyst based on Nature's example is, however, still a long term goal. It must also be pointed out that MMO is by far not a perfect system. When O_2 is used as the oxidant, it requires the presence of a reductant, NADH. Alternatively, a reduced form of O_2, for example H_2O_2, is generally used in model experiments. In practical applications however, this would most probably imply the use of hydrogen as a reducing agent, making the process less attractive. Ideally, methane should be oxidized by only O_2 to form methanol.

Methanol from Biomass

Although reserves of methane from natural gas are still very large, they are nevertheless finite and diminishing. The exploitation of unconventional natural gas resources and methane hydrates could significantly increase the amounts of methane available to mankind. Innovative processes continue to be explored for the transformation of methane into easy-to-handle liquid fuels, primarily methanol, but this will not solve the problem of increasing CO_2 concentrations in the atmosphere and its detrimental effect on the global climate. Even if methane does release less CO_2 than coal or petroleum, its increased utilization to fill the world's growing energy needs will still produce large quantities of CO_2. Thus, new ways are needed to fulfill mankind's ever bigger appetite for energy, without adversely affecting our environment. In the long term therefore, methanol will have to be produced from sources that release only minimal quantities of CO_2, or even no CO_2. In this respect, the use of biomass is one possibility, although natural resources will be unable to cover all our needs.

Biomass is referred to as any type of plant or animal material – that is, materials produced by life forms. This includes wood and wood wastes, agricultural crops and their waste byproducts, municipal solid waste, animal waste, and aquatic plants and algae. Methanol was originally made from wood (hence the name wood alcohol), but due to its inefficiency and the advent of synthesis via syn-gas, this route was rapidly abandoned in the first half of the 20th century. Increased oil and natural gas prices, dependence on foreign countries for energy supply, growing concerns about CO_2-induced climate changes as well as the depletion of easily accessible fossil fuel resources, have prompted many to reconsider a wider and more extensive use of biomass to fulfill our energy demands. Since biomass itself is both bulky and heterogeneous, one way to achieve this goal would be to convert it into a convenient liquid fuel, namely methanol. The modern production of methanol from biomass is very different, however, and is much more efficient than the methods used a century ago. Basically, not only wood but also any organic (i.e., carbon-containing) material obtained from living systems can be used in the process. Technologies to transform biomass into methanol are generally similar to those used to produce methanol from coal. They imply the gasification of biomass to syn-gas, followed by the synthesis of methanol with the same processes employed in plants based on fossil fuels.

Before the gasification step, the biomass feedstock is usually dried and pulverized to yield particles of a uniform size and with moisture content no higher than 15–20% for optimal results. Gasification is a thermochemical process that converts biomass at high temperature into a gas mixture containing hydrogen, carbon monoxide, carbon dioxide, and water vapor. The pretreated biomass is sent to the gasifier where it is mixed, generally under pressure, with oxygen and water. There, a part of the biomass is burned with oxygen to generate the heat necessary for gasification. The combustion gases (CO_2 and water) react with the rest of the biomass to produce carbon monoxide and hydrogen. With biomass being used as the heating fuel, no external heat source is necessary. The production of syn-gas from biomass in a single-step operation by partial oxidation is very attractive, but it has experienced technical problems. Gasification of biomass feedstocks is therefore generally a two-step process. In the first stage, called "pyrolysis" or "destructive distillation", the dried biomass is heated to temperatures ranging from 400 to 600 °C in an atmosphere too deficient in oxygen to allow complete combustion. The pyrolysis gas obtained consists of carbon monoxide, hydrogen, methane, volatile tars, carbon dioxide, and water. The residue, which is about 10–25% of the original fuel mass, is charcoal. In the second stage of gasification, called "char conversion", charcoal residue from the pyrolysis step reacts with oxygen at temperatures from 1300 to 1500 °C, producing carbon monoxide. Before being sent to the methanol production unit, the syn-gas obtained must be purified. Compared to most coals, the advantage of biomass is its much lower sulfur content (0.05–0.20 wt.%), whilst heavy metals impurities (mercury, arsenic, etc.) are present only in insignificant quantities.

Whilst the production of methanol is possible on a small scale, large-scale methanol plants are preferable because they induce substantial economies of scale, thereby lowering the cost of production and increasing efficiency. The quantities of biomass needed to feed a world-scale 2500 tonnes per day methanol plant are, however, very large (in the order of 1.5 million tonnes per year). To deliver such amounts, biomass would have to be collected over vast areas of land [209]. The transportation of bulky biomass products with low energy density over long distances is not economical, and consequently its transformation to an easy-to-handle and store liquid intermediate through fast pyrolysis has been proposed. In fast pyrolysis, small biomass particles are heated very rapidly to 400–600 °C at atmospheric pressure to yield oxygenated hydrocarbon gases. The generated gases are then rapidly quenched to avoid their decomposition by cracking. The black liquid obtained, called "biocrude" [210,211] (because of its resemblance to crude oil), has a wide range of possible applications. It can be processed into a replacement for fuel oil and used directly in furnaces for heat or electricity production. By altering its chemical composition, biocrude could also be combined and processed alongside petroleum crude in modified refineries. Although further development is required, biocrude holds promise as a domestically available synthetic alternative to petroleum. Besides biocrude, which is obtained in 70–80% yield, fast pyrolysis also produces some combustible gases and char, a fraction of which is used to drive the process. A fraction of the char could also be finely ground and added to the biocrude to form a slurry; this, like biocrude, might be pumped, stored and transported via road, rail, pipelines and tankers in much the same way as crude oil today, facilitating greatly the handling of biomass feedstock. Fast pyrolysis is a relatively simple, low-pressure and moderate-temperature process which is applicable on both large and small scales, though having been introduced only recently, further development is required for its commercial application. In contrast, the production of syn-gas from liquid feedstocks from diverse fossil origins is a well-established process which is used in numerous plants worldwide. The production of methanol in world-scale plants (>2500 tonnes per day) through syn-gas obtained from biocrude generated in delocalized plants is therefore a feasible and a promising option.

All biomass materials can be gasified for methanol production (Fig. 12.7). The efficiency of the process will, however, depend upon the nature and quality of the feedstocks. Low-moisture materials such as wood and its byproducts and herbaceous plants and crop residues are the most suitable. Using wood and forest residues, efficiencies as high as 50–55% have been recorded. Several demonstration projects are under way in different parts of the world. In Japan for example, Mitsubishi Heavy Industries is operating a pilot plant for the production of methanol from biomass [212]. Various feedstocks such as Italian ryegrass, rice straw, rice bran and sorghum have been successfully tested, with the best yields having been obtained with sawdust and rice bran. In the United States, the Hynol process originally developed at the Brookhaven National Laboratory was tested on a pilot scale [213], and was successful in converting materials such as woodchips into methanol. In Europe, the paper industry has turned its attention to black liquor gasification as a possible source for methanol [214]. Black liquor is a pulp-rich slurry obtained as byproduct of the Kraft pulping operation used for paper

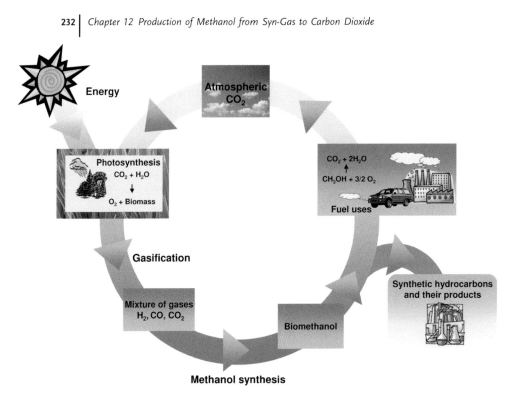

Figure 12.7 The carbon neutral cycle of biomethanol production and uses.

production, and contains about half of the organic material that was originally in the wood. It has the great advantage of being already partially processed and in a pumpable, liquid form. With the gasification process, 65–75% of the biomass energy contained in black liquor can be converted to methanol. Worldwide, the pulp and paper industry currently produces about 170 million tonnes of black liquor each year. The United States alone, could potentially produce 28 millions tonnes of methanol per year from black liquor. Clearly, since the United States consumption of gasoline and diesel fuels totals the equivalent of more than 1000 million tonnes of methanol, this would replace only a small fraction of the petroleum-derived fuels. In smaller countries with an important paper industry and limited population (e.g., Finland and Sweden), methanol from black liquor could replace a substantial part of the motor fuel demand (50% and 28%, respectively). Municipal solid waste (about 240 million tonnes are produced each year in the United States) is another possible feedstock for methanol – implying that landfills could actually become energy fields of the future.

Although waste products from wood processing, agricultural residues and by-products, as well as solid municipal waste, represent suitable feedstocks for methanol production in the near term, the quantities which can be generated from these resources are limited. In the long term, the growing demand for bio-methanol will necessitate a larger and reliable source of raw biomass material. Thus, crops selected specifically for energy purposes would have to be cultivated on a large scale if a significant amount of methanol were to be produced from biomass resources. Suitable "energy crops" are being identified, and the most promising – essentially fast-growing grasses and trees – are being field tested. Depending upon the climate, the species giving the best results are logically different. The potential impact of the large-scale culture of energy-devoted plants on ecosystem, soil erosion, water quality and wildlife is also being assessed. With regard to wood, fast-growing species such as poplar, sycamore, willow, eucalyptus and silver maple have been proposed [215] for cultivation in "short rotation" tree farms in the United States and Europe. In conventional silviculture for plywood or paper production, trees are grown for 20 to 50 years, or more. The term "short rotation" means that fast-growing trees planted at close spacing are harvested after only four to 10 years in order to maximize productivity since, after harvesting, new trees can be planted. Most of the trees under consideration, however, have the ability to resprout from cut stumps in a process called "coppicing", thus eliminating the need to replant after each harvest. Such intensive tree culture has shown to produce generally between 4 and 10 tonnes of oven-dry wood per hectare and year under temperate climates. Higher rates can be obtained in tropical climates with appropriate species, such as eucalyptus in Florida, Hawaii and South America. Besides wood, energy crops such as switchgrass, sorghum or sugar cane have been identified as having a high efficiency in converting solar energy into biomass. To contain the cost of planting, growing and harvesting energy crops, highest productivity is expected. In most cases, this means that prime agricultural land would have to be used for their production. In large countries with a low population density such as the United States, Australia or Brazil, a significant portion of idle food-crop lands, pasture range and forest land could be used for energy crops. However, in more densely populated areas such as Western Europe, Japan or China, most of the arable land is already in use for food production. Unless surplus land is available, it is therefore highly likely that the land needed for energy crops will directly compete with food and fiber production.

As with any form of plant life, high and sustainable productivity for energy crops over the years will only be maintained if the necessary nutrients for the plant's growth are present in the soil. In intensive agriculture, as practiced in most developed countries, this implies the massive use of fertilizers. Nitrogen fertilizers in particular, obtained from NH_3 derived from natural gas, if used for energy crops, could partially offset the benefits expected from the production of fuels from biomass. With nitrogen fertilizers on cultivated soils and animal wastes being responsible for 65–80% of the total emissions of N_2O (a powerful greenhouse gas, 296-fold more damaging than CO_2 [216]), a judicious choice of energy crops requiring minimum input of fertilizers is therefore necessary. Crop rotation, which is a widely accepted practice in agriculture to fight soil impoverishment, should also be applied. Although environmental risks must still be carefully assessed, biotechnology and genetic modifications are also expected to provide additional gains in crop yields.

From an industrial viewpoint, the production of energy crops in vast monocultures including only one species would be preferable. Such types of culture are, however, generally more prone to epidemics of pests and diseases, requiring frequent applications of pesticides and fungicides in order to keep them free from potentially disastrous infestations that might wipe out the entire plantation. In order to avoid this situation, several types or species of energy crops could be planted instead of only one; this would also allow a higher degree of biodiversity. At the same time, the introduction of nitrogen-fixing trees such a alder or acacia in fast-growing poplar and eucalyptus plantations could also help to reduce fertilizer requirements [117].

In avoiding the use of prime farmland, energy crops capable of being grown on marginal or degraded lands have also attracted much attention as a means of adding value to vast areas unsuitable for most other agricultural purposes. The mesquite tree and its sugar-rich pods for example, already thrives in the United States on tens of millions of hectares of semi-desert land which, without proper irrigation, would otherwise be of little value for the production of food crops or as pasture land. Due to its extensive root system, which also helps to maintain the soil in place and to protect the land from erosion, the mesquite tree is able to reach far into the ground to capture water. Also present in the root systems are bacterial nodules that fix the nitrogen contained in air and provide not only the mesquite tree but also its neighboring plants with the nutrients necessary for growth. Although the productivity of plants such as the mesquite trees is less than for fast-growing poplar or eucalyptus, they have the advantage of growing, without irrigation needs, on vast areas of land that would otherwise be of limited value.

In Australia, the National Science Agency devised a computer model to show that 30 million hectares of fast-growing trees planted over the next 50 years could produce enough methanol to replace 90% of the oil requirements of the country's 25 million people by 2050. It would, by the same time, create numerous jobs in rural areas and help to reduce the impact of salt that otherwise would be brought to the surface by rising water tables.

The European Union, with a population of over 450 million after its expansion to 25 member states, has much less potential for energy crops cultivation. Of the almost 400 million hectares total land area, about 170 million hectares are currently used for food production, while another 160 million hectares are covered by forests. Most of the remaining land, is classified as unsuitable for the production of biomass (mountains, deserts, degraded lands, urban areas, etc.). In order to remain self-sufficient, only a small fraction of the land currently used for food production could be converted to energy crops. Forests are another potential source for large-scale methanol production from biomass in Europe. They could be used directly for wood production or be cleared to enable agriculture, but due to their importance for the environmental balance and biodiversity, transforming a large proportion of the forests into vast, highly standardized and mechanized tree plantations would run contrary to the sustainable use of natural resources. When considering possible feedstocks, biomass is thus not expected to represent more than 10–15% of Europe's total energy demand in the future.

In those developing countries already struggling to produce the necessary food for their growing population, the production capacity for energy crops is clearly limited.

Methanol from Biogas

Most mammals (including humans), as well as termites, produce a flammable gas termed "biogas" when they digest their food. Biogas is also generated in wetlands, swamps and bogs where large amounts of rotting vegetation may accumulate. Biogas is formed when anaerobic bacteria break down organic material in the absence of oxygen. Anaerobic bacteria are some of the oldest forms of life on Earth, having evolved before the photosynthetic processes of green plants were able to release large quantities of oxygen into the primitive atmosphere.

Biogas is in fact a waste product of those microorganisms, composed mainly of methane and CO_2 in variable proportions, and trace levels of other elements such as hydrogen sulfide (H_2S, rotten egg smell), hydrogen, or carbon monoxide. The biological process for biogas generation by anaerobic methanogenic bacteria, termed "methanogenesis", was discovered in 1776 by Alessandro Volta, and has been used since the 19th century in so-called "anaerobic digestion reactors". The process was used quite extensively in Europe when energy supplies were reduced during and after World War II. Interest in biogas was then revived after the oil crisis of the 1970s. Most common feedstocks are animal dung, sewage sludge and wastewater or municipal organic waste. The use of anaerobic digesters to treat industrial waste water has also rapidly gained popularity during the past decade. Wastewaters treated include those from food, paper and pulp, fiber, meat, milk, brewing and pharmaceutical plants. In Brazil, anaerobic digesters are used to treat the waste liquid (called vinasse) which is co-produced during ethanol manufacture from sugar-cane. Anaerobic digestion is thus an important method of reducing the impact of waste disposal on the environment [217].

Anaerobic digester can be constructed from concrete, steel, bricks, or plastic, they may be shaped like silos, basins or ponds, and they may be placed either underground or on the surface. In large-scale operations, organic materials are constantly fed into a digestion chamber, and the gas produced is allowed to exit through a gas outlet placed on the top of the reactor. The sludge left behind in the digester, rich in nutrients, can be used as a soil fertilizer or conditioner.

The gas produced in anaerobic digesters, depending on the feedstock and effectiveness of the process, consists of 50–70% methane with the remainder being mostly CO_2. At present, the gas is mainly used to produce electricity or heat. Compressed methane obtained from biogas (basically the same as CNG) is also used as an alternative transportation fuel in light- and heavy-duty vehicles, notably in Brazil. Today, Sweden runs some 800 buses on biogas and has recently unveiled a small biogas train powered by two biogas bus engines. After the removal of impurities (especially hydrogen sulfide), biogas could also be used directly for the production of methanol in the same way as natural gas. A fraction of the CO_2 already present in the gas could react with the excess hydrogen generated during methane steam reforming in order to form more methanol. Using this technology, the manure of 250 000 pigs is currently planned to be used in Utah to produce some 30 000 L of methanol per day.

Another source for biogas production is landfills which, worldwide, is the dominant method for municipal solid waste disposal. As in anaerobic digesters, the organic material present in the landfills is decomposed in the absence of oxygen by anaerobic bacteria. The resultant gas comprises 50–60% methane, 40–50% carbon monoxide, and traces of H_2S as well as other volatile organic compounds. The landfill gas, which for safety reasons is otherwise vented to the atmosphere or is flared, can also be collected through elaborate underground piping for use as a fuel or feedstock for methanol production.

In the United States, 240 million tonnes of municipal solid waste were produced in 2003. Of this, 30% was recycled or composted, 14% incinerated, and 56% landfilled. From the more than 1700 landfills currently in operation, close to 400 have operational landfill gas utilization units in place [218]. Most of the gas produced is used to generate electricity or produce heat. The production of methanol from landfill gas is also possible, and a plant using this technology, with a capacity of 15 000 tonnes per year, is planned to be constructed in Ohio. A molten carbonate fuel cell (MCFC) that can operate on methanol from municipal organic waste of the city of Berlin, Germany, has also been in use since 2004 (Fig. 12.8).

It is important to note, however, that in 2004 the combined production of electricity from all landfill gas energy projects presently installed in the United States was only 9 billion kWh [218] – similar to the output of a single large-scale nuclear reactor, or 0.2% of the national electricity production! While landfill gas utilization provides a good means of controlling methane greenhouse gas emissions from municipal solid waste, it has a very limited capacity for energy or methanol supply.

Figure 12.8 The HotModule from MTU: The first bi-fuel molten carbonate fuel cell (MCFC) went into service at Vattenfall Europe AG in Berlin in September 2004. It is fuelled with natural gas or methanol or any mixture of these two. The methanol used is derived from waste generated locally. (Courtesy: MTU Friedrichshafen GmbH 2004.)

Aquaculture

Water, in the form of oceans, lakes, rivers and wetlands, covers more than 70% of our planet. Thus, although much less attention has been given to aquatic plants and organisms for the production of energy, those biomass sources could have considerable potential.

Water Plants

In freshwater, plants growing partially submerged, such as water hyacinth and cattail, have been identified as promising feedstock for energy applications. Water hyacinth, which is native to South America, is a free-floating herb which forms dense mats on the surface of the water. Introduced intentionally or by accident in most tropical and subtropical parts of the world, it has become a floating nightmare for the affected countries. In Africa it infests every major river and almost every lake. In the United States, it flourishes in countless bodies of water throughout the South, from California to Virginia (Fig. 12.9). Dense and rapidly expanding growth of water hyacinth can clog canals and water intakes, restrict navigation along rivers and lakes, and exclude native vegetation. In some areas it must be continually cleared from vital waterways. It has a very high annual productivity of 30–80 tonnes dry biomass per hectare.

Cattail (Fig. 12.10), which thrives in marshland and lakes throughout the planet, is also viewed as a good candidate for energy production, because it can

Figure 12.9 Water hyacinth infesting waterways in Southern United States.

Figure 12.10 Cattail.

yield annually up to 40 tonnes of biomass per hectare. Once collected, the water hyacinth, cattail or any other suitable water plant could be processed by anaerobic conversion to form methane and subsequently methanol.

Algae

Found in both marine and freshwater environment, algae have also been studied as potential sources of energy. Macroalgae – more commonly known as seaweed or kelp – are fast-growing plants that can reach a considerable size (up to 60 m in length) [219]. A few experimental seaweed farms were built along the coast of southern California in the 1970s, but due to difficult weather conditions and rough waters, the open-sea project was rapidly abandoned. More protected from the elements, near-shore farming of *Macrocystis algae* gave better results, and yearly biomass production of more than 30 tonnes per hectare were reported (Fig. 12.11).

Another form of algae – microalgae – are microscopic and the most primitive form of photosynthetic organisms (Fig. 12.12). Whilst their mechanism of photo-

Figure 12.11 Kelp forest. (Courtesy: Ocean Brite Systems).

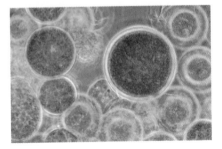

Figure 12.12 Microalgae.

synthesis is similar to that of higher plants, the algae are generally more efficient in converting solar energy to biomass, because of their simple cellular structure and the fact that they grow in aqueous suspension surrounded by water, CO_2 and other nutrients necessary for their growth. Some species of microalgae are very rich in oil, and this can account for up to 50% of their mass. It has been claimed that microalgae are capable of producing 30 times more oil per hectare than terrestrial oilseed crops. This prompted the National Renewable Energy Laboratory (NREL) from the U.S. Department of Energy (DOE) to study these organisms for the manufacture of biodiesel. Experimental production of microalgae was conducted in open, shallow ponds located in regions with high sun exposure in California, Hawaii, and New Mexico. Unlike higher, more complex organisms, microalgae reproduce very rapidly and can be harvested continuously as they are produced. To maintain high growth rates, the ponds were constantly fed with CO_2, water, and other nutrients needed by the algae. For large-scale production in "algae farms", CO_2 could be obtained from a number of sources, especially fossil fuel-burning power plants. The carbon contained in fossil fuel could thus be used twice, reducing global emissions of CO_2. This would provide a practical approach for the recycling of CO_2 into a useable fuel. Japan, which is extremely dependent on energy imports, is actively pursuing this approach to extend its domestic energy production. Ultimately it will, however, only postpone the release of CO_2 into the atmosphere. To be sustainable in the long term, algae should use the CO_2 contained naturally in air and dissolved in water. This will probably considerably lower the yield of biomass production, requiring vast areas of land to produce significant amounts of energy. One of simplest way to use the biomass generated from macro- or microalgae is to produce methane. Because of the high water content of algae, gasification requiring prior drying of the material is not attractive, and anaerobic digestion – which is anyway conducted in water – is therefore more adaptable. The methane generated can than be used in the same way as natural gas to produce methanol. More research is needed, however, to provide a better understanding of the biology of aquatic organisms and to identify the best species and growing conditions for optimal conversion of solar energy into biomass. Genetic engineering could also be used to improve the characteristics of such plants and algae. So far, however, the production of biomass from aquatic plants has proven too expensive to compete with fossil fuels, or even plants grown on land.

Methanol from Carbon Dioxide

When hydrocarbons are burned they produce CO_2 and water. The great challenge is to reverse this process and to produce – efficiently and economically – hydrocarbon fuels and materials from CO_2 and water. Of course Nature, in its process of photosynthesis, captures CO_2 by green plants from the air and converts it with water, using the Sun's energy and chlorophyll as the catalyst, into new plant life. Thus, plant life replenishes itself by recycling atmospheric CO_2. The difficulty is that the conversion of plant life into fossil fuels is a very long process, taking many million years. As we cannot wait that long, we must develop our own chemical recycling processes to achieve it within the required very short time scale. The most promising approach is to convert CO_2 chemically by catalytic or electrochemical hydrogenation to methanol, and subsequently to hydrocarbons. Chemists have known for more than 80 years how to produce methanol from CO_2 and H_2. In fact, some of the earliest methanol plants operating in the 1920 to 1930s were commonly using CO_2 and hydrogen, generally obtained as byproducts of others processes, for methanol production. Efficient catalysts based on metals and their oxides (notably copper and zinc) have been developed for this reaction. These catalysts are very similar to those used presently in the industry for methanol production via syn-gas. In view of the understanding of the mechanism for methanol synthesis from syn-gas, this is not really unexpected. It is now well established, that methanol is most probably formed almost exclusively by hydrogenation of CO_2 contained in syn-gas on the catalyst's surface. In order to be converted to methanol, the CO present in the syn-gas must first undergo a water gas shift reaction to form CO_2 and H_2. The formed CO_2 then reacts with hydrogen to yield methanol [183].

Although further improvements are needed, many methods of producing methanol from CO_2 and H_2 are already known [220, 221, 221a, 221b], and there should be no major technological barrier for their application on a large commercial scale. The limiting factor for the large-scale use of such a process is the availability of the feedstocks, namely CO_2 and H_2. CO_2 can be obtained relatively easily in large amounts from various exhaust sources such as those of fossil fuel-burning power plants and various industrial plants. Eventually, even the CO_2 contained in air can be separated and chemically recycled to methanol and desired synthetic hydrocarbons and their products. Hydrogen is presently mainly produced from non-renewable fossil fuel-based syn-gas, and thus its availability is limited. In view of our diminishing fossil fuel resources, this is not the way of the future. In addition, the generation of syn-gas from fossil fuels releases CO_2, which contributes further to the greenhouse effect.

To address this problem, while fossil fuels are still available improved routes to produce hydrogen from them have been investigated. In a process called "Carnol", which was developed at the Brookhaven National Laboratory, hydrogen and solid carbon are produced by the thermal decomposition of methane [222]. The generated hydrogen is then reacted with CO_2 recovered from fossil fuel-burning power plants, industrial flue gases or the atmosphere to produce methanol. Overall, the net emission of CO_2 from this process is close to zero, because CO_2 released when methanol is used as a fuel was recycled form existing emission sources. The solid carbon formed as a stable byproduct can be handled and stored much more easily than gaseous CO_2, and be disposed of or used as a commodity material, for example in soil conditioning or as a filler for road construction. Clearly, the process depends on the availability of methane:

Methane decomposition: $CH_4 \longrightarrow C + 2H_2$

Methanol synthesis: $CO_2 + 3H_2 \longrightarrow CH_3OH + H_2O$

Overall reaction: $3CH_4 + 2CO_2 \longrightarrow 2CH_3OH + 2H_2O + 3C$

Thermal decomposition of methane occurs when methane is heated at high temperature in the absence of air. To obtain reasonable conversion rates under industrial conditions, temperatures above 800 °C are required. This process has been used for many years, not for the production of hydrogen but of carbon black used in tires and as a pigment for inks and paints. For the primary generation of hydrogen, different reactor designs have been proposed. Attention has recently been focused on reactors operating with a molten metal bath, such as molten tin or copper heated to about 900 °C, into which methane gas is introduced.

Being an endothermic reaction, thermal decomposition of methane requires about 18 kcal mol^{-1} to produce 2 moles of hydrogen from methane, or 9 kcal of energy for the production of 1 mol of hydrogen. As a comparison, this is less than the highly endothermic methane steam reforming reaction, which requires about 49.2 kcal mol^{-1} to produce 4 moles of hydrogen, or 12.3 kcal mol^{-1} hydrogen. It should be clear that, whereas methane steam reforming produces four molecules of hydrogen for every molecule of methane used, methane decomposition, like partial oxidation of methane, yields only two. On the other hand, methane's thermal decomposition byproduct (i.e., carbon) can be easily handled, stored and used without much further treatment. The suppression of CO_2 emissions generated by methane steam reforming or partial oxidation is more difficult and energy-intensive. The CO_2 must first be collected, concentrated and finally transported to a suitable sequestration site; this may be distant from the production site and require an extensive pipeline network. Methane thermal decomposition process is still in its infancy, and will require considerable further investigation to mold it into a mature and efficient technology for commercial application. In a carbon-constrained world which is aiming to reduce greenhouse gas emissions, methane thermal decomposition could be a useful alternative for the generation of hydrogen and methanol, avoiding the problems associated with CO_2 emission and sequestration. It would allow mankind to use extensively the world's remaining natural gas resources, including unconventional sources such as the huge methane hydrates deposits under the ocean floors and arctic tundra (given that an effective means of collecting them is found) with limited effects on the Earth's atmosphere.

Ultimately, however, all of these resources are finite and non-renewable. Over time, they will become depleted or economically too prohibitive to exploit. In the long term, therefore, the large-scale, cost-effective production of hydrogen by electrolysis of water is the key to the development of CO_2 to methanol processes, and thus to the methanol economy. The electrolysis of water to produce hydrogen is a well-developed and straightforward process, and is achieved by applying an electric current between electrodes inserted into water with some electrolytes present. Hydrogen evolves at the cathode, and oxygen at the anode. The electricity needed for the process can be provided by any form of energy. At present, a large part of the electricity produced is derived from fossil fuel, but in the future, in order to be sustainable and environmentally sound, the electric power required for large-scale electrolysis of water should be obtained from atomic energy (fission and later fusion, if proven technically and commercially feasible) and any renewable energy source, including hydro, solar, wind, geothermal, wave and tides. Such production methods of hydrogen, avoiding the emission of CO_2 to the atmosphere, are discussed in Chapter 9.

Although the catalytic reduction of CO_2 with hydrogen is feasible, one drawback is that the process requires 3 moles of hydrogen for every mole of CO_2. As the generation of the required hydrogen is highly energy-consuming, the production of methanol from CO_2 was often considered economically less attractive than its formation from carbon monoxide and hydrogen.

Olah and Prakash have recently found ways of overcoming some of the difficulties of converting CO_2 into methanol. Using electrochemical or photochemical reduction of CO_2, besides methanol, formaldehyde and formic acid can be obtained in high selectivity and good conversions. The formic acid and formaldehyde in a subsequent step, not unlike the previously discussed oxidative conversion of methane, can then be converted to methanol, with formic acid providing the required hydrogen [223, 223a, 223b, 223c].

$$HCHO \ + \ HCO_2H \ \longrightarrow \ CH_3OH \ + \ CO_2$$

Carbon Dioxide from Industrial Flue Gases

The CO_2 necessary for the production of methanol can be obtained from various emissions and, eventually, from the CO_2 content of the atmosphere. Presently, worldwide, more than 20 billion tonnes of CO_2 related to human activities are released into the atmosphere every year. Globally, CO_2 emissions from electricity generation, industries, the transportation sector and heating or cooling of buildings are further increasing. They all contribute to the increase in CO_2 levels in the atmosphere, from 270 ppm before the beginning of the industrial era to 370 ppm today. The resulting greenhouse effect could have a detrimental effect on the Earth's global climate and our ecological systems. For the coming decades, fossil fuels will foreseeably continue to provide the largest share of humanity's energy needs. The Kyoto agreement, although not yet approved by all countries, limits the CO_2 emissions considerably. To reduce CO_2 emissions, mitigation technologies must be developed and enforced. In this respect, more energy-efficient technologies and conservation can help, but this will not be sufficient to stop the global increase of CO_2 emissions. In order to make a significant reduction in emissions, the recovery of CO_2 from fossil fuel-burning sources will be necessary. CO_2 can be obtained most readily from concentrated sources such as flue gases from fossil fuel-burning power plants, containing typically between 10–15% CO_2 by volume. Many industrial exhausts, including those from cement, iron, steel and aluminum factories, also contain considerable concentrations of CO_2. The recovery of CO_2, though not yet employed on a large scale, is a well-known procedure, with the separation of CO_2 being carried out with either liquid-phase absorbents or selective membranes.

Because they are large and concentrated sources of CO_2, most proposed methods for CO_2 recovery and disposal are currently focused on power plants and heavy industries. For such sources, on-site capture of CO_2 offers the most sensible and cost-effective approach. On the other hand, more than half of the CO_2 emissions are the result of small distributed sources such as home and office heating, and most importantly the transportation sector. Whilst collecting CO_2 from vehicles onboard may be technically possible, it would be economically prohibitive. The CO_2 storage system to be added to the vehicle would also be a difficult problem. Moreover, CO_2 once collected, would have to be transported to a sequestration or recycling site, requiring the construction of a massive and expensive infrastructure. Capturing CO_2 onboard airplanes is even less feasible, because of the added weight involved. In homes and offices, producing limited amounts of CO_2, the collection and transportation of CO_2 would also require an extensive and costly infrastructure. While these dispersed CO_2 emissions can probably presently not be addressed, they represent a preponderant part of the global CO_2 emissions and their importance can not be ignored in the long term.

Carbon Dioxide from the Atmosphere

To deal with small and dispersed CO_2 emitters and to avoid the need to develop and construct a huge CO_2-collecting infrastructure, CO_2 could be captured from the atmosphere – an approach that has already been proposed by some in the past [224–227]. The atmosphere could thus serve as an effective means of transporting CO_2 emissions to the site of capture. This would make the CO_2 collection independent of CO_2 sources, and CO_2 could be captured from any source – small or large, static or mobile. With the concentration of CO_2 in air being at equilibrium all around the world, CO_2 extraction facilities could be located anywhere, but to allow for any subsequent methanol synthesis they should ideally be placed close to the hydrogen production sites. As mixing and equilibration of air is relatively rapid, local depletion in CO_2 is not likely to pose any problem. If this were not the case, emissions from power plants would cause much higher local concentration of CO_2 near the plants, which is not the case.

Despite the low concentration of CO_2 of only 0.037% in the atmosphere, Nature routinely recycles CO_2 by photosynthesis in plants, trees, and algaes to produce carbohydrates, cellulose and lipids, and eventually new plant life, while simultaneously releasing oxygen and thereby sustaining life on Earth. Following Nature's example, mankind will be able to capture excess CO_2 from air and to recycle it to generate hydrocarbons and their products. CO_2 can be captured from the atmosphere using basic absorbents such as calcium hydroxide ($Ca(OH)_2$) or potassium hydroxide (KOH) which react with CO_2 to form calcium carbonate ($CaCO_3$) and potassium carbonate (K_2CO_3), respectively [228]. Due to its low CO_2 content, large volumes of air should be contacted with the sorbent material, and this could be achieved with minimum energy input, preferably using natural air convection. After capture, CO_2 would be recovered from the sorbent by desorption, through heating, vacuum or electrochemically. Calcium carbonate, for example, is thermally calcinated to release CO_2. CO_2 absorption is an exothermic reaction, which liberates heat, and is readily achieved by simply contacting CO_2 with an adequate base. The energy-demanding step is the endothermic desorption, requiring energy to regenerate the base and to recover CO_2. Calcium carbonate or sodium carbonate, requiring high energy input for recovery are therefore probably not the ideal candidates for CO_2 capture from air, and other bases might be more appropriate for this application. Research into this area, though still in its relatively early phases of development, should determine the best absorbents and technologies to remove CO_2 from air, with the lowest possible energy input. For example, when using KOH as an absorbent, it has been shown that the electrolysis of K_2CO_3 in water could efficiently produce not only CO_2 but also H_2 with relatively modest energy input [229]. With further developments and improvements, CO_2 capture from the atmosphere, which has already been described as technically feasible, will also become economically viable [228].

Among the various advantages of CO_2 extraction from air is the fact that CO_2 capture, in being independent of CO_2 sources, allows more CO_2 to be captured than is actually emitted from man-made activities. This means that this technology could allow mankind not only to stabilize CO_2 levels, but also eventually to lower them.

The removal of a significant share of CO_2 from industrial emissions and capture of CO_2 from the atmosphere would make huge amounts of CO_2 available. As proposed presently, the captured CO_2 could be stored/sequestered in depleted gas- and oil- fields, deep aquifers, underground cavities, or at the bottom of the seas. This approach, however, does not provide a permanent solution, nor does it assist in mankind's future needs for fuels, hydrocarbons, and their products. The recycling of CO_2 via its chemical reduction with hydrogen to produce methanol (i.e., the "Methanol Economy") is, therefore, an attractive alternative. As fossil fuels become more scarce, the capture and recycling of atmospheric CO_2 would become and remain feasible for the production of methanol, together with synthetic hydrocarbons and associated products. The hydrogen required would be obtained by the electrolysis of seawater (an unlimited resource), while also releasing oxygen. The electrical power required would be provided by atomic energy, and/or by any suitable alternate energy source. Upon their combustion, methanol and the synthetic hydrocarbons produced would be transformed back to CO_2 and water, thereby closing the methanol cycle. This would constitute mankind's artificial version of Nature's CO_2 recycling via photosynthesis. By using this approach, there would be no need for any drastic change in the nature of our energy storage and transportation systems, as would be required by switching to the hydrogen economy, while also providing synthetic hydrocarbons (Fig. 12.13). Furthermore, as CO_2 is available to everybody on Earth, it would liberate us from the reliance on diminishing and non-renewable fossil fuels and all the geopolitical instability associated with them.

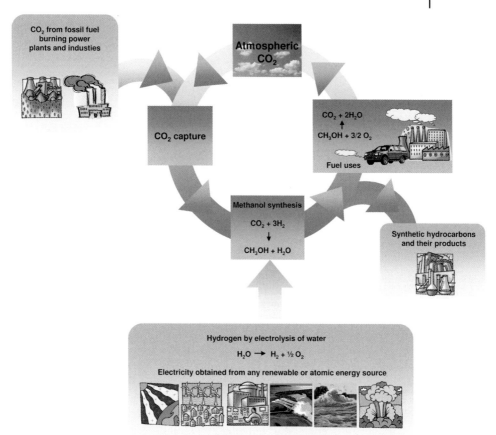

Figure 12.13 CO_2 recycling for methanol and synthetic hydrocarbons production.

Chapter 13
Methanol-Based Chemicals, Synthetic Hydrocarbons and Materials

Methanol-Based Chemical Products and Materials

Today, methanol is one of the most important feedstocks for the chemical industry. Most of the 32 million tonnes of methanol produced yearly are used for the production of a large variety of chemical products and materials, including such basic chemicals as formaldehyde, acetic acid and methyl-*tert*-butyl ether (MTBE; which is, however, phased out in most of the United States), as well as various polymers, paints, adhesives, construction materials and others. In processes for the production of basic chemicals, raw material feedstocks constitute typically up to 60–70% of the manufacturing costs (Fig. 13.1). The cost of feedstock therefore plays a significant economic role. In the past, for example, acetic acid was predominantly produced from ethylene using the Wacker process, but during the early 1970s Monsanto introduced a process which, by carbonylation of methanol using a Wilkinsons rhodium-phosphine catalyst and iodide (in the form of HI, CH$_3$I or I$_2$), produces acetic acid with a conversion and a selectivity close to 100%. Because of the high efficiency of this process and the lower cost of methanol compared to ethylene, most of the new acetic acid plants built worldwide since then are based on this technology [230]. Taking advantage of its lower cost, methanol is also considered as a potential feedstock for other processes currently utilizing ethylene. Rhodium-based catalysts have been found to promote the reductive carbonylation of methanol to acetaldehyde, with selectivities close to 90%. With the addition of ruthenium as a co-catalyst, the further reduction of acetaldehyde to ethanol is possible, providing a new catalytic route for the direct conversion of methanol into ethanol. The possibility of producing ethylene glycol via methanol oxidative coupling instead of the usual process using ethylene as a feedstock is also pursued. Significant advances have also been made on the synthesis of ethylene glycol from dimethyl ether, obtained by methanol dehydration. In addition to acetic acid, acetaldehyde, ethanol and ethylene glycol, other large-volume chemicals produced currently from ethylene or propylene such as styrene and ethylbenzene may be also manufactured from methanol in the future.

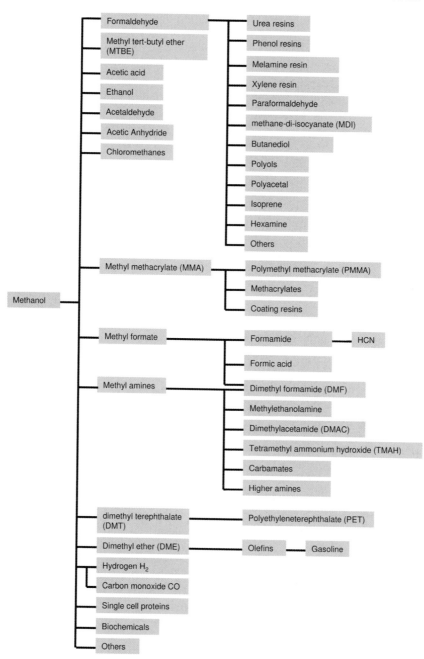

Figure 13.1 Methanol-derived chemical products and materials.

Methanol Conversion to Olefins and Synthetic Hydrocarbons

While methanol can replace light olefins (ethylene and propylene) in some applications, these compounds will remain indispensable building blocks for synthetic hydrocarbons and other products. Ethylene and propylene are by far the two largest volume chemicals produced by the petrochemical industry. In 2004, about 105 million tonnes of ethylene and 60 million tonnes propylene were consumed worldwide. They are important starting materials in the production of plastics, fibers and chemical intermediates such as ethylene oxide, ethylene dichloride, propylene oxide, acrylonitrile and others. The demand for light olefins however, is primarily driven by the polyolefin production. Today, almost 60% of ethylene and propylene is consumed in the manufacture of polyethylene (low-density polyethylene LDPE, high-density polyethylene HDPE, etc.) and polypropylene resins. The majority of light olefins are currently produced by the petrochemical industry as a byproduct of steam cracking and fluid catalytic cracking (FCC) of naphtha and other gas liquids. The demand for propylene is growing at a faster rate than that for ethylene; about 6% per year compared to 4% per year for ethylene. The reason for the higher growth in propylene demand is the increasing popularity of polypropylene, which is being substituted for many other materials and more expensive polymers, especially for automobile parts. Because of the high cost of transporting light olefins by sea, and the proximity of downstream markets, they are mainly produced in North America and Western Europe. With lower feedstock costs, a large part of the new production capacity for light olefins, however, will be installed in the Middle East. Through steam cracking, light hydrocarbons, especially ethane, produced in association with crude oil or from natural gas in that region, yields predominantly ethylene. Given the higher growth rate in propylene demand, this could lead to supply problems. Steam cracking and fluid catalytic cracking will not be able to cover the expected demand for propylene in the coming decade. Thus, the balance will have to be supplied by other sources including propane dehydrogenation, metathesis, olefin cracking and significantly new methanol to olefins (MTO) technologies. MTO can also provide a part of the demand for ethylene. Because methanol is presently mainly produced from natural gas (through syn-gas), it can decrease the dependence of ethylene on petroleum feedstocks. Considering the very large market for ethylene and propylene, these applications will also substantially increase methanol demand, calling for the construction of numerous mega-methanol plants each able to produce between 1 and 3.5 million tonnes of methanol per year. With further new technologies to produce methanol – most significantly via the hydrogenative conversion of CO_2 – the methanol to olefins processes already in commercial development will gain increased significance, allowing the production of these light olefins from non-fossil sources and their subsequent conversion to synthetic hydrocarbons and their various products.

More than a century ago, LeBel and Greene first reported the formation of gaseous saturated hydrocarbons (and some hexamethylbenzene) by adding methanol dropwise to "hot" zinc chloride. Under high pressure, significant amounts of light hydrocarbons were formed when methanol or dimethyl ether was reacted over zinc chloride at about 400 °C. It was later also reported that when methanol was reacted with zinc iodide at 200 °C, a mixture of C_4–C_{13} hydrocarbons containing almost 50% 2,2,3-trimethylbutane (triptane, an excellent high-octane jet fuel) was obtained [191]. Other catalysts, including phosphorus pentoxide, polyphosphoric acid, and later tantalum pentafluoride and other superacid systems, have also been reported for the synthesis of hydrocarbons from methanol. Most of the described catalysts, however, deactivate rapidly. It was only during the 1970s that researchers at the Mobil Oil Company discovered that an acidic zeolite called ZSM-5 was able to catalyze the practical conversion of methanol to both olefins (MTO process) and hydrocarbons in the gasoline range (MTG process) [231–233].

Methanol to Olefin (MTO) Process

The methanol to olefin technology, or MTO, was developed as a two-step process, which first converts natural gas via syn-gas to methanol, followed by its transformation to light olefins. The driving force for the development of this technology was to utilize natural gas sources far from major consumer centers.

The conversion of methanol to olefins proceed through the pathway:

$$2\ CH_3OH \underset{+\ H_2O}{\overset{-\ H_2O}{\rightleftharpoons}} CH_3OCH_3 \xrightarrow{-\ H_2O} \begin{array}{l} \text{Ethylene \& Propylene} \\ H_2C{=}CH_2\ \&\ H_2C{=}CH{-}CH_3 \end{array}$$

The initial step is the dehydration of methanol to dimethyl ether (DME), which then reacts further to form ethylene and propylene. In the process, small amounts of butenes, higher olefins, alkanes and some aromatics are also produced.

Besides the synthetic aluminosilicate zeolite (ZSM-5) catalysts, numerous other catalysts were also studied. UOP developed silicoaluminophosphate (SAPO) molecular sieves such as SAPO-34 and SAPO-17, which have demonstrated high activity and selectivity for the MTO process. Both type of zeolites have defined three-dimensional crystalline structures. They are microporous solids permeated with channels and cages of very specific size. The many different zeolite catalysts differ by their chemical composition and the size and structure of these channels and cages, which have molecular dimensions ranging from 3 to 13 Å. The catalytic sites responsible for the catalyst's activity are placed in the pores and channels of these catalysts. Access to these sites will therefore be limited to chemical reagents small enough to enter the zeolites pores and channels. At the same time, the size of the reaction products is also governed by space constraints imposed by the catalyst's structure. As a result, zeolites can be highly shape-selective catalysts. ZSM-5, for example has pore openings of some 5.5 Å, allowing much faster diffusion of *para*-xylene than *meta*- or *ortho*-xylene with a larger molecular size. With a pore size of only 3.8 Å, SAPO-34 used in the MTO process, allows effective control of the size of the olefins that emerge from the catalyst. Larger olefins diffuse out at a lower rate, making smaller olefins such as ethylene and propylene the predominant products.

Independent from zeolites, bi-functional supported acid–base catalysts, such as tungsten oxide over alumina (WO_3/Al_2O_3) were found by Olah and coworkers during the 1980s also to be active for the conversion of methanol to ethylene and propylene at temperatures between 250 and 350 °C [234,235], and subsequently to hydrocarbons. These heterogeneous bi-functional catalysts catalyze the reaction despite the fact that they lack the well-defined three-dimensional structure of shape-selective zeolite catalysts.

Based on SAPO-34, an MTO process has been developed jointly by UOP and Norsk Hydro [236, 237]. This process converts methanol in more than 80% selectivity to ethylene and propylene. In addition, about 10% is converted to butenes, which are also valuable starting materials for a variety of products. Depending on the operating conditions, the propylene to ethylene weight ratio can be modified from 0.77 to 1.33. This allows considerable flexibility and adaptation to changing market conditions. The technology has been extensively tested in a demonstration plant in Norway, and more than ten years of development have now been completed. Currently, the UOP/Hydro MTO process is commercialized in Nigeria and units are being considered in other locations. A methanol plant, built close to the capital city of Lagos and with a capacity of 2.5 million tons per year will be the largest in the world. The methanol produced will then be converted to ethylene and propylene to produce 400 000 tonnes per year of polyethylene and 400 000 tonnes per year of polypropylene. This will be the first commercial large-scale demonstration of the MTO technology.

Lurgi has also developed an MTO process [238] which, unlike the UOP/Hydro technology, is designed to yield mostly propylene. It is thus described as a methanol to propylene (MTP) process. In a first step, methanol is dehydrated over a slightly acidic catalyst to produce DME. The DME is then reacted over a ZSM-5-based catalyst under moderate pressure (1.3–1.6 bar) and temperatures between 420 °C and 490 °C to form light olefins. The achievable overall yield of propylene is above 70%. The process has been demonstrated at Statoil's Tjeldbergodden methanol plant in Norway, and is now ready for commercialization. Propylene obtained from this unit with a purity of 99.7% has also been successfully polymerized to polypropylene, showing the feasibility of producing polypropylene directly from methanol obtained from natural gas.

Mobil (now ExxonMobil), which pioneered the MTO technology using the company's ZSM-5 catalyst (ZSM meaning literally Zeolite Synthesized by Mobil) also demonstrated this technology on a 100 barrel-per-day scale in Wesseling, Germany. In addition, Mobil also developed the subsequent Olefins to Gasoline and Distillate (MOGD) process. In this process, which was originally developed as a refinery process, the olefins from the MTO unit are oligomerized over a ZSM-5 catalyst to yield, with selectivity greater than 95%, hydrocarbons in the gasoline and/or distillate range. Depending on the reaction conditions, the ratio between gasoline and distillate can be varied considerably, allowing a significant flexibility in production. When operated at relatively low temperatures and high pressures (200–300 °C, 20–105 bar), called the distillate mode, the products are higher molecular weight olefins which, after being subjected to hydrogenation, produce fuels including diesel and premium quality jet fuels. Changing the operating conditions to higher temperatures and lower pressures leads to the formation of products of lower molecular weight but with higher aromatic content (i.e., high-octane gasoline). Besides being a good catalyst for the MTO and MOGD processes, ZSM-5 was also found to catalyze the direct conversion of methanol to hydrocarbons in the gasoline range. This led to the development of the Methanol to Gasoline process (MTG).

Methanol to Gasoline (MTG) Process

The MTG process was conceived and developed in response to the energy crisis of the 1970s. It was the first major new route to synthetic hydrocarbons since the introduction of the Fischer–Tropsch process before World War II, and provided an alternative pathway for the production of high-octane gasoline from coal or natural gas. Discovered by accident by a research team at Mobil, the MTG process was actually developed before the MTO process. In fact, the MTO process can be considered as a modified MTG process designed to produce mainly olefins instead of gasoline. For the MTG reaction, medium-pore zeolites with considerable acidity are the most suitable catalysts, with ZSM-5 recognized as the most selective and stable. Because of the defined structure and geometry of their pores, channels and cavities, they are shape-selective catalysts, able to control product selectivity depending on their molecular size and shape. Over this catalyst, methanol is first dehydrated to an equilibrium mixture of DME, methanol and water. This mixture is then converted to light olefins, primarily ethylene and propylene. Once these small olefins are formed, they can undergo further transformations to higher olefins, C_3–C_6 alkanes and C_6–C_{10} aromatics.

$$2\ CH_3OH \underset{+H_2O}{\overset{-H_2O}{\rightleftharpoons}} CH_3OCH_3 \xrightarrow{-H_2O} \text{light olefins} \longrightarrow \begin{array}{l} \text{alkanes} \\ \text{higher olefins} \\ \text{aromatics} \end{array}$$

Due to the shape selectivity of ZSM-5, heavier hydrocarbons containing more than 10 carbon atoms are practically not produced in this process. This is fortuitous, as C_{10} is also the usual limit for conventional gasoline. At the same time, this process produces aromatics (toluene, xylenes, trimethylbenzene, etc.), providing also a route to aromatic hydrocarbons. Depending on the catalyst and conditions used, the product distribution can be modified if desired. Using zeolites with larger pores, channels and cavities, such as ZSM-12 or Mordenite for example, leads to products with higher molecular weight.

Bi-functional acid–base catalysts such as tungsten oxide (WO_3) supported on alumina active for the conversion of methanol to light olefins can also catalyze their further reaction to a mixture of higher hydrocarbons including alkanes, alkenes as well as aromatic compounds.

In 1979, the New Zealand government selected the MTG process developed by Mobil [233] for the conversion of natural gas from the large off-shore Maui field into gasoline. The New Zealand plant began operations in 1986, producing about 600 000 tonnes of gasoline per year, supplying one-third of New Zealand's gasoline needs. Methanol was produced using the ICI low-pressure methanol process in two production units, each capable of producing 2200 tonnes of methanol per day. Crude methanol from these units was fed directly to the MTG section, where it was first converted over an alumina catalyst to an equilibrium mixture of methanol, DME and water. This mixture was then transferred to the gasoline synthesis reactor where it was reacted over ZSM-5 at 350–400 °C and 20 bar [232]. The crude gasoline was then treated to remove minimal amounts of heavier components. Without further distillation or refining necessary, the high-octane gasoline obtained can be blended directly with the general gasoline pool.

As mentioned earlier, the MTG process was developed in response to the energy crisis of the 1970s, which saw the price of oil and its products rise dramatically. However, as the oil price fell again during the 1980s, dipping briefly under $10 in 1986, the commercial interest for MTG dropped accordingly. Gasoline production at the New Zealand plant ceased because it was now cheaper to use low-cost gasoline derived from petroleum than to produce it from natural gas via methanol. Methanol production itself is, however, still in operation, providing methanol at a competitive cost. Increasing oil prices experienced during the past few years will most probably revive the interest in MTG.

Methanol-Based Proteins

Methanol can also serve as the source for single-cell protein production. Single-cell proteins (SCP) refers to proteins produced by a variety of microorganisms degrading hydrocarbon substrates while gaining energy [108, 113]. The protein content depends on the type of microorganism; bacteria, yeast, mold, etc. The use of microorganisms in human alimentation has been practiced since ancient times, in the form of yeasts used in brewing and baking, and bacteria in cultured dairy products such as cheese, yogurt, and sour cream. In modern times, the possibility of using proteins produced by microorganisms for animal and human alimentation first emerged in Germany during World War I. The main development of SCP, however, began during the 1950s when the petroleum industry realized that, by the degradation of hydrocarbons, some microorganisms could produce high-quality proteins that were suitable for animal feed or as a nutritional source for human food. Because of the possibility of contamination and build-up of carcinogenic compounds in animals fed with SCP produced from oil products, however, methanol was chosen as a substitute. In contrast with petroleum-derived feeds, methanol is non-carcinogenic, forms a homogeneous solution with the aqueous nutrient salt solutions used in the process, and can be readily separated from the protein products after their formation. A number of companies, including Shell, Mitsubishi, Hoechst, Phillips Petroleum and ICI, have studied the bacterial fermentation process. For some time, ICI operated a commercial plant in Billingham, England, producing 70 000 tonnes of SCP from 100 000 tonnes of methanol per year. The bacteria obtained had a very high protein content, with the dried cells containing up to 80% protein (much higher than other types of food such as fish and soybean). As the structure of the bacteria was highly complex, a wide range of amino acids – including in particular aspartic and glutamic acid, alanine, leucine and lysine – was obtained. The overall quality of SCP produced from methanol by the ICI process was very high, and the product sold as an animal feed under the name Pruteen.

Besides serving as a medium for microorganism culture, methanol was also found to increase substantially the growth rate in a variety of plants [108]. In plants with C3 metabolism – for which the first product of photosynthesis is a three-carbon sugar – significantly higher photosynthetic productivity has been observed, especially in regions with high light intensity such as the South-Western United States. C3 plants include for example sunflower, watermelon, tomato, strawberry, lettuce, and eggplant. Methanol sprayed onto the plants is rapidly absorbed by the foliage and metabolized to CO_2, sugars, amino acids and other structural components. It is thus used as a concentrated source of carbon in place of CO_2 (1 mL of liquid methanol contains about as much carbon as 2 000 000 mL of air).

Methanol was also found to be an economic and effective means of inhibiting the process of photorespiration; this describes the plant uptake of oxygen, which is competing with CO_2 uptake for photosynthesis. Oxygen assimilation results in the breakdown of sugars, reversing the photosynthetic process. When exposed to stressful conditions such as high light intensity and high temperatures, the stomata (tiny pores used by the plant to absorb atmospheric CO_2) close, reducing the uptake of CO_2, and this results in increased photorespiration. This can stop plant growth for several hours during the hottest period of day. The control of photorespiration is, therefore, key to enhancing the photosynthetic yield of plants. Further studies on the effects of methanol on the complex growth mechanism of plants are needed before large-scale applications can be envisioned. Methanol, however, has a good potential to effect significant improvements in crop productivity.

With an increasing world population, agricultural production might encounter difficulties in providing sufficient protein for food and animal feed. The "Methanol Economy" could, therefore, also supplement essential protein needs through SCP.

Outlook

With decreasing oil and gas reserves, it is inevitable that synthetic hydrocarbons must play a major role as energy sources. Since methanol can be produced not only from still-available carbonaceous sources of biomass (e.g., coal, natural gas) but also from concentrated industrial CO_2 emissions and eventually from the air itself, methanol-based chemicals and products – and particularly synthetic hydrocarbons available through MTG and MTO processes – will assume increasing importance in replacing decreasing oil- and gas-based resources.

Chapter 14
Future Perspectives

All fossil fuels are mixtures of hydrocarbons, which contain varying ratios of carbon and hydrogen. Upon their combustion, carbon is converted into CO_2 and hydrogen into water. Consequently, when burned these fuels are irreversibly used up. The increases in the CO_2 content of the atmosphere which result from human activities and the excessive combustion of fossil fuel is considered to be a major man-made cause of global warming.

Much has been said of the extent of our available oil and gas resources (see Chapters 4 and 5). Although our coal reserves may last for another two or three centuries, the mining of coal (except in areas suited to surface strip mining) involves difficult and dangerous labor, hazards and environmental difficulties. Moreover, our oil and gas reserves, although readily accessible, will not last much longer than the end of the 21st century – even taking into consideration new discoveries, improved technologies and unconventional sources.

Besides accessible petroleum oil and natural gas (and coal) resources, we have additional unconventional hydrocarbon sources such as heavy oil deposits in Venezuela, oil shale in various geological formations, including the U.S. Rocky Mountains, and vast tar sand deposits in Alberta. The hydrates of methane, as are found under the Siberian tundra and along the continental shelves of the oceans, represent significant resources for natural gas for the future. They too will all eventually be exploited, although the difficulties and costs involved are immense.

Besides the size of the reserves, as discussed, one must also consider the expanding world population, which currently exceeds six billion and will most likely attain eight to ten billion by the end of the 21st century. The consequences of increasing consumption of our reserves, due to improving standards of living in the world, as well as increased demands in fast-developing countries such as China and India, are clear. Potential oil reserves estimated at one to two trillion barrels or some 135 to 270 billion tonnes must be considered when taking into account these factors. The best present estimates of our readily accessible oil reserves imply that they would last for no more than 70 years at the current rate of consumption. Natural gas reserves are somewhat larger, and may last for another 80 to 100 years. New discoveries and improved recovery methods can extend these estimates, while increased oil consumption will place more pressure on

Beyond Oil and Gas: The Methanol Economy. G. A. Olah, A. Goeppert, G. K. S. Prakash
Copyright © 2006 WILEY-VCH Verlag GmbH & Co. KGaA, Weinheim
ISBN 3-527-31275-7

our reserves. In any case, mankind must begin to prepare itself for the future, and find new sources and solutions.

In order to satisfy mankind's ever-increasing energy needs, the use of all feasible alternative energy sources will be necessary in the future. Hydro and geothermal energy are already well used where Nature makes them feasible, but no further major new suitable locations are expected to be found in developed countries. The energy of the Sun, wind, waves and tides of the seas all have great potential and are increasingly being exploited, though their large-scale use as substitutes for fossil fuels is not expected to have significant effect on our energy picture in the foreseeable future.

Perhaps the greatest technological achievement of the 20th century has been mankind's ability to harness the energy of the atom. Regretfully, as this was first achieved in the building of the atom bomb, public opinion has in subsequent years increasingly turned against this energy source, even when considering only its peaceful uses. During the past few decades, relatively few new atomic power plants have been constructed (and none in the United States). There is even strong sentiment in some countries to close them down altogether, whilst other countries such as France depend on them for some 80% of their electricity needs. Much progress has been made to limit the use of atomic energy only to peaceful uses and to improve safety aspects, including radioactive waste storage and disposal. Our society, which was able to build the atom bomb, can – and will – solve these problems. The decline of the atomic energy industry in most industrialized countries is most regrettable and shortsighted. Whether or not one likes atomic energy, it is for the foreseeable future the most feasible and massive energy source available to mankind. Of course it should be made even safer and more effective, solving the problem of reuse and storage of radioactive wastes as well as developing new improved reactor designs and the use of atomic fusion. Conservation, as well as use of alternate energy sources, are most desirable, but these alone cannot solve our enormous appetite for energy. Nonetheless, there have been recent hopeful signs that the generally negative perception of atomic energy is slowly changing.

Despite their non-renewable nature and diminishing resources, fossil fuels – and particularly oil and gas – will maintain their leading role as long as they are readily available. A vast infrastructure exists for their transport and distribution. As transportation fuels for our cars, trucks and airplanes they are used in the form of their convenient products (gasoline, diesel fuel or as compressed natural gas). Natural gas and heating oil are essential for heating our homes and offices, and to provide energy for industry. Oil and gas are also the raw materials for chemical products and materials essential for our everyday life. However, the bulk of the fossil fuels are still utilized in power plants to generate electricity. From whatever source electricity is generated, its storage on a large scale is still unresolved (batteries, for example, are inefficient and bulky), and it is therefore necessary to find, besides new energy sources, efficient ways for energy storage and distribution. We also need to develop new and efficient ways to produce synthetic hydrocarbons and their varied products from non-fossil fuel sources. It is ironic that, in knowing fully that we will need to produce synthetic hydrocarbons at a significant cost and major technological efforts, we continue to burn much of the still-existing natural fossil fuel resources in order to provide energy.

One approach which has been proposed and much discussed recently is the use of hydrogen as a clean fuel (the so-called "hydrogen economy"; see Chapter 9). Free hydrogen is however not a natural energy source on Earth as it is incompatible with the high oxygen content of our atmosphere. Whilst it is indeed "clean burning" (forming only water), its generation from its bound compounds (hydrocarbons, water) is a highly energy-consuming process which is at present far from clean as hydrogen is mainly produced by reforming of natural gas, oil or coal (i.e., fossil fuels) to syn-gas (a mixture of CO and H_2). The CO generated is oxidized to CO_2, or it can be further used in the water gas shift reaction to yield more hydrogen. Overall, however, one-fourth of the energy of fossil fuels is lost as heat. Hydrogen can also be produced by the electrolysis of water, a process which does not produce CO_2 and does not necessarily involve a fossil fuel source. Our oceans represent an inexhaustible source of water that can be split by electricity (or by other means) to produce hydrogen and oxygen. Hydrogen is, however, not a convenient energy storage medium, as its storage, transportation and distribution are both difficult and costly. The handling of extremely volatile and potentially explosive hydrogen gas necessitates high-pressure equipment, the use of special materials to minimize diffusion and leakage, and extensive safety precautions to prevent explosions. In addition to these difficulties, the essential infrastructure for the development of the hydrogen economy is yet to be developed, at extreme cost.

The "Methanol Economy" and its Advantages

One feasible alternative to the hydrogen economy – what we now call the "Methanol Economy" – has been proposed and discussed to great extent in this book (Fig. 14.1). Methanol is a convenient, oxygenated liquid hydrocarbon that at present is prepared from fossil fuel-based syn gas. As discussed earlier, however, new methods are currently under development for its production by the direct oxidative conversion of still-existing large natural gas (methane) sources, or by the hydrogenative conversion of exhausts of fossil fuel-burning power plants and other industrial plants, which are rich in CO_2. Eventually, it will be possible chemically to recycle atmospheric CO_2 itself via hydrogenative conversion to methanol. The required hydrogen will be obtained from water (an inexhaustible resource), using any energy source – atomic or renewable energy. In this way, extremely volatile hydrogen gas will be conveniently and safely stored by converting it, with CO_2, into liquid methanol.

Methanol represents not only a convenient and safe means of storing energy but, when combined with easily derived dimethyl ether (DME), this pair of compounds represent excellent fuels in their own right. Methanol and DME can be blended with gasoline/diesel and used in internal combustion engines, as well as in electricity generators. Methanol is particularly efficient, when used in the

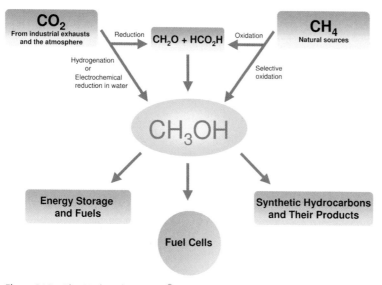

Figure 14.1 The Methanol Economy®.

direct methanol fuel cell (DMFC) (see Chapter 11), where methanol is oxidized directly with air to CO_2 and water while producing electricity. In addition to its many uses for diverse chemical products and materials, methanol can also be readily converted to ethylene and/or propylene (the MTO process), which then can be used to produce synthetic hydrocarbons and their products, that presently are obtained from oil and gas.

$$CH_3OH \longrightarrow CH_2=CH_2 \text{ and/or} \longrightarrow \text{hydrocarbons}$$
$$CH_3CH=CH_2$$

Although today, methanol is prepared exclusively from fossil fuel-based syn-gas, while natural gas remains available it would seem reasonable to convert it directly into methanol, without first going through the syn-gas stage. This developing technology would not only greatly simplify its production but also extend its availability.

Methanol can also be obtained from CO_2 by catalytic reduction with hydrogen, or by electrochemical reduction in water. The emissions of fossil fuel-burning power plants and chemical plants contain high concentrations of readily isolable CO_2. As these large amounts of CO_2, when released, contribute greatly to global warming, it is now generally agreed that they must be captured and disposed of. Rather than simply sequestering them, however, their chemical recycling to methanol seems a most feasible approach. Water could provide the required hydrogen for converting CO_2 to methanol using any energy source (alternatives include atomic, photochemical and even bacterial conversions). In the longer perspective, however, fossil fuel-derived excessive CO_2 generation will not be problematic, as our non-renewable fossil fuels will last at most for some centuries. At the same time, the recycling of atmospheric CO_2 itself via its reductive hydrogenation to methanol will offer an inexhaustible carbon source for fuels, synthetic hydrocarbons and their products.

As the CO_2 content of the atmosphere is low (0.037%), new and efficient ways for the separation of CO_2 are needed. Today, selective absorption and other separation methods are making significant advances to allow the separation of atmospheric CO_2 from the air on a practical scale. Methanol produced efficiently on a large scale from atmospheric CO_2 and hydrogen from water will therefore be able to replace oil and gas both as a convenient way to store energy, as a suitable fuel and chemical raw material for synthetic hydrocarbons and their varied products (including polymers and even proteins). Thus, the "Methanol Economy" will eventually liberate mankind from reliance on diminishing and non-renewable fossil fuels.

Nature itself recycles CO_2 in the photosynthetic processes conducted by plants (using water and energy from the Sun). The subsequent formation of fossil fuels from plant life is however, a very slow process requiring hundred of millions of years (although rapid bacterial conversion is a possibility). Hence, in a way the "Methanol Economy" supplements and greatly accelerates Nature's own recycling processes.

Ultimately, the majority of our energy on Earth comes from the Sun. It is considered that the Sun will last for at least 4.5 billion years, during which time the possibilities for mankind to devise more efficient ways of harnessing its energy are limitless. Whilst we cannot even begin to imagine the advances that will be made by future generations, our present discussions have been limited to what might be achieved in the foreseeable future, based on our present and developing knowledge base.

Our conclusion for the future is optimistic. Humankind is an ingenious species which always seems to find ways of overcoming adversities and challenges. As history teaches us, however, humankind's reaction to significant major problems and challenges usually comes only when a crisis is already upon us. Many believe that the problem of our oil and gas reserves is not yet at the crisis stage, and that we do not yet need to unduly worry about them. Past pessimistic predictions concerning our diminishing fossil fuel resources have always turned out to be "false alarms", and this is substantially true considering the short time spans to which they were applied. However, with regard to the longer range, the outlook is different. We must face the fact that our Nature-given non-renewable fossil fuel resources are finite and diminishing, whilst worldwide both the population and consumption is growing. We need to find new solutions, if we wish to continue our lives at a comparable or even higher standard of living. We need to start developing new solutions now – while we still have the time and resources to do it in an orderly fashion.

As rightly argued, one way of extending our oil and gas reserves is their better and more economical use, conservation measures and the introduction of new efficient technologies, particularly in the transportation area where oil-based gasoline and diesel fuel continue to be used primarily. Fuel savings, together with more efficient vehicles (such as using hybrid propulsion systems combining internal combustion engines with onboard generated electricity driven electric motors) can reduce gasoline and diesel fuel use and extend their availability. Fuel cells based not only on hydrogen, but also directly on methanol (DMFC) can provide cars with great fuel efficiency. The wide use of hydrogen for energy storage and as a fuel (the so-called "hydrogen economy") – except perhaps for larger static installations – is considered less feasible, as volatile hydrogen gas is extremely difficult to handle. This would necessitate not only the development of an entirely new and expensive infrastructure but also the recognition and control of serious safety hazards.

The proposed "Methanol Economy" represents a feasible new approach which extends beyond the era of abundant and cheap oil and gas. We hope that this book will further raise interest in the study of such an economy, together with its development and applications. It is not suggested, however, that this is the only approach to be followed – or is even necessarily the most feasible in all aspects. Rather, mankind will need to rely on all possible solutions available. We believe however, that the "Methanol Economy" is feasible and warrants extensive further study, development and evaluation.

Further Reading and Information

General Information on Energy

- Schobert H. H., *Energy and Society, an Introduction*, Taylor and Francis, New York, **2002**.
- Smil V., *Energy at the Crossroads, Global Perspectives and Uncertainties*, MIT Press, Cambridge, **2003**.
- Smil V., *Energy in World History*, Westview Press, Boulder, Colorado, **1994**.
- Bent R., Orr L., Baker R., *Energy. Science, Policy, and the Pursuit of Sustainability*, Island Press, Washington, DC, **2002**.
- *Energy Technologies for the 21st Century*, International Energy Agency, Paris, **1997**.
- *Key World Energy Statistics 2004*, International Energy Agency, Paris, **2004**.
- *World Energy Outlook 2001: Insights*, International Energy Agency, Paris, **2001**.
- *World Energy Outlook 2004*, International Energy Agency, Paris, **2004**.
- *Annual Energy Outlook 2005 with Projections to 2025*, Energy Information Agency, Washington, DC, **2005**, available at http://www.eia.doe.gov/oiaf/aeo.
- World Energy Council (WEC): http://www.worldenergy.org/wec-geis/
- International Energy Agency (IEA): http://www.iea.org/
- U.S. Energy Information Administration (EIA): http://www.eia.doe.gov/
- BP Statistical Review of World Energy: http://www.bp.com/statisticalreview
- U.S. Department of Energy (DOE): http://www.energy.gov/

Beyond Oil and Gas: The Methanol Economy. G. A. Olah, A. Goeppert, G. K. S. Prakash
Copyright © 2006 WILEY-VCH Verlag GmbH & Co. KGaA, Weinheim
ISBN 3-527-31275-7

Coal

To learn more about the history of coal:
- Freese B., *Coal, a Human History*, Perseus Publishing, Cambridge, **2003**.

General information about coal, including statistics on production, uses and impact on the environment:
- Coal information 2005, International Energy Agency, Paris, **2005**.
- World Coal Institute: http://www.worldcoal.org/
- National Mining Association (U.S.): http://www.nma.org/

Clean coal technology

- NETL clean coal technology: http://www.netl.doe.gov/cctc/
- IEA clean coal center: http://www.iea-coal.org.uk/site/ieaccc/home

Oil and Natural Gas

To learn more about the history of oil:
- Black B., *Petrolia, the Landscape of America's First Oil Boom*, The Johns Hopkins University Press, Baltimore, **2000**.
- Yergin D., *The Prize*, Simon & Schuster, New York **1991**.

To learn more about discovery and exploitation of oil and natural gas:
- Conaway C. F., *The Petroleum Industry. A Nontechnical Guide*, PennWell, Tulsa, Oklahoma, **1999**.
- Hyne N. J., *Nontechnical Guide to Petroleum Geology, Exploration, Drilling and Production*, PennWell, Tulsa, Oklahoma, **1995**.
- Campbell C. J., *The Coming Oil Crisis*, Multi-science Publishing, Brentwood, England, **1988**.
- Deffeyes K. S., *Hubbert's Peak, the Impending World Oil Shortage*, Princeton University Press, Princeton, **2001**.
- Leffler W. L., Pattarozzi R., Sterling G., *Deepwater Petroleum Exploitation & Production: A Nontechnical Guide*, PennWell, Tulsa, Oklahoma, **2003**.
- *All About Petroleum*, brochure from American Petroleum Institute, available at http://api-ec.api.org
- American Petroleum Institute (API): http://www.api.org

Oil and natural gas statistics on production, reserves and uses.
- *Oil Information 2005*, International energy Agency, Paris, **2005**.

- *Natural Gas Information 2005*, International Energy Agency, Paris, **2005**.

Liquefied Natural Gas (LNG)

- The Center for LNG: http://www.lngfacts.org/
- Department of Energy, information on LNG: http://www.fossil. energy.gov/programs/oilgas/storage/lng/feature/index.html

Unconventional Oil and Gas Resources

Tar Sands

- *Oil Sands Technology Roadmap. Unlocking the Potential*, Alberta Chamber or Resources, Edmonton, Alberta, **2004**, available at http://www.acr-alberta.com/.
- *World Energy Council, information about bitumen and extra-heavy oil*: http://www.worldenergy.org/wec-geis/publications/reports/ ser/bitumen/bitumen.asp

Oil Shale

- Is Oil Shale America's Answer to Peak-Oil challenge, *Oil & Gas Journal*, August 9, **2004**.
- *Strategic Significance of America's Oil Shale Resource*, Naval Petroleum and Oil Shale Reserves & Department of Energy (U.S.), Washington, DC, **2004**.
- *World Energy Council, information about oil shale*: http:// www.worldenergy.org/wec-geis/publications/reports/ser/shale/ shale.asp
- Loucks R. A., *Shale Oil: Tapping the Treasure*, Xlibris, **2002**.

Coalbed Methane, Tight Gas Sands and Shale Gas

- Canadian Society for Unconventional Gas: http://www.csug.ca/
- Fischer P. A., Unconventional gas resources fill the gap in future supplies, *World Oil Magazine*, August **2004**, vol. 225, available at http://worldoil.com.
- Perry K. F., Cleary M. P., Curtis J. B., *New Technology for Tight Gas Sands*, 17th World energy Congress, Houston, September 13-18, **1998**, available at http://www.worldenergy.org/wec-geis.
- Garbutt D., *Unconventional Gas*, Schlumberger white paper, **2004**, available at http://www.oilfield.slb.com.

Methane Hydrate

To learn more about methane hydrate:
- The National Energy Technology Laboratory (NETL, U.S.):
  http://www.netl.doe.gov/scngo/NaturalGas/hydrates/
- Department of Energy (U.S.), methane hydrate program:
  http://www.fe.doe.gov/programs/oilgas/hydrates/index.html

Diminishing Oil and Natural Gas Resources, Production Peak and Shortage

- Appenzeller T., The End of Cheap Oil, *National Geographic Magazine*, June, **2004**, p. 80.
- Campbell C. J., *The Coming Oil Crisis*, Multi-science Publishing, Brentwood, England, **1988**.
- Deffeyes K. S., *Hubbert's Peak, the Impending World Oil Shortage*, Princeton University Press, Princeton, **2001**.
- Deffeyes K. S., *Beyond Oil, the View from Hubbert's Peak*, Hill and Wang, New York, **2005**.
- Heinberg R., *The party's Over. Oil, War and the Fate of Industrial Societies*, New Society Publishers, Gabriola Island, Canada, **2003**.
- Odell P. R., Rosing K. E., *The Future of Oil. World Oil Resources and Use*, Kogan Page, London, **1983**.
- Roberts P., *The End of Oil*, Houghton Mifflin Company, New York, **2004**.
- Campbell C. J., Laherrère J. H., The End of Cheap Oil, *Scientific American*, March, **1998**, 78.
- Goodstein D., *Out of Oil: The End of the Age of Oil*, W. W. Norton & Company, New York, **2004**.
- Ruppert M. C., *Crossing the Rubicon: The Decline of the American Empire at the End of the Age of Oil*, New Society Publishers, **2004**.
- Bentley R. W., Oil & Gas Depletion: an Overview, *Energy Policy*, **2002**, 30, 189.
- Odell P. R., *Why Carbon Fuels Will Dominate the 21st Century's Global Energy Economy*, Multi-Science Publishing, Brentwood, England, **2004**.
- Huber P., Mills M. P., *The Bottomless Well. The Twilight of Fuel, The Virtue of Waste, and Why we Will Never Run Out of Energy*, Basic Books, Cambridge, **2005**.
- Maugeri L., Oil: Never Cry Wolf. Why the Petroleum Age Is Far from Over, *Science*, **2004**, vol. 304, p. 1115.

On the Internet

- see wikipedia under "Hubbert peak theory" and numerous references therein: http://en.wikipedia.org/
- Association for the study of peak oil & gas: http://peakoil.net/
- Other sites for information about oil peak: http://www.hubbertpeak.com/, http://www.peakoil.com/ and http://www.oilscenarios.info/

Hydrocarbons and their Products

- Olah G. A., Molnár Á., *Hydrocarbon Chemistry*, 2nd edn., John Wiley & Sons, Hoboken, New Jersey, **2003**.
- Weissermel K., Arpe H.-J., *Industrial Organic Chemistry*, 4th edn., Wiley-VCH, Weinheim, Germany, **2003**.

To learn more about petrochemistry and chemistry:
- http://www.petrochemistry.net/
- http://www.americanchemistry.com/

To learn more about plastics:
- American Plastic Council: http://www.plastics.org/
- Plastic Europe, Association of plastic manufacturers: http://www.plasticseurope.org

Climate Change

- *Third Assessment Report: Climate Change 2001*, Intergovernmental Panel on Climate Change (IPCC), **2001**, available at http://www.ipcc.ch/
- Johansen B. E., *The Global Warming Desk Reference*, Greenwood Press, Westport, Connecticut, **2002**.
- Kelly R. C., *The Carbon Conundrum. Global Warming and Energy Policy in the Third Millennium*, CountryWatch, Houston, **2002**.
- Leggett J., *The Carbon War. Global Warming and the End of the Oil Era*, Routledge, New York, **2001**.
- Crichton M., *State of Fear*, HarperCollins Publisher, New York, **2004**.
- *Beyond Kyoto. Energy Dynamics and Climate stabilisation*, International Energy Agency (IEA), Paris, 2002, available at http://www.iea.org/
- Intergovernmental Panel on Climate Change (IPCC): http://www.ipcc.ch/
- U.S. Environmental Protection Agency (EPA) global warming Site: http://yosemite.epa.gov/oar/globalwarming.nsf/

CO₂ Capture and Storage

- Socolow R.H., Can we Bury Global Warming?, *Scientific American*, July 2005, p. 49.
- Herzog H., Eliasson B., Kaarstad O., Capturing Greenhouse Gases, *Scientific American*, February 2000, p. 72.
- *Prospects for CO2 Capture and Storage*, International Energy Agency (IEA), Paris, 2004.
- Anderson S., Newell R., Prospects for Carbon Capture and Storage Technologies, *Annual Review of Environment and Resources*, 2004, 29, p. 109.
- IPCC Special Report on Carbon Dioxide Capture and Storage, 2005, available at http://www.ipcc.ch/
- Johnson J., Putting a Lid on carbon Dioxide, *Chemical & Engineering News*, December 20, 2004, 82, p. 51.
- IEA Greenhouse Gas Research & Development Program: http://www.ieagreen.org.uk/
- Department of Energy (U.S.), Office of Fossil Energy, Carbon Sequestration R&D: http://www.fe.doe.gov/programs/sequestration/
- Princeton University Carbon Mitigation initiative: http://www.princeton.edu/~cmi/
- CO₂ Capture Project: http://www.co2captureproject.org/

Renewable Energies

General Information about Renewable Energies

- Cassedy E. S., *Prospects for Sustainable Energy. A Critical Assessment*, Cambridge University Press, Cambridge, 2000.
- Sørensen B., *Renewable Energy. Its Physics, Engineering, Use, Environmental Impacts, Economy and Planning Aspects*, 2nd edn., Academic Press, London, 2000.
- *Renewables Information 2005*, International Energy Agency (IEA), Paris, 2005.
- Renewables for Power Generation: Status and Prospects, International Energy Agency (IEA), Paris, 2003.
- Renewable Energy Outlook, in *World Energy Outlook 2004*, International Energy Agency (IEA), Paris, 2004, p. 225.
- Global Renewable Energy Supply Outlook, in *World Energy Outlook 2001*, International energy Agency (IEA), Paris, 2001.
- *Renewable Energy Technology Characterization*, Electric Power Research Institute (EPRI) and U.S. Department of Energy (DOE), 1997.

- Berinstein P., *Alternative Energy. Facts, Statistics, and Issues*, Oryx Press, Westport, Connecticut, **2001**.
- Patel M. R., *Wind and Solar Power Systems*, CRC Press, Boca Raton, Florida, **1999**.
- European Union websites for renewable energies: http://europa.eu.int/comm/energy_transport/atlas/htmlu/renewables.html; http://europa.eu.int/comm/research/energy/nn/nn_rt/article_1075_en.htm
- U.S. Department of Energy (DOE), Energy Efficiency and Renewable Energy (EERE) website: http://www.eere.energy.gov/
- *Renewable Energy Journal*, available for free from http://www.energies-renouvelables.org/

Hydropower

- International Energy Agency (IEA), Hydropower technologies: http://www.ieahydro.org/
- World Commission on Dams: http://www.dams.org/
- International Commission on Large Dams: http://www.icold-cigb.net/

Geothermal

- International Energy Agency (IEA), Geothermal energy: http://www.iea-gia.org/
- International Geothermal Association: http://iga.igg.cnr.it/index.php
- World Bank, Geothermal energy information: http://www.worldbank.org/html/fpd/energy/geothermal/
- Geo-Heat Center, Oregon Institute of Technology: http://geoheat.oit.edu/

Wind

- Archer C. L., Jacobson m. Z., Evaluation of Global Wind Power, *Journal of Geophysical Research*, **2005**, vol. 110, p. D12110.
- International Energy Agency (IEA), Wind Energy Systems: http://www.ieawind.org/
- U.S. Department of Energy, Energy efficiency and Renewable Energy (EERE), wind technology program: http://www.eere.energy.gov/windandhydro/
- Global Wind Energy Council: http://www.gwec.net/
- European Wind Energy Association: http://www.ewea.org/
- American Wind Energy Association: http://www.awea.org/

Solar Energy

- *A Vision for Photovoltaic Technology*, European Communities, **2005**, available from: http://europa.eu.int/comm/research/energy/photovoltaics/intro
- Komp R. J., *Practical Photovoltaic. Electricity from Solar Cells*, Aatec Publications, Ann Arbor, **1995**.
- U.S. Department of Energy, Energy Efficiency and Renewable Energy (EERE), solar energy topics: http://www.eere.energy.gov/RE/solar.html
- *Basic Research Needs for Solar Energy Utilization. Report of the Basic Energy Sciences Workshop on Solar Energy Utilization*, Office of Science, U.S. Department of Energy, **2005**, available at http://www.sc.doe.gov/bes/reports/files/SEU_rpt.pdf.
- *Solar Energy Technologies Program: Multi-Year Technical Plan 2003– 2007 and Beyond*, Energy Efficiency and Renewable Energy (EERE) U.S. DOE, **2004**.
- International Energy Agency (IEA), Photovoltaic power program: http://www.iea-pvps.org/

Solar thermal for electricity production:
- International Energy Agency (IEA), Concentrated solar power for electricity generation: http://www.solarpaces.org/

Solar for heat production:
- International Energy Agency (IEA), Solar heating & cooling program: http://www.iea-shc.org/

Biomass

- *Industrial Uses of Biomass Energy. The Example of Brazil*, (Eds.: F. Rosillo-Calle, S. V. Bajay, H. Rothman), Taylor & Francis, London, **2000**.
- Rothman H., Greenshields R., Rosillo Callé F., *Energy from Alcohol, the Brazilian Experience*, University Press of Kentucky, Lexington, Kentucky, **1983**.
- Schobert H. H., *Renewable Energy from Biomass, in Energy and Society, an Introduction*, Taylor and Francis, New York, **2002**, p. 531.
- *Biomass as a Feedstock for a Bioenergy and Bioproducts Industry: The Technical Feasibility of a Billion-Ton Annual Supply*, U.S. Department of Agriculture (USDA) and U.S. Department of Energy (DOE), **2005**.
- International Energy Agency (IEA), bioenergy website: http://www.ieabioenergy.com/

- European Union website for biomass energy:
  http://europa.eu.int/comm/energy_transport/atlas/htmlu/
  renewables.html
- U.S. Department of Energy, National Renewable Energy Laboratory (NREL) Biomass Research: http://www.nrel.gov/biomass/
  and Energy Efficiency and Renewable Energy (EERE), Biomass
  Program: http://www.eere.energy.gov/biomass/

Ocean Energy

- General information about ocean energy form International
  Energy Agency (IEA), Ocean energy systems:
  http://www.iea-oceans.org/

Tidal and Current Power

- European Union website for tidal energy: http://europa.eu.int/
  comm/energy_transport/atlas/htmlu/tidal.html
- Johnson J., Power From Moving Water, *Chemical & Engineering
  News*, October 4, **2004**, p. 23.

Wave Energy

- Electric Power Research Institute (EPRI) studied the potential of
  offshore devices to produce electricity from waves. Reports avail-
  able from http://my.epri.com include:
  - Previsic M., Bedard R., Hagerman G., *E21 EPRI Assessment.
    Offshore Wave Energy Conversion Devices*, **2004**.
  - Bedard R., Hagerman G., Previsic M., et al., *Offshore Wave
    Power Feasibility Demonstration Project. Project Definition Study*.
    Final Summary Report, 2005.
- European Wave Energy Network: http://www.wave-energy.net/
- European Union website for wave energy: http://europa.eu.int/
  comm/energy_transport/atlas/htmlu/wave.html
- Ocean Power Delivery Ltd, energy production with the Pelamis
  device: http://www.oceanpd.com/

Ocean Thermal Energy

- U.S. National Renewable Energy Laboratory (NREL), Ocean
  Thermal Energy Conversion (OTEC) website:
  http://www.nrel.gov/otec/
- Sea Solar Power International, OTEC technology:
  http://www.seasolarpower.com/

Nuclear Energy

- *Nuclear Energy Today*, Nuclear Energy Agency (NEA), OECD Publication, Paris, **2003**, Available from: http://www.nea.fr/html/pub/nuclearenergytoday/welcome.html.
- Morris R. C., *The Environmental Case for Nuclear Power. Economical, Medical and political Considerations*, Paragon House, St. Paul, Minnesota, **2000**.
- Ramsey C. B., Modarres M., *Commercial nuclear Power. Assuring Safety for the future*, John Wiley & Sons, New York, **1998**.
- *The Economic Future of Nuclear power*, A study Conducted at the University of Chicago, **2004**, Available from: http://www.anl.gov/Special_Reports/NuclEconSumAug04.pdf. This study demonstrates that future nuclear power plants in the United States can be competitive with either natural gas or coal.
- *The Future of Nuclear Power*. An Interdisciplinary MIT Study, MIT, Cambridge, **2003**, Available from: http://web.mit.edu/nuclear-power/
- *A technology Roadmap for Generation IV Nuclear Systems*, US DOE Nuclear Energy Research Advisory Committee and the Generation IV International Forum, **2002** , Available from: http://gen-iv.ne.doe.gov/
- *Uranium 2003: Resources, Production and Demand*, OECD Nuclear Energy Agency and the International Atomic Energy agency, **2004**.
- World Nuclear Association: http://www.world-nuclear.org/
- OECD Nuclear Energy Agency (NEA): http://www.nea.fr/welcome.html
- Nuclear Energy Institute (NEI): http://www.nei.org/
- International Atomic Energy Agency: http://www.iaea.org/
- AREVA, worldwide leader in Nuclear Energy: http://www.areva.com/
- Commisariat à l'Energie Atomique (CEA): http://www.cea.fr/
- Generation IV Nuclear Energy Systems: http://gen-iv.ne.doe.gov/

Nuclear Fusion

- International Thermonuclear Experimental Reactor (ITER) to be constructed in Cadarache, France: http://www.iter.org/
- International Energy Agency (IEA), fusion section: http://www.iea.org/textbase/techno/technologies/index_fusion.asp
- *The Sun on the Earth*, Clefs CEA 49, **2004**, Available online from: http://www.cea.fr/gb/publications/Clefs49/contents.htm.

Hydrogen

- Romm J. J., *The Hype about Hydrogen. Fact and Fiction in the Race to Save the Climate*, Island Press, Washington, DC, **2004**.
- Hoffmann P., *Tomorrow's Energy. Hydrogen, Fuel Cells, and the Prospects for a Cleaner Planet*, The MIT Press, Cambridge, **2002**.
- Rifkin J., *The Hydrogen Economy*, Tarcher/Putnam, New York, **2002**.
- Sperling D., Cannon J., *The Hydrogen Energy Transition: Moving Toward the Post Petroleum Age in Transportation*, Elsevier Academic Press, **2004**.
- *The Hydrogen Economy: Opportunities, Costs, Barriers and R&D Needs*, National Research Council and National Academy Engineering, The National Academic Press, Washington, DC, **2004**.
- *Hydrogen and Other Alternative Fuels for Air and Ground Transportation*, (Ed.: H. W. Pohl), John Wiley & Sons, Chichester, England, **1995**.
- Bossel U., Eliasson B., Taylor G., *The Future of the Hydrogen Economy: Bright or Bleak?*, **2003**, Available from http://www.efcf.com/reports/
- Toward a Hydrogen Economy, editorial and special issue: *Science*, **2004**, vol. 305, p. 957.
- Wald M. L., Questions about a Hydrogen Economy, *Scientific American*, May, **2004**, p. 66.
- U.S. Department of Energy Hydrogen Program: http://www.hydrogen.energy.gov/
- U.S. Federal government's central source of information on R&D activities related to hydrogen and fuel cells: http://www.hydrogen.gov/
- International Energy Agency (IEA) Hydrogen Program: http://www.ieahia.org/
- National Hydrogen Association (U.S.): http://www.hydrogenus.com/
- Hydrogen and fuel cell information system: http://www.hyweb.de
- European Hydrogen Association: http://www.h2euro.org/

Fuel Cells

- *Fuel Cell Handbook*, 7th edn., U.S. DOE, National Energy Technology Laboratory (NREL), **2004**.
- Koppel T., *Powering the Future. The Ballard Fuel Cell and the Race to Change the World*, John Wiley & Sons Canada, Toronto, **1999**.
- Online fuel cell information resources: http://www.fuelcells.org/
- World Fuel Cell Council: http://fuelcellworld.org/

- Fuel Cell Today: http://www.fuelcelltoday.com/
- U.S. Department of Energy, Energy Efficiency and Renewable Energy (EERE), fuel cells: http://www.eere.energy.gov/hydrogenandfuelcells/

Methanol and the Methanol Economy

- *Methanol Production and Use*, (Eds.: W.-H. Cheng, H. H. Kung), Marcel Dekker, New York, **1994**.
- Asinger F., *Methanol, Chemie- und Energierohstoff. Die Mobilisation der Kohle*, Springer-Verlag, Heidelberg, **1987**.
- Perry J. H., Perry C. P., *Methanol, Bridge to a Renewable Energy Future*, University Press of America, Lanham, Maryland, **1990**.
- Bernton H., Kovarik W., Sklar S., *The Forbidden Fuel. Power Alcohol in the Twentieth Century*, Boyd Griffin, New York, **1982**.
- Dovring F., *Farming for Fuel*, Praeger, New York, **1988**.
- Gray Jr. C. L., Alson J. A., *Moving America to Methanol*, The University of Michigan Press, Ann Arbor, **1985**.
- *Methanol as an Alternative Fuel Choice: An Assessment*, (Ed.: W. L. Kohl), The Johns Hopkins University, Washington, DC, **1990**.
- Supp E., *How to Produce Methanol from Coal*, Springer-Verlag, Berlin, **1990**.
- Pavone A., *Mega Methanol Plants*, Report No. 43D, Process Economics Program, SRI Consulting, Menlo Park, California, **2003**.
- Lee S., *Methanol Synthesis Technology*, CRC Press, Boca Raton, Florida, **1990**.
- Fiedler E., Grossmann G., Kersebohm D. B., et al., Methanol, in *Ullmann's Encyclopedia of Industrial Chemistry*, Vol. 21, 6th edn., Wiley-VCH, Weinheim, Germany, **2003**, p. 611.
- Hansen J. B., Methanol Synthesis, in *Handbook of Heterogeneous Catalysis*, Vol. 4 (Eds.: G. Ertl, H. Knözinger, J. Weitkamp), Wiley-VCH, Weinheim, Germany, **1997**, p. 1856.
- Weissermel K., Arpe H.-J., *Industrial Organic Chemistry*, 4th edn., Wiley-VCH, Weinheim, Germany, **2003**, p. 30.
- Olah G. A., Molnár Á. *Hydrocarbon Chemistry*, 2nd edn., John Wiley & Sons, Hoboken, New Jersey, **2003**.
- Olah G. A., *Electrophilic Methane Conversion, Accounts of Chemical Research*, **1987**, vol. 20, p. 422.
- Olah G. A., Oil and Hydrocarbons in the 21st Century, in *Chemical Research - 2000 and Beyond: Challenges and Vision* (Ed.: P. Barkan), American Chemical Society, Washington DC, Oxford University Press, Oxford, **1998**.
- Olah G. A., The Methanol Economy, *Chemical & Engineering News*, September 22, **2003**, p. 5.

- Olah G. A., Beyond Oil and Gas: The Methanol Economy, *Angewandte Chemie Int. Ed.*, **2005**, vol. 44, p. 2636.
- Olah G. A., Prakash G. K. S., *Recycling of Carbon Dioxide into Methyl Alcohol and Related Oxygenates for Hydrocarbons*, US Patent 5,928,806, **1999**.
- Reed T. B., Lerner R. M., Methanol: A versatile Fuel for Immediate Use, *Science*, **1973**, vol. 182, p. 1299.
- Stiles A. B., Methanol, Past, Present, and Speculation on the Future, *AI ChE Journal*, **1977**, vol. 23, p. 362.
- *Beyond the Internal Combustion Engine: The Promise of Methanol Fuel Cell Vehicles*, Brochure published by the American Methanol Institute, Available from: http://www.methanol.org/.
- *Methanol in our Lives*, Brochure by Methanex, illustrating the presence of methanol in many products and materials of our daily lives, Available from: http://www.methanex.com.
- *Evaluation of the Fate and Transport of Methanol in the Environment*, prepared by Malcolm Pirnie, Inc. for the Methanol Institute, **1999**, Available from http://www.methanol.org/
- American Methanol Institute: http://www.methanol.org/
- Methanex methanol producer website: http://www.methanex.com
- Methanol Fuel Cell Alliance (MFCA): http://www.methanolresearch.com/

Methanol to Hydrocarbons

- Special issue covering methanol to hydrocarbons technologies and processes, in *Microporous and Mesoporous Materials*, Vol. 29 (1-2) (Eds.: M. Stocker, J. Weitkamp), **1999**.
- Chang C.D., Methanol to Hydrocarbons, in *Handbook of Heterogeneous Catalysis*, Vol. 4 (Eds.: G. Ertl, H. Knözinger, J. Weitkamp), Wiley-VCH, Weinheim, Germany, **1997**, p. 1894.
- Chang C.D., Methanol to Gasoline and Olefins, in *Methanol Production and Use* (Eds.: W.-H. Cheng, H. H. Kung), Marcel Dekker, New York, **1994**, p. 133.

Direct Methanol Fuel Cells (DMFC)

- Apanel G., Johnson E., Direct Methanol Fuel Cells: Ready to go Commercial?, *Fuel Cells Bulletin*, November, **2004**, p. 12.
- McGrath K. M., Prakash G. K. S., Olah G. A., Direct Methanol Fuel Cells, *Journal of Industrial and Engineering Chemistry*, **2004**, vol. 10, p. 1063.

- Surampudi S., Narayanan S. R., Vamos E., et al., Advances in Direct Oxidation Methanol Fuel Cells, *Journal of Power Sources*, **1994**, vol. 47, p. 377.

Dimethyl ether (DME)

- International DME Association: http://www.vs.ag/ida/
- For current information on DME:
  http://www.greencarcongress.com/dme/
- Haldor Topsoe DME information: http://www.haldortopsoe.com/

Other Reading

- Lomborg B., *The Skeptical Environmentalist. Measuring the Real State of the World*, Cambridge University press, Cambridge, **2001**.
- Prof. Olah's Nobel Lecture: Olah G. A., Carbocations and Their Role in Chemistry, *Angewandte Chemie Int. Ed.*, **1995**, vol. 34, p. 1393.

Transportation

- *Automotive Fuels for the Future: The Search for Alternatives*, International Energy Agency (IEA), **1999**.
- *Energy Technologies for a Sustainable Future: Transport*, International Energy Agency (IEA), **2004**.
- *Well-to-Wheels Analysis of Future Automotive Fuels and Powertrains in the European Context*, Tank-to-Wheels Report, Version 1, Concawe, European Council for Automotive R&D and European Commission Joint Research Centre, **2003**.

References

1. Olah G. A., *Methanol Economy* (trademark) No 78/692,647.
2. *World Energy Outlook* 2004, International Energy Agency, Paris, **2004**.
3. World Energy Council (WEC):, http://www.worldenergy.org/wec-geis/.
4. Smil V., *Energy at the Crossroads, Global Perspectives and Uncertainties*, MIT Press, Cambridge, **2003**.
5. *World Energy Outlook 2001: Insights*, International Energy Agency, Paris, **2001**.
6. *Energy Technologies for the 21st Century*, International Energy Agency, Paris, **1997**.
7. *FutureGen. Integrated Hydrogen, Electric Power Production and Carbon Sequestration Research Initiative. Report to the Congress*, Department of Energy (US), **2004**, available at http://www.energy.gov.
8. Getting to 'Clean Coal', *Chemical Engineering News*, February 23, **2004**.
9. Black B., *Petrolia, The Landscape of America's First Oil Boom*, The Johns Hopkins University Press, Baltimore, **2000**.
10. *All About Petroleum*, brochure from American Petroleum Institute, available at http://api-ec.api.org.
11. Johnson J., LNG Weighs Anchor, *Chemical Engineering News*, April 25, **2005**, p. 19.
12. Hightower M., Gritzo L., Anay L.-H., et al., *Guidance on Risk Analysis and Safety Implications of Large Liquefied Natural Gas (LNG) Spill over Water*, Sandia National Laboratories, **2004**.
13. Campbell C. J., Laherrère J. H., The End of Cheap Oil, *Scientific American*, March, **1998**, p. 78.
14. *Key World Energy Statistics 2004*, International Energy Agency (IEA).
15. World Coal Institute.
16. *Annual Energy Outlook 2005 With Projections to 2025*, Energy Information Administration (EIA), US DOE, Washington, **2005**.
17. *BP Statistical Review of World Energy*, BP, **2005**, available online at www.bp.com/statisticalreview.
18. U.S. Bureau of Transportation Statistics.
19. Oil Tech Inc.: http://www.oiltechinc.com/.
20. Snyder R. E., Oil Shale back in the Picture, *WorldOil Magazine online*, August, **2004**, vol. 225.
21. *IPCC Third Assessment Report: Climate Change 2001: The Scientific Basis*, (Eds.: J. T. Houghton, Y. Ding, D. J. Griggs, M. Noguer), Cambridge University Press, Cambridge, U.K., **2001**.
22. Sasol: http://www.sasol.com/.
23. Gold R., In Qatar, Oil Firms Make Huge Bet an Alternative Fuel, *The Wall Street Journal*, February 15, **2005**.
24. *U.S. Geological Survey World Petroleum Assessment 2000*, USGS, Denver, Colorado, **2000**, available at: http://pubs.usgs.gov/dds/dds-060/.
25. Hubbert M. K., *Nuclear Energy and the Fossil Fuels, American Petroleum Institute Drilling and Production Practice*, Proceedings of the spring Meeting, San Antonio, March 7-9, **1956**.
26. Campbell C. J., *The Coming Oil Crisis*, Multi-science Publishing, Brentwood, England, **1988**.

Beyond Oil and Gas: The Methanol Economy. G. A. Olah, A. Goeppert, G. K. S. Prakash
Copyright © 2006 WILEY-VCH Verlag GmbH & Co. KGaA, Weinheim
ISBN 3-527-31275-7

27. Deffeyes K. S., *Hubbert's Peak, the Impending World Oil Shortage*, Princeton University Press, Princeton, **2001**.

28. Hirsch R. L., Bezdek R., Wendling R., *Peaking of World Oil Production: Impact, Mitigation & Risk Management*, prepared for the U.S. DOE's National Energy Technology Laboratory (NETL) by Science Applications International Corporation (SAIC). **2005**.

29. Bentley R. W., Oil & Gas Depletion: an Overview, *Energy Policy*, **2002**, vol. 30, p. 189.

30. Donnely J. K., Pendergast D. R., *Nuclear Energy in Industry: Application to Oil Production, Climate Change and Energy Options Symposium*, Canadian Nuclear Society, Ottawa, Canada, November 17–19, **1999**.

31. *Nuclear Power an Attractive Option for Tar Sands: Alberta Chamber Report, Nuclear Canada*, February 10, **2004**, p. 2.

32. *Oil Sands Technology Roadmap. Unlocking the Potential*, Alberta Chamber of Resources, Edmonton, Alberta, **2004**, Available at http://www.acr-alberta.com/.

33. Odell P. R., Rosing K. E., *The Future of Oil. World Oil Resources and Use*, Kogan Page, London, **1983**.

34. Maugeri L., Oil: Never Cry Wolf. Why the Petroleum Age Is Far from Over, *Science*, **2004**, vol. 304, p. 1115.

35. Greene D. L., Hopson J. L., Li J., *Running Out of and Into Oil: Analyzing Global Oil Depletion and Transition Through 2050*, prepared by Oak Ridge National Laboratory for the U.S. DOE, **2003**.

36. Laherrere J., *Natural Gas Future Supply*, International Institute for Applied Systems Analysis (IIASA) International Energy Workshop, June 22-24, Paris, **2004**, available at: http://www.hubbertpeak.com/laherrere/.

37. Rogner H. H., An Assessment of World Hydrocarbon resources, *Annual Review of Energy and the Environment*, **1997**, vol. 22, p. 217.

38. Odell P. R., Fossil Fuel Resources in the 21st Century, *Financial Times Energy*, London, **1999**.

39. Odell P. R., *Why Carbon Fuels will Dominate the 21st Century's Global Energy Economy*, Multi-Science Publishing, Brentwood, U.K., **2004**.

40. Gold T., *The Deep Hot Biosphere*, Copernicus Press, New York, **1999**.

41. *IPCC Third Assessment Report: Climate Change* 2001, Cambridge University Press, Cambridge, U.K., **2001**.

42. Arrhenius S., On the Influence of Carbonic Acid in the Air Upon the Temperature of the Ground, *Philosophical Magazine*, **1896**, vol. 41, p. 237.

43. Solanki S. K., Usoskin I. G., Kromer B., et al., Unusual Activity of the Sun During Recent Decades Compared to the Previous 11,000 years, *Nature*, **2004**, vol. 431, p. 1084.

44. Essenhigh R. H., Does CO_2 Really Drive Global Warming?, *Chemical Innovation*, May, **2001**, p. 44.

45. Idso S., B., *Biological Consequences of Increased Concentrations of Atmospheric CO_2*, in Global Warming: The Science and the Politics (Ed.: L. Jones), The Fraser Institute, Vancouver, **1997**.

46. Socolow R. H., Can we Bury Global Warming?, *Scientific American*, July **2005**, p. 49.

47. Johnson J., Putting a Lid on Carbon Dioxide, *Chemical Engineering News*, December 20, **2004**.

48. World Commission on Dams: http://www.dams.org/.

49. Wonder of the World Databank, http://www.pbs.org.

50. European Deep Geothermal Energy Programme: http://www.soultz.net/.

51. Global Wind Energy Council (GWEC): http://www.gwec.net/.

52. *Wind Energy: The Facts. An Analysis of Wind Energy in the EU-25*, (Ed.: H. Chandler), European Wind Energy Association (EWEA), available at: http://www.ewea.org/, **2003**.

53. *Solar Energy Technologies Program: Multi-Year Technical Plan 2003-2007 and Beyond*, Energy Efficiency and Renewable Energy (EERE) U.S. DOE, **2004**.

54. IEA Photovoltaic Power Systems Programme. http://www.iea-pvps.org/.

55. Sayigh A., Spotlight on PV Energy: As Commercialisation Grows, Solar Needs Attention, in *Renewable Energy 2003*, An official publication of the World

Renewable Energy Network, UNESCO, 2003.

56. *Renewable Energy Technology Characterization,* Electric Power Research Institute (EPRI) and U.S. DOE, **1997**.

57. Port O., Power from the Sunbaked Desert: Solar Generator May Be the Hot Source of Plentiful Electricity, *Business-Week,* September 12, **2005**, p. 76.

58. Stirling Energy Systems: http://www.stirlingenergy.com/.

59. Einav A., Solar Energy Research and Development Achievements in Israel and Their Practical Significance, *Journal of Solar Energy Engineering,* **2004**, vol. 126, p. 921.

60. *Industrial Uses of Biomass Energy. The Example of Brazil,* (Eds.: F. Rosillo-Calle, S. V. Bajay, H. Rothman), Taylor & Francis, London, **2000**.

61. Buarque de Hollanda J., Dougals Poole A., *Sugarcane as an Energy Source in Brazil,* Instituto Nacional de Eficiencia Energetica.

62. São Paulo Sugarcane Agroindustry Union (UNICA): http://www.unica.com.br/.

63. Dickerson M., Homegrown Fuel Supply Helps Brazil Breathe Easy, *Los Angeles Times,* **2005**.

64. Shapouri H., Duffield J. A., Wang M., *The Energy Balance of Corn Ethanol: An Update,* U.S. Department of Agriculture, Washington, DC, **2002**.

65. Hess G., Ethanol Wins Big in Energy Policy, *Chemical Engineering News,* September 12, **2005**.

66. Patzek T. W., Thermodynamics of the Corn-Ethanol Biofuel Cycle, *Critical Reviews in Plant Sciences,* **2004**, vol. 23, p. 519.

67. Pimentel D., Ethanol Fuels: Energy Balance, Economics, and Environmental Impacts are Negative, *Natural Resources Research,* **2003**, vol. 12, p. 127.

68. Pimentel D., Patzek T. W., Ethanol Production Using Corn, Switchgrass, and Wood; Biodiesel Production Using Soybean and Sunflower, *Natural Resources Research,* **2005**, vol. 14, p. 65.

69. Bensaïd B., *Road Transport Fuels in Europe: the Explosion of Demand for Diesel Fuel, Panorama 2005,* Institut Francais du Petrole (IFP), **2005**.

70. *Biofuel Barometer, Systemes Solaire,* June, **2005**, vol. 167.

71. *Energy Technologies for a Sustainable Future: Transport,* International Energy Agency (IEA), Paris, **2004**.

72. Pontes T., António F., *Ocean Energies: Resources and Utilisation,* 18th World Energy Conference, Buenos Aires, Argentina, October 21-25, **2001**.

73. European Union, Atlas Project: http://europa.eu.int/comm/energy_transport/atlas/.

74. Hammerfest Strøm AS: http://www.e-tidevannsenergi.com/.

75. Johnson J., Power From Moving Water, *Chemical & Engineering News,* October 4, **2004**, p. 23.

76. Clément A., McCullen P., Falcão A., et al., Wave Energy in Europe: Current Status and Perspectives, *Renewable and Sustainable Energy Reviews,* **2002**, vol. 6, p. 405.

77. Ocean Power Delivery Ltd.: http://www.oceanpd.com.

78. Cowan G. A., A Natural Fission Reactor, *Scientific American,* July, **1976**, p. 36.

79. *A Technology Roadmap for Generation IV Nuclear Systems,* US DOE Nuclear Energy Research Advisory Committee and the Generation IV International Forum, **2002**, Available from: http://gen-iv.ne.doe.gov/.

80. Claude B., *Superphénix, le Nucléaire à la Française,* l'Harmattan, Paris, **1999**.

81. Argonne National Laboratory-West. http://www.anlw.anl.gov/.

82. *Uranium 2003: Resources, Production and Demand,* OECD Nuclear Energy Agency and the International Atomic Energy agency, **2004**.

83. Lidsky L. M., Miller M. M., Nuclear Power and Energy Security: A Revised Strategy for Japan, *Science and Global Security,* **2002**, vol. 10, p. 127.

84. *Chernobyl's Legacy: Health, Environmental and Socio-economic Impacts and Recommendations to the Governments of Belarus, the Russian Federation and Ukraine,* UN, **2005**.

85. Gabbard A., Coal Combustion: Nuclear Resource or Danger?, *ORNL Review,* vol. 26, **1993**, available at: http://www.ornl.

gov/info/ornlreview/rev26-34/text/colmain.html.

86. Yucca Mountain Standards, EPA. http://www.epa.gov/radiation/yucca/.

87. Nuclear Energy Institute. http://www.nei.org/.

88. Morris R. C., *The Environmental Case for Nuclear Power. Economical, Medical and Political Considerations*, Paragon House, St. Paul, Minnesota, **2000**.

89. Hoffmann P., *Tomorrow's Energy. Hydrogen, Fuel Cells, and the Prospects for a Cleaner Planet*, The MIT Press, Cambridge, **2002**.

90. *Hydrogen and Other Alternative Fuels for Air and Ground Transportation*, (Ed.: H. W. Pohl), John Wiley & Sons, Chichester, England, **1995**.

91. *The Hydrogen Economy: Opportunities, Costs, Barriers and R&D Needs*, National Research Council and National Academy Engineering, The National Academic Press, Washington, DC, **2004**.

92. U.S. Department of Energy. http://www.energy.gov/.

93. Toward a Hydrogen Economy, editorial and special issue: *Science*, **2004**, vol. 305, p. 957.

94. Milne T. A., Elam C. C., Evans R. J., *Hydrogen from Biomass: State of the Art and Research Challenges*, IEA/H2/TR-02/001, International Energy Agency (IEA).

95. Romm J. J., *The Hype about Hydrogen. Fact and Fiction in the Race to Save the Climate*, Island Press, Washington, DC, **2004**.

96. Bossel U., Eliasson B., Taylor G., *The Future of the Hydrogen Economy: Bright or Bleak?*, **2003**, Available from http://www.efcf.com/reports/.

97. Altmann M., Gaus S., Landinger H., et al., *Wasserstofferzeugung in Offshore Windparks 'Killer-Kriterien', Grobe Auslegung und Kostenabschaetzung*, Studie im Auftrag von GEO Gesellschaft fuer Energie Und Oekologie mbH, L-B-Systemtechnik GmbH, Ottobrunn, Germany, **2001**.

98. His S., *Panorama 2004: Hydrogen: An Energy Vector for the Future?*, Institut Francais du Petrole (IFP), **2004**, available at http://www.ifp.fr/IFP/en/aa.htm.

99. Bientôt l'Ère Hydrogène, *Alternatives Magazine*, **2004**, vol. 7, p. 8, available at http://www.areva.com/.

100. Source: BMW.

101. U.S. Department of Energy, Energy Efficiency and Renewable Energy (EERE).

102. Fairley P., Recharging the Power Grid, *Technology Review*, March, **2003**, p. 50.

103. VRB-ESS: The Great Leveller, *Modern Power Systems*, June, **2005**, p. 55.

104. Williams B., Hennesy T., Electric Oasis, *IEE Power Engineering*, February/March, **2005**, p. 28.

105. Wilks N., Whatever the Weather. Advances in Battery Technology Could Hold the Key to Successful Development of Alternative Sources of Energy, *Professional Engineering*, October 6, **2004**, p. 33.

106. Boyle R., *The Sceptical Chymist*, London, **1661**.

107. Stiles A. B., Methanol, Past, Present, and Speculation on the Future, *AIChE Journal*, **1977**, vol. 23, p. 362.

108. *Methanol Production and Use*, (Eds.: W.-H. Cheng, H. H. Kung), Marcel Dekker, New York, **1994**.

109. Fiedler E., Grossmann G., Kersebohm D. B., et al., Methanol, in *Ullmann's Encyclopedia of Industrial Chemistry*, Vol. 21, 6th ed., Wiley-VCH, Weinheim, Germany, **2003**, p. 611.

110. Fischer F., Tropsch H., Synthesis of Higher Members of the Aliphatic Series from Carbon Monoxide, *Berichte*, **1923**, vol. 56B, p. 2428.

111. Fischer F., Tropsch H., Direct Synthesis of Petroleum Hydrocarbons at Ordinary Pressure, *Berichte*, **1926**, vol. 59B, p. 830.

112. Edmonds W. J., *Synthetic Methanol Process, Commercial Solvents Corporation*, US Patent 1,875,714, **1932**.

113. Weissermel K., Arpe H.-J., *Industrial Organic Chemistry*, 4th edn, Wiley-VCH, Weinheim, Germany, **2003**.

114. Methanol, *Chemical Week*, June 22, **2005**, p. 33.

115. *Methanol in our Lives*, Brochure by methanol producer Methanex, illustrating the presence of methanol in many products and materials of our daily lives, available from: http://www.methanex.com.

116. *On the Road with Methanol: The present and Future Benefits of Methanol Fuel*, Prepared for the Methanol Institute, **1994**.

117. Bernton H., Kovarik W., Sklar S., *The Forbidden Fuel. Power Alcohol in the Twentieth Century*, Boyd Griffin, New York, **1982**.

118. Reed T. B., Lerner R. M., Methanol: A versatile Fuel for Immediate Use, *Science*, **1973**, vol. 182, p. 1299.

119. *Beyond the Internal Combustion Engine: The Promise of Methanol Fuel Cell Vehicles*, Brochure published by the American Methanol Institute, available from: http://www.methanol.org/.

120. Perry J. H., Perry C. P., *Methanol, Bridge to a Renewable Energy Future*, University Press of America, Lanham, Maryland, **1990**.

121. Moffat A. S., Methanol-Powered, *Science*, **1991**, vol. 251, p. 514.

122. *Alternative to Traditional Transportation Fuels 1998*, DOE/EIA-0585(98), Washington, DC.

123. *Alternative Fuels for Vehicles Fleet Demonstration Program Volume 3*, Technical Reports, New York State Energy Research and Development Authority, **1997**.

124. *Alternative Fuel: Transit Buses*. Final Results from the National Renewable Energy Laboratory Vehicle Evaluation Program, Produced for the US DOE, **1996**.

125. Ogawa T., Inoue N., Shikada T., et al., Direct Dimethyl Ether Synthesis, *Journal of Natural Gas Chemistry*, **2003**, vol. 12, p. 219.

126. Hirano M., Imai T., Yasutake T., et al., Dimethyl Ether Synthesis from Carbon Dioxide by Catalytic Hydrogenation (Part 2) Hybrid Catalyst Consisting of Methanol Synthesis and Methanol Dehydration Catalysts, *Journal of the Japan Petroleum Institute*, **2004**, vol. 47, p. 11.

127. Hansen J. B., Mikkelsen S.-E., *DME as a Transportation Fuel*, Project Carried out for the Danish Road Safety & Transport Agency and the Danish Environmental Protection agency, **2001**.

128. Volvo Bus Corporation Company Presentation, **2004**.

129. JFE Holdings, Inc, http://www.jfe-holdings.co.jp/en/dme/.

130. Basu A., Wainwright J. M., *DME as a Power Generation Fuel: Performance in Gas Turbines*, Presented at the PETROTECH-2001 Conference, New Delhi, India, January, **2001**.

131. Ohno Y., Omiya M., *Coal Conversion into Dimethyl Ether as an Innovative Clean Fuel*, Presented at the 12th International Conference on Coal Science, November, **2003**.

132. Pavone A., *Mega Methanol Plants, Report No. 43D*, Process Economics Program, SRI Consulting, Menlo Park, California, **2003**.

133. Ryu J. Y., Gelbein A. P., *Producing Dimethyl Carbonate from CO₂ and Methanol. A Green Chemistry Alternative to Phosgene as a Chemical Intermediate*, in 4th Annual Green Chemistry and Engineering Conference Proceedings, Washington, DC, June 27-29, **2000**, p. 33.

134. *Methanol Institute Comments to US DOE On-Board Fuel Processing Review Panel*, **2004**, available from: http://www.methanol.org.

135. See, *Fuel Cell Vehicles Chart (from Auto Manufacturers)*, http://www.fuelcells.org.

136. See, DaimlerChrysler, http://www.daimlerchrysler.com.

137. See, Ford Motor Company, http://www.ford.com/.

138. For more information see the Georgetown University website on fuel cell buses, http://fuelcellbus.georgetown.edu/.

139. *Methanol to Hydrogen Fueling Stations*, Methanol Institute, **2003**.

140. Japan Hydrogen & Fuel Cell Demonstration Project (JHFC), http://www.jhfc.jp.

141. Apanel G., Johnson E., Direct Methanol Fuel Cells - Ready to go Commercial?, *Fuel Cells Bulletin*, November, **2004**, p. 12.

142. Voss D., A Fuel Cell in your Phone, *Technology Review*, November, **2001**, p. 68.

143. Surampudi S., Narayanan S. R., Vamos E., et al., Advances in Direct Oxidation Methanol Fuel Cells, *Journal of Power Sources*, **1994**, vol. 47, p. 377.

144. Surampudi S., Narayanan S. R., Vamos E., et al., US Patent 5,599,638 (**1997**), US Patent 6,248,460 (**2001**), US Patent

6,740,434 (**2004**), US Patent 6,821,659 (**2004**).

145. Prakash G. K. S., Smart M. C., Wang Q.-J., et al., High Efficiency Direct Methanol Fuel Cell Based on Poly(styrenesulfonic) Acid (PSSA) - Poly(vinylidenefluoride) (PVDF) Composite Membranes, *Journal of Fluorine Chemistry*, **2004**, vol. 125, p. 1217.

146. Prakash G. K. S., Olah G. A., Smart M. C., et al., *Polymer Electrolyte Membranes for Use in Fuel Cells*, US Patent 6,444,343, **2002**.

147. Dillon R., Srinivasan S., Arico A. S., et al., International Activities in DMFC R&D: Status of Technologies and Potential Applications, *Journal of Power Sources*, **2004**, vol. 127, p. 112.

148. Jung D. H., Jo Y.-K., Jung J.-H., et al., *Proc. of the Fuel Cell Seminar*, Portland, Oregon, **2000**, p. 420.

149. Bostaph J., Koripella R., Fisher A., et al., in *Proc. of the 199th Meeting of Direct Methanol Fuel Cells*, Electrochemical Society, Washington, DC, March 25-29, **2001**.

150. Jung D. H., Jo Y.-K., Jung J.-H., et al., in *Proceedings of the Fuel Cell Seminar*, Portland, Oregon, **2000**, p. 420.

151. Kim D., Cho E. A., Hong S.-A., et al., Recent Progress in the Passive Direct Methanol Fuel Cell at KIST, *Journal of Power Sources*, **2004**, vol. 130, p. 172.

152. http://www.fuelcellstore.com.

153. Dolan G. A., In Search of the Perfect Clean-Fuel Options, *Hydrocarbon Processing*, March, **2002**, p. 1.

154. JuVOMe Presentation, Research Center Julich http://www.fz-juelich.de/portal/angebote/pressemitteilungen/scooter.

155. Yamaha Motor Co.: http://www.yamaha-motor.co.jp/motorshow/html/0003.html.

156. Geiger S., Jollie D., Report from the 2003 Fuel Cell Seminar, Miami, Fuel Cell Today, 14 November, **2003**, http://www.fuelcelltoday.com.

157. Vectrix Corp.: http://www.vectrixusa.com.

158. Olah G. A., Prakash G. K. S., *Recycling of Carbon Dioxide into Methyl Alcohol and Related Oxygenates for Hydrocarbons*, US Patent 5,928,806, **1999**.

159. Temchin J., *Analysis of Market Characteristics for Conversion of Liquid Fueled Turbines to Methanol*, Prepared for The Methanol Foundation and Methanex by Electrotek Concepts, **2003**.

160. *GE Position Paper: Feasibility of Methanol as Gas Turbine Fuel*, General Electric, **2001**.

161. http://www.fuelcelltoday.com/.

162. Stokes H., *Commercialization of a New Stove and Fuel System for Household Energy in Ethiopia Using Ethanol from Sugar Cane Residues and Methanol from Natural Gas*, Presented to the Ethiopian Society of Chemical Engineers ESChE at the Forum on "Alcohol as an Alternative Energy Resource for Household Use", October 30, **2004**.

163. Ebbeson B., Stokes H. C., Stokes C. A., *Methanol - The Other Alcohol: A Bridge to a Sustainable Clean Liquid Fuel*, **2000**.

164. *Methanol Refueling Station Costs*, Prepared by EA engineering, Science and Technology, Inc. for the American Methanol Foundation, **1999**, available from http://www.methanol.org.

165. Ashley S., On the Road to Fuel Cell Cars, *Scientific American*, March, **2005**, p. 62.

166. *Methanol Market Distribution Infrastructure in the United States*, Prepared by DeWitt & Company, Inc for the Methanol Institute, **2002**, available from: http://www.methanol.org.

167. *Methanol Fact Sheets*, American Methanol Institute, Washington, DC, **1993**.

168. Methanex website: http://www.methanex.com/.

169. *Methanol in Fuel Cell Vehicles: Human Toxicity and Risk Evaluation* (Revised), Prepared by Statoil, Norway, **2001**.

170. *Methanol Health Risk Fact Sheet*, Methanol Institute, available at http://www.methanol.org.

171. *Methanol Health Effects Fact Sheet*, Methanol Institute, available at http://www.methanol.org.

172. *Methanol Fuels and Fire Safety*, Fact Sheet OMS-8, EPA 400-F-92-010, US Environmental Protection Agency (EPA), Office of Mobile Sources, Washington, DC, **1994**.

173. *Clean Alternative Fuels: Methanol, Fact Sheet EPA 420-F-00-040*, US Environmental Protection Agency (EPA), Transportation and Air Quality, **2002**.

174. Pollutant Emissions from Georgetown University Methanol Powered Fuel Cell Buses: http://fuelcellbus.georgetown.edu/overview3.cfm.

175. *Methanol, Health and Safety Guide (HSG 105, 1997)*, International Programme on Chemical Safety (IPCS), http://www.inchem.org/.

176. *Wastewater Treatment with Methanol Denitrification, Fact Sheet*, Methanol Institute, available at http://www.methanol.org.

177. *Evaluation of the Fate and Transport of Methanol in the Environment*, prepared by Malcolm Pirnie, Inc. for the Methanol Institute, **1999**, available from http://www.methanol.org/.

178. Brown R., Methanol Market Quiet, *Chemical Market Reporter*, section 2, January 31, **2005**, p. 5.

179. Brown R., Methanol Pricing Steady as Supply Situation Changes, *Chemical Market Reporter*, section 2, October 4, **2004**, p. 19.

180. Plouchart G., Panorama 2005: *Energy Consumption in the Transportation Sector*, Institut Francais du Petrole (IFP), **2005**, available from http://www.ifp.fr/IFP/en/aa.htm.

181. *Commercial-Scale Demonstration of the Liquid Phase Methanol (LPMEOH™) Process: Final Report*, Prepared by Air Products Liquid Phase Conversion Company for the US DOE National Energy Technology Laboratory, **2003**.

182. Kochloefl K., Steam Reforming, in *Handbook of Heterogeneous Catalysis*, Vol. 4 (Eds.: G. Ertl, H. Knözinger, J. Weitkamp), Wiley-VCH, Weinheim, Germany, **1997**.

183. Hansen J. B., Methanol Synthesis, in *Handbook of Heterogeneous Catalysis*, Vol. 4 (Eds.: G. Ertl, H. Knözinger, J. Weitkamp), Wiley-VCH, Weinheim, Germany, **1997**, p. 1856.

184. Turek T., Trimm D. L., Cant N. W., The Catalytic Hydrogenolysis of Esters to Alcohols, *Catalysis Reviews – Science and Engineering*, **1994**, vol. 36, p. 645.

185. Christiansen J. A., *Method of Producing Methyl Alcohol From Alkyl Formate*, US Patent 1,302,011, **1919**.

186. Marchionna M., Lami M., Raspolli Galleti A. M., Synthesizing Methanol at Lower Temperature, *Chemtech*, April, **1997**, p. 27.

187. Crabtree R. H., Aspects of Methane Chemistry, *Chemical Reviews*, **1995**, vol. 95, p. 987.

188. Lunsford J. H., *Catalytic Conversion of Methane to more Useful Chemicals and Fuels: a Challenge for the 21st Century*, **2000**.

189. Otsuka K., Wang Y., Direct Conversion of Methane into Oxygenates, *Applied Catalysis A: General*, **2001**, vol. 222, p. 145.

190. Olah G. A., Electrophilic Methane Conversion, *Accounts of Chemical Research*, **1987**, vol. 20, p. 422.

191. Olah G. A., Molnár Á. *Hydrocarbon Chemistry*, 2nd ed., John Wiley & Sons, Hoboken, New Jersey, **2003**.

192. Olah G. A., Prakash G. K. S., *Efficient and Selective Conversion of Methane to Methanol*, US Provisional patent application No. 60 671-650, **2005**.

193. Periana R. A., Bhalla G., Tenn W. J., et al., Perspective on Some Challenges and Approaches for Developing the Next Generation of Selective, Low Temperature, Oxidation Catalysts for Alkane Hydroxylation Based on CH Activation Reaction, *Journal of Molecular Catalysis A*, **2004**, vol. 220, p. 7.

194. Periana R. A., Taube T. J., Evitt E. R., et al., A Mercury-Catalyzed, High-Yield System for the Oxidation of Methane to Methanol, *Science*, **1993**, vol. 259, p. 340.

195. De Vos D. E., Sels B. F., Gold Redox Catalysis for Selective Oxidation of Methane to Methanol, *Angewandte Chemie Int. Ed.*, **2005**, vol. 44, p. 30.

196. Jones C. J., Taube D., Ziatdinov V. R., et al., Selective Oxidation of Methane to Methanol Catalyzed, with C-H Activation, by Homogeneous, Cationic Gold, *Angewandte Chemie Int. Ed.*, **2004**, vol. 43, p. 4626.

197. Olah G. A., Gupta B., Farina M., et al., Selective Monohalogenation of Methane over Supported Acid or Platinum Metal Catalysts and Hydrolysis of Methyl Ha-

lides over Gamma-Alumina-Supported Metal Oxide/Hydroxide Catalysts. A Feasible Path for the Oxidative Conversion of Methane into Methyl Alcohol/Dimethyl Ether, *Journal of the American Chemical Society*, **1985**, vol. 107, p. 7097.

198. Olah G. A., *Methyl Halides and Methyl Alcohol from Methane*, US Patent 4,523,040, **1985**.

199. Olah G. A., Mo Y. K., Electrophilic Reaction at Single Bonds. XIII. Chlorination and Chlorolysis of Alkanes in SbF_5-Cl_2-SO_2ClF Solution at Low Temperature, *Journal of the American Chemical Society*, **1972**, vol. 94, p. 6864.

200. Olah G. A., Renner R., Schilling P., et al., Electrophilic Reactions at Single Bonds. XVII. SbF_5, $AlCl_3$, and $AgSbF_6$ Catalyzed Chlorination and Chlorolysis of Alkanes and Cycloalkanes, *Journal of the American Chemical Society*, **1973**, vol. 95, p. 7686.

201. Pan H. Y., Minet R. G., Benson S. W., et al., Process for Converting Hydrogen Chloride to Chlorine, *Industrial Engineering Chemical Research*, **1994**, vol. 33, p. 2996.

202. Mortensen M., Minet R. G., Tsotsis T. T., et al., The Development of Dual Fluidized-Bed reactor System for the Conversion of Hydrogen Chloride to Chlorine, *Chemical Engineering Science*, **1999**, vol. 54, p. 2131.

203. Lorkovic I., Noy M., Weiss M., et al., C_1 Coupling via Bromine Activation and Tandem Catalytic Condensation and Neutralization over CaO/Zeolite Composite, *Chemical Communications*, **2004**, p. 566.

204. Schweizer A. E., Jones M. E., Hickman D. A., *Oxidative Halogenation of C1 Hydrocarbons into Halogenated C1 Hydrocarbons and Integrated Processes Related Thereto*, US Patent 6,452,058, **2002**.

205. Periana R. A., Mirinov O., Taube D. J., et al., High Yield Conversion of Methane to Methyl Bisulfate Catalyzed by Iodine Cations, *Chemical Communications*, **2002**, p. 2376.

206. Baik M.-H., Newcomb M., Friesner R. A., et al., Mechanistic Studies on the Hydroxylation of Methane by Methane Monooxygenase, *Chemical Reviews*, **2003**, vol. 103, p. 2385.

207. Ayala M., Torres E., Enzymatic Activation of Alkanes: Constraints and Prospective, *Applied Catalysis A: General*, **2004**, vol. 272, p. 1.

208. Süss-Fink G., Stanislas S., Shul'pin G. B., et al., Catalytic Functionalization of Methane, *Applied Organometallic Chemistry*, **2000**, vol. 14, p. 623.

209. Hamelinck C. N., Faaij A. P. C., *Future Prospects for Production of Methanol and Hydrogen from Biomass*, University Utrecht, Copernicus Institute, The Netherlands, **2001**.

210. Swaaij W. P. M., Kersten S. R. A., van den Aarsen F. G., *Routes for Methanol from Biomass*, International Business Conference on Sustainable Industrial Developments, Delfzijl, The Netherlands, April, **2004**.

211. Henrich E., *Kraftstoff aus Stroh*, NRW Fachtagung "Was Tanken wir Morgen?", Oberhausen, November 25-26, **2002**.

212. Mitsubishi Heavy Industries, Ltd: http://www.mhi.co.jp/power/e_power/techno/biomass/.

213. Norbeck J. M., Johnson K., *The Hynol Process: A Promising Pathway for Renewable Production of Methanol*, College of Engineering, Center for Environmental research and Technology, University of California, Riverside, **2000**.

214. Ekbom T., Lindblom M., Berglin N., et al., *Cost-Competitive, Efficient Bio-Methanol Production from Biomass via Black Liquor Gasification*, Alterner Program of the European Union, **2003**.

215. Cassedy E. S., *Prospects for Sustainable Energy. A Critical Assessment*, Cambridge University Press, Cambridge, **2000**.

216. *Climate Change 2001, Mitigation*, IPCC Third Assessment Report.

217. Nyns E.-J., Methane, in *Ullmann's Encyclopedia of Industrial Chemistry*, Vol. 21, Wiley-VCH, Weinheim, Germany, **2003**, p. 599.

218. US Environmental Protection Agency (EPA), Municipal Solid Waste and Landfill Methane Outreach Program, http://www.epa.gov.

219. Sheehan J., Dunahay T., Benemann J., et al., *A Look Back at the U.S. Department of Energy's Aquatic Species Program – Biodiesel from Algae*, National Renewable Energy Laboratory (NREL), **1998**.

220. Xiaoding X., Moulijn J. A., Mitigation of CO_2 by Chemical Conversion: Plausible Chemical Reactions and Promising Products, *Energy & Fuels*, **1996**, vol. 10, p. 305.

221. Goehna H., Koenig P., Producing Methanol from CO_2, Chemtech, June, **1994**, p. 39.

221a. Saito M., R & D Activities in Japan on Methanol Synthesis from CO_2 and H_2, *Catalysis Surveys from Japan*, **1998**, Vol. 2, p. 175.

221b. Saito M., Murata K., Development of High Performance Cu/ZnO-Based Catalysts for Methanol Synthesis and Water-Gas Shift Reaction, *Catalysis Surveys from Asia*, **2004**, Vol. 8, p. 285.

222. Steinberg M., Fossil Fuel Decarbonization Technology for Mitigating Global Warming, *International Journal of Hydrogen Energy*, **1999**, vol. 24, p. 771.

223. Augustynski J., Sartoretti C. J., Kedzierzawski P., Electrochemical Conversion of Carbon Dioxide, in *Carbon Dioxide Recovery and Utilization* (Ed.: M. Aresta), Kluwer Academic Publisher, Dordrecht, The Netherlands, **2003**, p. 279.

223a. Bagotzky V. S., Osetrova N. V., Electrochemical Reduction of Carbon Dioxide, *Russian Journal of Electrochemistry*, **1995**, Vol. 31, p. 409.

223b. *Electrochemical and Electrocatalytic Reactions of Carbon Dioxide*, (Eds.: B. P. Sullivan, K. Krist, H. E. Guard), Elsevier, Amsterdam, **1993**.

223c. Olah G. A., Prakash G. K. S., *Efficient Selective Conversion of Carbon Dioxide to Methanol*, US Provisional patent application 60 671-651, **2005**.

224. Specht M., Bandi A., Herstellung von Fluessigen Kraftstoffen aus Atmosphaerischem Kohlendioxid, in *Forshungsverbund Sonnenenergie, Themen 1994-1995, Energiespeicherung*, **1995**, p. 41.

225. Asinger F., *Methanol, Chemie- und Energierohstoff. Die Mobilisation der Kohle*, Springer-Verlag, Heidelberg, **1987**.

226. Pasel J., Peters R., Specht M., Methanol Herstellung und Einsatz als Energietraeger fuer Brennstoffzellen, in *Forshungsverbund Sonnenenergie, Themen 1999-2000: Zukunftstechnologie Breenstoffzelle*, Berlin, Germany, **2000**, p. 46.

227. Specht M., Bandi A., 'The Methanol Cycle' – *Sustainable Supply of Liquid Fuels*, Center for Solar Energy and Hydrogen Research (ZSW), Stuttgart, Germany.

228. Lackner K. S., Ziock H.-J., Grimes P., The Case for Carbon Dioxide Extraction from Air, *SourceBook*, **1999**, vol. 57, p. 6.

229. Schuler S. S., Constantinescu M., Coupled CO_2 recovery from the Atmosphere and Water Electrolysis: Feasibility of a New Process for Hydrogen Storage, *International Journal of Hydrogen Energy*, **1995**, vol. 20, p. 653.

230. Rabo J. A., Catalysis: Past, Present and Future, in *Proceedings of the 10th International Congress on Catalysis*, Budapest, Hungary, July 19-24. Studies in Surface Science and Catalysis, 75, p. 1-30, **1993**.

231. Chang C. D., Methanol to Gasoline and Olefins, in *Methanol Production and Use* (Eds.: W.-H. Cheng, H. H. Kung), Marcel Dekker, New York, **1994**, p. 133.

232. Chang C. D., Methanol to Hydrocarbons, in *Handbook of Heterogeneous Catalysis*, Vol. 4 (Eds.: G. Ertl, H. Knözinger, J. Weitkamp), Wiley-VCH, Weinheim, Germany, **1997**, p. 1894.

233. Special issue covering methanol to hydrocarbons technologies and processes, in *Microporous and Mesoporous Materials*, Vol. 29 (1-2) (Eds.: M. Stocker, J. Weitkamp), **1999**.

234. Olah G. A., Doggweiler H., Felberg J. D., et al., Onium Ylide Chemistry. 1. Bifunctional Acid-Base-Catalyzed Conversion of Heterosubstituted Methanes into Ethylene and derived Hydrocarbons. The Onium Ylide Mechanism of the C_1 to C_2 Conversion, *Journal of the American Chemical Society*, **1984**, vol. 106, p. 2143.

235. Olah G. A., *Bifunctional Acid-Base Catalyzed Conversion of Heterosubstituted Methanes into Olefins*, US Patent 4,373,109, **1983**.

236. Andersen J., Bakas S., Foley T., et al., MTO: Meeting the Needs for Ethylene and Propylene Production, presented at ERTC Petrochemical Conference, Paris, France, March 3-5, **2003**.

237. UOP: http://www.uop.com/.

238. Lurgi: http://www.lurgi.de.

Index

a

acetic acid   175, 246
acid rain   32
advanced gas-cooled reactor (AGR)   116
Agrol   177
air composition   8
Air Products   214
al-Burgan   22
al-Ghawar   22
algae   238
alkaline fuel cell   156, 160
alkylation   21, 64
alternative energy   85
ammonia synthesis   68, 173
anthracite   29
Antizol®   202
aquaculture   237
Aramco   36
Archimedes   100
Arrhenius, Svante   8, 74
aspartame   203
Aswan dam   90
Athabaska   37, 55
atmospheric aerosol   78
atomic bomb   115
atomic energy   3, 5, 55, 81, 84, 255
atomic energy, see also nuclear energy   111
atomic pile   114
automobile   20
autothermal reforming   217

b

Bacon, Francis T.   156
bagasse   106
bakelite   66
Bank of America   179
BASF (Badische Anilin und Soda Fabrik)   174, 178
Benz, Carl   20

Bequerel, Edmond   98
Big Bang   133
biocrude   231
biodiesel   104, 107, 186
biofuel   104, 108
– cell   195
biogas   235
biomass   103, 230
– co-firing   103
– electricity   103
– gasification   103, 141, 230
– pyrolysis   230
biomass energy   103, 131
– cost   104
bituminous coal   29
black liquor   232
boiling water reactor (BWR)   115
Bolshevik Revolution   105, 177
Bosch, Carl   68, 173
Boyle, Robert   173
breeder reactor   118
bromine   227
Brookhaven National Laboratory   231
Bunsen burner   24
Bunsen, Robert   24

c

California's Geysers   91
CANDU (Canada deuterium uranium) reactor   116
carbon dioxide, see also CO_2
– from industrial flue gas   242
– from the atmosphere   243
Carnol process   240
Carnot efficiency of heat engine   156
catalytic cracking   21
cattail   237
cetane rating   182
charcoal   13

Beyond Oil and Gas: The Methanol Economy. G. A. Olah, A. Goeppert, G. K. S. Prakash
Copyright © 2006 WILEY-VCH Verlag GmbH & Co. KGaA, Weinheim
ISBN 3-527-31275-7

chemicals from methanol   175, 246
Chernobyl   120, 123
chlorination   226
chlorine   227
chlorofluorocarbon (CFC)   78
– greenhouse gas   76
clathrate   47
clean coal technologies   16
climate change   72, 208
– mitigation   81
CNG (compressed natural gas)   183, 235
CO_2
– atmospheric   8
– atmospheric concentration   74, 76, 80
– capture   81, 243-244
– emission   8, 76, 79, 208, 211, 229, 242
– greenhouse gas   74
– recycling   7-8, 208, 244, 256
– reduction   9, 196, 208, 239-244
– sequestration   82, 141, 244
– Sleipner platform   82
coal   28
– Atmospheric Fluidized Bed Combustion
  (AFBC)   16
– Britain   11
– clean coal technologies   16
– electricity production   29
– emission   15, 32, 127
– formation   11
– FutureGen   16, 140
– gasification   173, 218
– Integrated Gasification Combined Cycle
  (IGCC)   16
– mining   12, 31
– Pressurized Fluidized Bed Combustion
  (PFBC)   16
– production   29
– R/P ratio   15
– reserve   15, 29
– resource   29
– SO_2 emissions   31
– structure   62
coalbed methane   46
co-firing   103
Cold Lake   37, 55
Commisariat à l'Energie Atomique (CEA)
  122, 145
composition of air   8
compressed hydrogen   148, 152
cracking   21
cytochrome P-450   229

d
DaimlerChrysler   188, 194
Daimler, Gottlieb   20
deuterium   129
diesel   61, 63
– engine   182
– fuel   45, 182
diet soda   203
dimethoxymethane   195
dimethyl carbonate (DMC)   186
dimethyl ether (DME)   176, 195
– as a diesel fuel   182, 185
– in transportation   184 f.
– production   186
– property   184-185
– synthesis   184
Direct Methanol Fuel Cell, see DMFC
Discol   177
DME, see dimethyl ether
DMFC (Direct Methanol Fuel Cell)   9, 170,
  191, 257
– efficiency   192
– for mobile application   193
Drake, Edwin   18

e
earth temperature variation   73
Einstein, Albert   98, 128
electrical energy storage   166
electricity generation, industrial countries   4
energy carrier   135
energy crop   106, 108, 233
energy storage   170
enhanced oil recovery (EOR)   54
EPA (Environmental Protection Agency)   126,
  204
Erren, Rudolf   136
ethanol   6, 69, 104, 171, 195
– as a fuel   177
– Brazil   105, 178
– England   177
– flexible fuel vehicle (FFV)   106, 178
– from cellulose   107
– from corn   106
– from sugar cane   104, 178
– Germany   177
– in transportation   177, 183
– Soviet Union   177
– tax subsidy   106
– United States   106, 177
ethylene   65, 170, 248
– from methanol   248

eucalyptus   233
ExxonMobil   36

f

FC-me   194
Fermi, Enrico   114
fertilizer   68-69, 90, 108, 233
Fischer-Tropsch   6, 45, 57, 69, 174, 178
flexible fuel vehicle (FFV)   106, 178, 180, 198
fomepizol   202
Ford, Henry   20, 179, 188
formaldehyde   175, 195, 220, 222, 246
formic acid   195, 222
fossil fuel
– definition   2
– reserve   27
– resource   27
– use   27
fuel   170
fuel cell   9, 195
– alkaline   156, 160
– bus   189, 205
– direct methanol   191
– efficiency   156
– history   155
– hydrogen-based   159
– molten carbonate   160, 197, 236
– phosphoric acid   159, 189, 197
– proton exchange membrane   137, 158, 189
– regenerative   9, 165, 196
– solid oxide   161, 197
– vehicle   188
fusion reaction   130
fusion reactor   129
FutureGen   16, 140

g

gasification   103, 141, 173, 218, 230
gasoline   20, 61, 63
gasoline reforming   150
gas-to-liquid (GTL)   6, 45
Generation IV International Forum (GIF)   118
Georgetown University   189, 205
geothermal electricity   91
geothermal energy   85, 91
– cost   93
– heat   93
– hot dry rock   93
global carbon cycle   75
global warming   41, 72
Global Warming Potential (GWP)   77

Gold, Thomas   58
Greene   248
greenhouse effect   74
greenhouse gas   74
– chlorofluorocarbon   74, 78
– CO_2 74
– methane   41, 74
– moisture   79
– nitrous oxide   74, 76
Green River   39
Grove, William R.   9, 155
GTL (gas-to-liquid) facilities   45
– Qatar   45

h

H_2O_2 196, 224, 229
Haber, Fritz   68, 173
Haldor Topsoe AIS   184, 209, 213
hard coal   29
Hart, William   24
Hoover dam   88
Hubbert, M. King   52
Hubbert's peak   49, 53, 56
hybrid vehicle   155, 164, 187
hydrocarbon   60
– definition   60
– from methanol   170, 246, 257
hydrogen   168
– bus   163
– compressed   148, 152
– cryogenic storage   147
– discovery   133
– distribution   150
    – by pipeline   152
    – by truck   151
– economy   133, 136, 168, 256
– energy   135
– flammability   153
– fuel   135, 146, 148, 154
– fueling station   198
– in ICE   154
– liquefaction   147
– liquid   137, 147, 151, 154
– metal hybride storage   149
– property   133
– safety   153, 198
– storage   145
– use   138
hydrogen production   138, 146
– centralized   151
– cost   141, 143
– decentralized   151
– from biomass   141

– from fossil fuel 140
– from geothermal energy 144
– from hydropower 143
– from nuclear energy 144
– from solar energy 143
– from water electrolysis 142
– from wind energy 143
– source 139
hydropower 85, 87
Hynol process 231

i

ICE (internal combustion engine) 177,
  180, 187, 205
ICI (Imperial Chemical Industries) 174,
  213, 252
industrial revolution 13
insolation 97
Integrated Gasification Combined Cycle
  (IGCC) 140
International Panel on Climate Change
  (IPCC) 72, 79
iodine 227
isomerization 21
Itaipú hydroelectric plant 89
ITER (International Thermonuclear
  Experimental Reactor) 129

j

Jet Propulsion Laboratory (JPL) 9, 191, 193
Johnson Matthey Fuel Cells 190
Joint European Torus (JET) 129
Julich Research Center 194
JuMOVe 194

k

kelp 238
kerosene 19, 63
Korea Institute of Science and Technology
  (KIST) 193
Kyoto Protocol 83, 85, 242

l

landfill 236
Larderello 91
Lavoisier, Antoine 135
Le Chatelier's principle 212
LeBel, Joseph Achille 248
lignite 29
liquefied natural gas, see LNG
liquid hydrogen 137, 147, 151, 154
liquid phase methanol process 214
LNG 25, 43, 170

– accident 25
– infrastructure 43
– production 44
– transport 44
Los Alamos National Labs 193
LPG (liquid petroleum gas) 63, 185
Lurgi 209, 213, 250

m

M85 179-181, 198
M100 180, 198
macroalgae 238
Manchester 14
Manhattan project 115
Maracaibo sedimentary basin 37
marine current 109
Mars, temperature 74
Massachusetts Institute of Technology
  (MIT) 178
Mauna Loa 75
Maybach, Wilhelm 20
mega methanol plant 200, 209, 248
mesquite tree 234
metal hydrides 149
methane 60
– atmospheric concentration 76
– autothermal reforming 140, 217
– catalytic gas-phase oxidation 221
– decomposition 241
– from coalbed 46
– from tight sands and shales 47
– greenhouse gas 41, 76
– halogenation 226-227
– hydrate 47, 58, 208, 241
– monooxygenase (MMO) 228
– oxidation 216, 221-225, 227
– partial oxidation 140, 216
– steam reforming 140, 215
Methanex 199
methanogenesis 235
methanol 57
– and the environment 206
– as a fuel 169-170, 173, 178, 182-183
– as an energy carrier 171, 173
– Australia 234
– cetane rating 182
– Chile 209
– climate change 208
– demand 210
– denitrification process 207
– distribution 197
– energy content 180, 182, 211
– energy storage 170-171

– Europe 234
– fire 203
– flexible fuel vehicle (FFV) 180, 198
– for cooking 197
– for heat generation 196
– for static power generation 196
– history 173
– Indianapolis 500 179
– in transportation 179-180, 197-198
– New Zealand 252
– octane rating 180
– poisoning 202
– property 173, 180
– purification 214
– Qatar 209
– reforming 150, 187, 190
– refueling station 198
– refueling station cost 198
– safety 201, 203
– Saudi Arabia 209
– sewage treatment plant 206
– spill 203, 207
– storage 197, 207
– to enhance photosynthetic productivity 253
– toxicity 201, 206
– Trinidad and Tobago 209
– United States 209, 234
– use 175, 201, 246
methanol-based protein 252
Methanol Economy 7, 168, 171-172, 178, 256
methanol powered vehicles 179
– emission 205
methanol production 173, 209
– capacity 210
– from biogas 235
– from biomass 107, 208, 229
– from carbon dioxide (CO_2) 171, 208, 239, 244, 257
– from coal 174, 211
– from direct oxidation of methane 171
– from fossil fuel 228
– from landfill gas 236
– from liquid-phase oxidation of methane 224
– from methane 208, 217
– from natural gas 174, 213
– from recycling of atmospheric CO_2 170, 243
– from selective oxidation of methane 221
– from solid municipal waste 232
– from syn-gas 174, 212-213
– from water plant 238
– from wood 173, 231, 233
– high pressure synthesis 174, 213
– low pressure synthesis 175, 213
methanol reactions
– biodiesel production 107, 186
– hydrogen production 189
– to chemicals 246
– to hydrocarbon 170, 246, 257
methanol synthesis
– methyl formate 219
– microbial or photochemical conversion of methane 228
– mono-halogenated methane 226
methanol to gasoline (MTG) 251
methanol to hydrogen (MTH) 189
methanol to olefin (MTO) 170, 248
methanol to propylene (MTP) 250
methanotroph 228
4-methylpyrazole 202
methyl alcohol, see methanol 173
methyl bromide 226
methyl chloride 226
methyl formate 219, 223
methyl hydrogen sulfate 224, 227
methyl iodide 227
microalgae 238
micro-hydro system 90
Middle-East 36
Mitsubishi Gas Chemical 190
mixed oxides fuel (MOx) 117
Mobil Oil 249
Mojave Desert 101
molten carbonate fuel cell (MCFC) 160, 197, 236
molten salt reactor 119
Monsanto 246
Montreal Protocol 78, 83
Motorola 193
MTBE (methyl-*t*-butyl ether) 175, 178, 195, 200, 207, 209, 246

n

National Alcohol Program (PNA) 178
National Fire Protection Association (NFPA) 198
National Renewable Energy Laboratory (NREL) 239
natural gas 23, 39
– abiotic origin 59
– CO_2 emission 41
– consumption 39
– electricity generation 40

– formation 18
– liquefaction 43
– nature 23
– off-shore pipelines 44
– pipeline 24
– production 40
– R/P ratio 42, 58
– reserve 5, 41, 51, 58
– resource 58
natural nuclear reactor 115
Necar 5 188
Newcomen, Thomas 13
New Zealand 252
nitrous oxide
– atmospheric concentration 77
– emissions 82
– for methane oxidation to methanol 223
– greenhouse gas 76
non-conventional oil source 55
nuclear
– chain reaction 114
– economics 121
– electricity 121
– energy 111
– first reactor 114
– fission 113
– fuel reprocessing 117, 125
– hydrogen production 144
– plant 112
– plant safety 121
– power plant emission 127
– production cost 121
– reactor 111
– reactor under construction 120
– spent fuel 125
– waste 125
nuclear fusion 128
– hydrogen production 131
nylon 66

o
ocean energy 108
Ocean Thermal Energy Conversion (OTEC) 110
oil 32, 60, 62
– cracking 64
– depletion 4, 51
– distillation 63
– exploitation 21
– exploration 22
– extraction 22
– formation 18
– history 18

– in transportation 33
– Middle-East 34, 36
– non-conventional 55
– price 35, 51
– production 32
– properties 60
– quality 65
– R/P ratio 42, 51
– recovery factor 54
– refining 21, 63
– reserve 4, 28, 34, 36, 51
– transport 22
– ultimate recovery 52
– use 63
oil consumption
– China 5
– United States 5
– world 4, 34
oil price shock 35
oil shale 38
– Autun, France 38
– Estonia 39
– Scotland 38
– United States 39
– Sweden 38
Oklo, Gabon 115
Olah, George 171, 222, 224, 226-227, 242, 250
olefins to gasoline and distillate (MOGD) 250
on-board reforming 150, 155, 187
OPEC 28, 35, 51
Oracle of Delphi 23
Orinoco belt 37, 55
ozone layer hole 83

p
parabolic trough 100
Pelamis 110
PEM fuel cell (Proton Exchange Membrane fuel cell) 137, 158, 189
– for transportation 162
– platinum catalyst 163
– price 163
petrochemical 33, 62, 65
petroleum, see oil
Phénix 118
phosphoric acid fuel cell (PAFC) 159, 189, 197
photovoltaic 98
plastic 66
plexiglas 66
plutonium 239 116